Compendium of Plant Genomes

Series Editor

Chittaranjan Kole, Raja Ramanna Fellow, Government of India, ICAR-National Research Center on Plant Biotechnology, Pusa, New Delhi, India

Whole-genome sequencing is at the cutting edge of life sciences in the new millennium. Since the first genome sequencing of the model plant Arabidopsis thaliana in 2000, whole genomes of about 100 plant species have been sequenced and genome sequences of several other plants are in the pipeline. Research publications on these genome initiatives are scattered on dedicated web sites and in journals with all too brief descriptions. The individual volumes elucidate the background history of the national and international genome initiatives; public and private partners involved; strategies and genomic resources and tools utilized; enumeration on the sequences and their assembly; repetitive sequences; gene annotation and genome duplication. In addition, synteny with other sequences, comparison of gene families and most importantly potential of the genome sequence information for gene pool characterization and genetic improvement of crop plants are described.

Interested in editing a volume on a crop or model plant?

Please contact Prof. C. Kole, Series Editor, at ckoleorg@gmail.com

More information about this series at http://www.springer.com/series/11805

Ramakrishnan M. Nair ·
Roland Schafleitner ·
Suk-Ha Lee
Editors

The Mungbean Genome

Springer

Editors
Ramakrishnan M. Nair
World Vegetable Center
ICRISAT Campus
Hyderabad, Telangana, India

Roland Schafleitner
World Vegetable Center
Shanhua, Tainan, Taiwan

Suk-Ha Lee
College of Agriculture and Life Sciences
Seoul National University
Seoul, Korea (Republic of)

ISSN 2199-4781 ISSN 2199-479X (electronic)
Compendium of Plant Genomes
ISBN 978-3-030-20007-7 ISBN 978-3-030-20008-4 (eBook)
https://doi.org/10.1007/978-3-030-20008-4

This Springer imprint is published by the registered company Springer Nature Switzerland AG
The registered company address is: Gewerbestrasse 11, 6330 Cham, Switzerland

This book series is dedicated to my wife Phullara, and our children Sourav, and Devleena
Chittaranjan Kole

Preface to the Series

Genome sequencing has emerged as the leading discipline in the plant sciences coinciding with the start of the new century. For much of the twentieth century, plant geneticists were only successful in delineating putative chromosomal location, function, and changes in genes indirectly through the use of a number of "markers" physically linked to them. These included visible or morphological, cytological, protein, and molecular or DNA markers. Among them, the first DNA marker, the RFLPs, introduced a revolutionary change in plant genetics and breeding in the mid-1980s, mainly because of their infinite number and thus potential to cover maximum chromosomal regions, phenotypic neutrality, absence of epistasis, and codominant nature. An array of other hybridization-based markers, PCR-based markers, and markers based on both facilitated construction of genetic linkage maps, mapping of genes controlling simply inherited traits, and even gene clusters (QTLs) controlling polygenic traits in a large number of model and crop plants. During this period, a number of new mapping populations beyond F2 were utilized and a number of computer programs were developed for map construction, mapping of genes, and mapping of polygenic clusters or QTLs. Molecular markers were also used in the studies of evolution and phylogenetic relationship, genetic diversity, DNA fingerprinting, and map-based cloning. Markers tightly linked to the genes were used in crop improvement employing the so-called marker-assisted selection. These strategies of molecular genetic mapping and molecular breeding made a spectacular impact during the last one and a half decades of the twentieth century. But still they remained "indirect" approaches for elucidation and utilization of plant genomes since much of the chromosomes remained unknown and the complete chemical depiction of them was yet to be unraveled.

Physical mapping of genomes was the obvious consequence that facilitated the development of the "genomic resources" including BAC and YAC libraries to develop physical maps in some plant genomes. Subsequently, integrated genetic–physical maps were also developed in many plants. This led to the concept of structural genomics. Later on, emphasis was laid on EST and transcriptome analysis to decipher the function of the active gene sequences leading to another concept defined as functional genomics. The advent of techniques of bacteriophage gene and DNA sequencing in the 1970s was extended to facilitate sequencing of these genomic resources in the last decade of the twentieth century.

As expected, sequencing of chromosomal regions would have led to too much data to store, characterize, and utilize with the-then available computer software could handle. But the development of information technology made the life of biologists easier by leading to a swift and sweet marriage of biology and informatics, and a new subject was born—bioinformatics.

Thus, the evolution of the concepts, strategies, and tools of sequencing and bioinformatics reinforced the subject of genomics—structural and functional. Today, genome sequencing has traveled much beyond biology and involves biophysics, biochemistry, and bioinformatics!

Thanks to the efforts of both public and private agencies, genome sequencing strategies are evolving very fast, leading to cheaper, quicker, and automated techniques right from clone-by-clone and whole-genome shotgun approaches to a succession of second-generation sequencing methods. The development of software of different generations facilitated this genome sequencing. At the same time, newer concepts and strategies were emerging to handle sequencing of the complex genomes, particularly the polyploids.

It became a reality to chemically—and so directly—define plant genomes, popularly called whole-genome sequencing or simply genome sequencing.

The history of plant genome sequencing will always cite the sequencing of the genome of the model plant Arabidopsis thaliana in 2000 that was followed by sequencing the genome of the crop and model plant rice in 2002. Since then, the number of sequenced genomes of higher plants has been increasing exponentially, mainly due to the development of cheaper and quicker genomic techniques and, most importantly, the development of collaborative platforms such as national and international consortia involving partners from public and/or private agencies.

As I write this preface for the first volume of the new series "Compendium of Plant Genomes," a net search tells me that complete or nearly complete whole-genome sequencing of 45 crop plants, eight crop and model plants, eight model plants, 15 crop progenitors and relatives, and 3 basal plants is accomplished, the majority of which are in the public domain. This means that we nowadays know many of our model and crop plants chemically, i.e., directly, and we may depict them and utilize them precisely better than ever. Genome sequencing has covered all groups of crop plants. Hence, information on the precise depiction of plant genomes and the scope of their utilization are growing rapidly every day. However, the information is scattered in research articles and review papers in journals and dedicated Web pages of the consortia and databases. There is no compilation of plant genomes and the opportunity of using the information in sequence-assisted breeding or further genomic studies. This is the underlying rationale for starting this book series, with each volume dedicated to a particular plant.

Plant genome science has emerged as an important subject in academia, and the present compendium of plant genomes will be highly useful to both students and teaching faculties. Most importantly, research scientists involved in genomics research will have access to systematic deliberations on the plant genomes of their interest. Elucidation of plant genomes is of interest not only for the geneticists and breeders, but also for practitioners of an array of plant science disciplines, such as taxonomy, evolution, cytology, physiology, pathology, entomology, nematology, crop production, biochemistry, and obviously bioinformatics. It must be mentioned that information regarding each plant genome is ever-growing. The contents of the volumes of this compendium are, therefore, focusing on the basic aspects of the genomes and their utility. They include information on the academic and/or economic importance of the plants, description of their genomes from a molecular genetic and cytogenetic point of view, and the genomic resources developed. Detailed deliberations focus on the background history of the national and international genome initiatives, public and private partners involved, strategies and genomic resources and tools utilized, enumeration on the sequences and their assembly, repetitive sequences, gene annotation, and genome duplication. In addition, synteny with other sequences, comparison of gene families, and, most importantly, the potential of the genome sequence information for gene pool characterization through genotyping by sequencing (GBS) and genetic improvement of crop plants have been described. As expected, there is a lot of variation of these topics in the volumes based on the information available on the crop, model, or reference plants.

I must confess that as the series editor, it has been a daunting task for me to work on such a huge and broad knowledge base that spans so many diverse plant species. However, pioneering scientists with lifetime experience and expertise on the particular crops did excellent jobs editing the respective volumes. I myself have been a small science worker on plant genomes since the mid-1980s and that provided me the opportunity to personally know several stalwarts of plant genomics from all over the globe. Most, if not all, of the volume editors are my longtime friends and colleagues. It has been highly comfortable and enriching for me to work with them on this book series. To be honest, while working on this series I have been and will remain a student first, a science worker second, and a series editor last. And I must express my gratitude to the volume editors and the chapter authors for providing me the opportunity to work with them on this compendium.

I also wish to mention here my thanks and gratitude to the Springer staff particularly Dr. Christina Eckey and Dr. Jutta Lindenborn for the earlier set of volumes and presently Ing. Zuzana Bernhart for all their timely help and support.

I always had to set aside additional hours to edit books beside my professional and personal commitments—hours I could and should have given to my wife, Phullara, and our kids, Sourav and Devleena. I must mention that they not only allowed me the freedom to take away those hours from them but also offered their support in the editing job itself. I am really not sure whether my dedication of this compendium to them will suffice to do justice to their sacrifices for the interest of science and the science community.

New Delhi, India Chittaranjan Kole

Foreword

Mungbean is a crop with great potential for improving the livelihood of millions of smallholder farmers in tropical developing countries. Its short duration makes it an attractive option for intensifying farming systems: the crop can usually fit within existing rotations, it diversifies the diet of farmers consuming it, and provide soil benefits (disease control and added nitrogen) to the following cereal crop. The product is in high demand and attracts a high price, making the crop profitable for farmers who manage to protect it against the biotic stresses affecting it in many places. The crop has also recently become an important summer crop in the industrial farming systems of Northeast Australia.

Consistent with its mandate of investing in research benefiting developing countries and Australia, the Australian Centre for International Agricultural Research (ACIAR) is supporting mungbean research in South Asia, South East Asia and Eastern Africa. Research on farming system intensification includes mungbean trials and demonstration. The sustainability and ongoing increase in mungbean production depends on continually improving the potential of new varieties available to farmers. Under the leadership of the World Vegetable Center, the International Mungbean Improvement Network (IMIN) has been supported by ACIAR since 2016 to provide breeding programs in Bangladesh, India, Myanmar and Australia tools and resources to conduct modern mungbean pre-breeding and breeding, and to collect results from a global multilocation trial network. The detailed phenotypic characterization by the Network of a carefully assembled Mini-Core Collection will provide breeders with new sources of useful traits. The Mini Core Collection has been genotyped at high density, thereby allowing the identification of molecular markers for the traits discovered, and facilitating breeding for these traits. IMIN welcomes other partners who would like to contribute to the Network by sharing multilocation trial data and evaluating the common germplasm.

As a small crop, mungbean is only recently benefiting from advances in genetics and genomics made over the past 10–20 years. Molecular markers, low- cost and high-throughput genotyping tools, and the corresponding data storage, management and analysis methods and skills, are facilitating and accelerating the discovery, and then the application, of genes and Quantitative Trait Loci (QTL) for important traits.

This book is a timely review and presentation of the genetic and genomic resources available to mungbean breeders. The authors have led IMIN since its inception and have contributed more than anyone else to the development and application of the modern resources which will in the future underpin the continual improvement of mungbean. ACIAR is proud to support their effort.

Eric Huttner
Research Program Manager, Crops
Australian Centre for International Agricultural Research
Canberra, Australia

Preface

Mungbean (*Vigna radiata* (L.) R. Wilczek var. *radiata*), also called green gram, is an important food and cash crop in the rice based farming systems of South and Southeast Asia, but is also grown in other parts of the world. Short crop duration, tolerance to heat, low input requirement and high global demand make mungbean an ideal rotation crop for smallholder farmers. It generates a triple benefit for its users: additional income, additional nutrient-rich food, and increased soil fertility. The book outlines the global status of mungbean and its economic importance. Mungbean collections maintained by different organizations and their utilization are described, especially for adaptation of mungbean to new environments. The progress in breeding for tolerance to biotic and abiotic stresses, improvement of the nutritional quality, as well as future challenges for mungbean breeding are discussed. The state of use of molecular markers and the potential of molecular breeding for mungbean improvement are reviewed. The mungbean genome sequence was published in 2014. How the mungbean genome compares with other *Vigna* species and genomic approaches tackling various breeding objectives is elaborated. Overall, the book aims at depicting the current status of mungbean breeding to promote research on this important tropical legume crop.

Hyderabad, India — Ramakrishnan M. Nair
Shanhua, Tainan, Taiwan — Roland Schafleitner
Seoul, Korea (Republic of) — Suk-Ha Lee

Acknowledgement The authors acknowledge the support from their respective organisations and also the Australian Centre for International Agricultural Research (ACIAR) for the support of the International Mungbean Improvement Network (IMIN).

Contents

1 Global Status and Economic Importance of Mungbean 1
Ramakrishnan Nair and Pepijn Schreinemachers

2 Mungbean Genetic Resources and Utilization 9
Gayacharan, Sunil Archak, Kavita Gupta, Veena Gupta, Vandana Tyagi and Kuldeep Singh

3 Taxonomy of Mungbean and Its Relatives 27
Yu Takahashi and Norihiko Tomooka

4 Classical Genetics and Traditional Breeding in Mungbean 43
Harsh Kumar Dikshit, Gyan Prakash Mishra, Prakit Somta, Tun Shwe, A. K. M. M. Alam, Tejinderjit S. Bains and Ramakrishnan M. Nair

5 Breeding Progress and Future Challenges: Biotic Stresses 55
Aditya Pratap, Col Douglas, Umashanker Prajapati, Gita Kumari, Abdul Rasheed War, Rakhi Tomar, Abhay K. Pandey and Sonali Dubey

6 Breeding Progress and Future Challenges: Abiotic Stresses 81
Col Douglas, Aditya Pratap, Bindumadhava Hanumantha Rao, B. Manu, Sonali Dubey, Parikshit Singh and Rakhi Tomar

7 Breeding Progress and Future Challenges—Nutritional Quality 97
Ramakrishnan Nair

8 Molecular Marker Resources and Their Application 107
Roland Schafleitner

9 Mungbean Genome and Synteny with Other Genomes 125
Yang Jae Kang and Jungmin Ha

10 Genomic Approaches to Biotic Stresses 133
Kularb Laosatit, Prakit Somta, Xin Chen and Peerasak Srinives

11 Genomic Approaches to Abiotic Stresses in Mungbean 169
Thomas J. Noble, Brett Williams, Thi My Linh Hoang, Sudipta Shekhar Das Bhowmik, Grace Zi Hao Tan and Sagadevan Mundree

12 Future Prospects and Challenges 181
Roland Schafleitner and Ramakrishnan M. Nair

Abbreviations

ACIAR	Australian Centre for International Agricultural Research
AFLP	Amplified Fragment Length Polymorphisms
AGES	Austrian Agency for Health and Food Safety
AGG	Australian Grains Genebank
AICRP	All India Co-ordinated Research Project
AVRDC	Asian Vegetable Research and Development Center
BARI	Bangladesh Agriculture Research Institute
BDA	Biological Diversity Act
CAAS	Chinese Academy of Agricultural Sciences
CAP	Cleaved Amplified Polymorphism
CBD	Convention on Biological Diversity
CLS	Cercospora Leaf Spot
CTSI	Cold Tolerance Scoring Index
CVRC	Central Variety Release Committee
CWR	Crop Wild Relatives
DAR	Department of Agriculture Research
DAS	Day After Sowing
DBT	Department of Biotechnology
DIP	Destructive Insects and Pests
DNA	Deoxyribo Nucleic Acid
FRS	Field Root Strucure
FSD	Fresh Seed Dormancy
GWAS	Genome-Wide Association Study
ICAR	Indian Council of Agricultural Research
IIPR	Indian Institute of Pulse Research
IMIN	International Mungbean Improvement Network
MABC	Marker Assisted Back Crossing
MAS	Marker Assisted Selection
MoEFCC	Ministry of Environment, Forest and Climate Change
MYIMV	Mungbean Yellow India Mosaic Virus
MYMD	Mungbean Yellow Mosaic Disease
NARS	Indian National Agricultural Research System
NBA	National Biodiversity Authority
NBPGR	National Bureau of Plant Genetic Resources
NCBI	National Center for Biotechnology Information
NGS	Next Generation Sequencing

NIAB	Nuclear Institute for Agriculture and Biology
PCR	Polymerase Chain Reaction
PGRC	Plant Germplasm Registration Committee
PGRFA	Plant Genetic Resource for Food and Agriculture
PGRs	Plant Genetic Resources
PHS	Pre-Harvest Sprouting
PM	Powdery Mildew
PQO	Plant Quarantine Order
QTLs	Quantitative Trait Loci
RAPD	Random Amplification of Polymorphic DNA
RFLP	Restriction Fragment Length Polymorphism
SCAR	Sequence Characterized Amplified Regions
SNPs	Single-Nucleotide Polymorphism
SSRs	Simple Sequence Repeats
STI	Stress Tolerance Index
STS	Sequence Tagged Sites
SVRCs	State Variety release committee
TIU	Trypsin Inhibitor Unit
ULCV	Urdbean Leaf Crinkle Virus
USDA	United States Department of Agriculture
WGS	Whole Genome Sequence
YMV	Yellow Mosaic Virus

Global Status and Economic Importance of Mungbean

1

Ramakrishnan Nair and Pepijn Schreinemachers

Abstract

This chapter provides an overview of the economic importance of mungbean globally and the status of mungbean improvement research. The global mungbean area is about 7.3 million ha, and the average yield is 721 kg/ha. India and Myanmar each account for 30% of global output of 5.3 million t. Other large producers are China, Indonesia, Thailand, Kenya, and Tanzania. The mungbean market is divided in four main segments by usage: dry grains (important in South Asia and Kenya), sprouts (important in East and Southeast Asia), transparent noodles/starch (important in East and Southeast Asia), and paste (important in East Asia). Mungbean research is under-resourced in most countries as it is considered a minor crop. There is a history of strong international collaboration in mungbean improvement research in Asia, which is particularly important for a minor crop like mungbean as no single country has the capacity to cover all aspects requiring research. The International Mungbean Improvement Network was established in 2016 to further this collaboration and is coordinated by the World Vegetable Center.

R. Nair (✉)
World Vegetable Center, South Asia, ICRISAT Campus, Patancheru, Hyderabad, Telangana, India
e-mail: ramakrishnan.nair@worldveg.org

P. Schreinemachers
World Vegetable Center, East and Southeast Asia, Bangkok, Thailand

1.1 Introduction

Mungbean (*Vigna radiata* (L.) R. Wilczek var. *radiata*), also known as green gram or moong, is an important food and cash crop in the rice-based farming systems of South and Southeast Asia, but is also produced in other parts of the world. The chief attractions of mungbean to farmers are the crop's short duration, low input requirement, suitability as a rotation crop in cereal-based systems, and good performance under heat and drought stress. In some parts of the world, such as in India and Pakistan, it is an important subsistence crop adding essential nutrients (especially protein, iron, and zinc) to the diets of farm families, whereas it is an income-generating crop in countries such as Myanmar, China, and Kenya.

Mungbean yields in most countries are relatively low, usually ranging from 0.5 to 1.5 t/ha. Genetic improvement of mungbean is an important part of the strategy to raise mungbean productivity. Genetic improvement of mungbean is mostly done by public sector organizations in the national agricultural research systems. The international mungbean program of the World Vegetable Center, previously known as AVRDC,

R. M. Nair et al. (eds.), *The Mungbean Genome*, Compendium of Plant Genomes,
https://doi.org/10.1007/978-3-030-20008-4_1

has made an important contribution to coordinate these programs, share mungbean germplasm between countries, and develop varieties with improved performance.

The objective of this chapter is to give a brief overview of the economic importance of mungbean globally and to relate this to mungbean breeding efforts. The chapter relies on secondary data, published studies, and our own observations from working on mungbean improvement across Africa and Asia. Accordingly, the remainder of this chapter is divided into three sections. We continue in the next section by first describing the main uses of mungbean as the market is broadly divided into four segments. We then describe the main mungbean-producing countries in the world, after which we give an overview of the mungbean improvement efforts in Asia.

1.2 Mungbean Market Segments and Varietal Requirements

Mungbean has different uses in different parts of the world, some of which require specific varieties. Each of the main uses and variety requirements is described in the following.

1.2.1 Use of Dry Grains

The use of dry grains in cooking is the most common use of mungbean globally. In South Asia (India, Pakistan, Bangladesh, and Nepal), mungbean grains are typically consumed as a *dahl* (a stew made of dried pulses). In Kenya, it is mostly eaten as a thick bean stew. In other countries, mungbean grains may be cooked with rice, cooked with sugar to make a sweet desert soup (e.g., in China), or grilled or roasted as a snack. When used for *dahl*, consumers often prefer mungbean in de-husked and split form as this reduces the cooking time. The grain segment represents a relatively low-value segment of the market as quality attributes are generally less important when grains are de-husked and split. However, when whole grains are used, then consumers may have particular preferences for grain size, seed coat luster (shiny or dull), and seed coat color (green or yellow). For instance, in the Indian sub-continent, consumers tend to prefer shiny green grains, while shiny yellow grains are preferred in Bangladesh, Sri Lanka, and some parts of India. Consumers in Indonesia, Taiwan, Kenya, and Tanzania prefer dull green grains. In eastern India, consumers prefer mungbean with a particular pleasant aroma (called *Sona mung*).

1.2.2 Bean Sprouts

Mungbean is highly suitable for producing vegetable sprouts, and this use is common in East Asia and Southeast Asia. Fresh mungbean spouts are a commonly used garnish for noodle soups in countries such as Thailand, Laos, and Vietnam. In Korea, sprouts are blanched and used as a side dish (called *sukju namul*) or used with ground mungbeans in stuffed pancakes (*bindaetteok*). In Indonesia, mungbean sprouts, locally known as *tauge*, are found in numerous dishes such as stir-fried mungbean sprouts with tofu (*tauge goreng*), and blanched mungbean sprouts in a vegetable salad served with peanut sauce (*gado gado*), in chicken soup (*soto ayam*) and stir-fried with various vegetables and eggs (*orak-arik*).

The sprouting segment represents the high-value segment of the market as the grains need to meet exacting quality attributes. Consumers prefer sprouts that are bright white and crisp, with short roots and small cotyledons and have a shelf life of at least seven days. Hence, the sprout segment requires different varieties from the dry grain segment. Countries such as China, Australia, Indonesia and to some extent Myanmar control this segment. There is also an increasing demand for mungbean sprouts in high-income countries in Europe and North America, but these continents strict standards regarding pesticide residues on grains used for sprout production, which makes it challenging for low-income countries to supply these markets.

1.2.3 Transparent Noodles and Starch Extraction

Mungbean starch produces a transparent noodle (vermicelli or glass noodle). These noodles are particularly common in China and many other countries influenced by Chinese cuisine. Grain quality is less important than in the sprout segment, but starch quality could be important. To our knowledge, there are no specialized mungbean varieties for noodle production, and this is something that can be taken up in the future.

1.2.4 Bean Paste and Sweets

Mungbean paste is used as an ingredient for various types of sweets in Asia. Mungbean paste mixed with butter and sugar is used to fill mooncakes in China and Taiwan. Mungbean cakes are an important snack and gift in Vietnamese culture. In Thailand, mungbean paste is turned into colorfully painted small imitations of fruits and vegetables, called *khanom luk chup*. In Andhra Pradesh, pancakes (called *pesarattu* dishes) are also made from mungbean paste.

1.2.5 Other Uses

Mungbean flour mixed with other flours such as wheat flour is used as a weaning food for children in some countries. Both Chinese and traditional Indian (Ayurvedic) medicine use mungbean for therapeutic purposes. High levels of total phenols and flavonoids and other compounds have been reported in mungbean grains and sprouts (Kim et al. 2012), which support its use as a preventive or/and therapeutic agent for human health. In China, certain health drinks (reducing body heat due to flavonoids, vitexin, and isovitexin) have been prepared from mungbean seed coats (Cao et al. 2011). Improvement in the knowledge of the food processing properties of mungbean may lead to the development of new value-added products from mungbean (Dahiya et al. 2015).

1.2.6 Animal Feed

Mungbean is utilized as a forage-cum-grain crop in countries like Egypt (Abd El-Salam et al. 2013), and mungbean crop residues are sometimes used as a livestock feed (Nair et al. 2019). The use of mungbean grains in the feed industry is marginal as the market price is high compared to other grains such as soybean. In some countries, mungbean is used as a green manure (Pataczek et al. 2018).

1.3 Mungbean Production Per Country

Mungbean is not separately listed in the statistical database of the Food and Agriculture Organization of the United Nations, and there is therefore a paucity of good internationally comparable data on mungbean. It is a minor crop in many countries and therefore not always included in national statistics. Available data show that mungbean in Asia, Africa, and Oceania is planted on over 7.3 million ha (Table 1.1). Average mungbean yields are lowest in South Asia and Africa and highest in Southeast Asia. The average yield is 721 kg/ha. Each of the main producers is introduced in the following.

1.3.1 India

India is the world's largest producer and consumer of mungbean. However, average mungbean yields in India are among the lowest in the world. The area of mungbean in India has increased over the past few years, and currently, it is estimated at 3.8 million ha. The country produces about 1.6 million t of mungbean. Among the states, Rajasthan (31%) followed by Maharashtra (11%) lead in both area and production. Mungbean is grown as a rotation crop in the rice and rice–wheat cropping systems. Intercropping is also practiced with sugarcane and maize. The government of India has tried to promote domestic production by limiting imports in years when the domestic

Table 1.1 Mungbean area, production, and yield for selected countries

Region/country	Area planted (1,000 ha)	Production (1,000 t)	Average yield (kg/ha)	Year of data
South Asia:	**4,182**	**1,880**	**450**	
• Bangladesh	175	181	1,030	2015
• India	3,828	1,600	420	2016
• Pakistan	179	99	730	2016
East Asia:	**788**	**852**	**1,081**	
• China	786	850	1,081	2008–2009
• Korea	2	2	1,000	1999
Southeast Asia:	**1,867**	**2,330**	**1,282**	
• Cambodia	53	68	1,275	2013
• Lao PDR	3	4	1,430	2017
• Myanmar	1,209	1,597	1,320	2016
• Thailand	275	320	1,164	2014–2015
• Indonesia	229	271	1,183	2015
• Philippines	41	34	811	2015
• Vietnam	89	142	1,600	2017
Central Asia:	**18**	**35**	**1,920**	
• Uzbekistan	18	35	1,920	2015
East Africa:	**554**	**255**	**460**	
• Tanzania	217	73	336	2017
• Kenya	302	149	493	2017
• Uganda	35	33	950	2017
Oceania:	**38**	**35**	**916**	
• Australia	38	35	916	2012–2013
Total	**7,287**	**5,258**	**721**	

Source National statistical agencies for all countries except China (USDA 2009); Thailand (USDA 2014); Korea (Lee 2003); Uzbekistan (Rani et al. 2018); and Australia (AMA 2014)

production is high, but allowing imports when there is a high shortfall in the market. This has created much uncertainty in the global mungbean market with prices rising and falling.

1.3.2 Myanmar

The mungbean area in Myanmar has increased rapidly from about 100,000 ha in 1990 to 1,000,000 ha in 1999, to about 1,200,000 ha now (Ministry of Agriculture 2016; Shanmugasundaram 2003). Mungbean is not widely consumed in Myanmar, but it is one of the country's most widely produced crops (second to rice). Myanmar is the world's largest exporter of mungbean. Most of it is exported to India, and it accounts for 70% of India's mungbean import. Mungbean yields in Myanmar, according to government statistics, are relatively high compared to most other countries. The adoption of improved mungbean varieties is 89%, and most of these are based on germplasm obtained from the World Vegetable Center (Schreinemachers et al. 2019). Mandalay, Sagaing, and Magway regions account for over 80% of the production area. Mungbean is popular among farmers in the legume-based farming systems of Sagaing, Magway, and Mandalay regions as well as the rice-based farming systems of the lower parts of

Myanmar such as Bago, Yangon, and Ayeyarwaddy regions. Further expansion of the mungbean area is limited by the high costs of harvesting due to labor shortages at harvest time. As in most Asian countries, the crop is still mostly harvested by hand and requires several times harvesting as the pods progressively mature. Mechanization of harvesting is therefore important to maintain farmer profits.

1.3.3 China

China is both a large producer and a large consumer of mungbean. The main production areas are in Inner Mongolia, Jilin, Anhui, and Henan provinces accounting for over 60% of the planted area (USDA 2009). Most of the country's production is consumed locally. In 2008, about 140,000 t was exported to other countries in East Asia, while about 80,000 t was imported (USDA 2009). Mungbean is used to produce transparent noodles (vermicelli), used to make bean sprouts, and processed into paste for a range of bakery products such as mooncakes.

1.3.4 South Asia

Mungbean is produced on a substantial area in Bangladesh and Pakistan, and both countries have a high adoption of improved mungbean varieties, most of which originate from their collaboration with the World Vegetable Center (Schreinemachers et al. 2019; Haque et al. 2014; Ali et al. 1997). High labor cost for manual harvesting is also a constraint to farmers in Bangladesh and Pakistan. Smaller areas of mungbean are found in Sri Lanka, Nepal, and Bhutan (not shown).

1.3.5 Southeast Asia

Indonesia and Thailand are the largest mungbean producers in Southeast Asia after Myanmar. The crop is also grown in the Philippines, Vietnam, Cambodia, and Laos. The mungbean prices have been favorable to farmers as market demand is high. High production costs are a challenge to farmers, and there is a need for mechanized harvesting in many countries to reduce costs.

1.3.6 Australia

Mungbean is an export crop for Australia, and most of it is produced in New South Wales and Queensland (AMA 2014). The planted area has seen a decline in recent years as a result of drought in spite of favorable prices. The strength of the Australian mungbean is in grain quality assurance and traceability, which allows it to capture high-value markets. Mungbean harvesting is fully mechanized in Australia. The mungbean industry in the country is well-organized, and there is a strong system of variety improvement in the public sector with royalties of improved varieties flowing into research.

1.3.7 East Asia

Mungbean production, and the production of pulses in general, has strongly declined in Japan, Korea, and Taiwan (Lee 2003; Chen 2003). Although the prices are favorable, the production costs are too high. These countries have become mungbean importers. Mungbean is used for producing vegetable sprouts, mungbean paste for bakery products, mungbean soup, and vermicelli production. Chen (2003) mentioned that sprouts are an important vegetable during the typhoon season in Taiwan when other vegetables are unavailable.

1.3.8 Africa

Mungbean in Africa is grown as a monocrop or as an intercrop with maize, sorghum, or pigeon pea. Mungbean-producing countries include Kenya, Tanzania, Ethiopia, Mozambique, and

Uganda. Kenya is probably the largest producer of mungbean in Africa (most of it produced in Eastern Province), but it is also its largest consumer and even imports mungbean in some years.

1.3.9 South America

Mungbean is produced in Brazil and Argentina, but production data are not available.

1.3.10 Central Asia

Mungbean was recently introduced in Uzbekistan, Kazakhstan, and Tajikistan as a catch crop suitable to grow in between two consecutive crops of wheat and cotton (Rani et al. 2018). The planted area in Uzbekistan is estimated to be about 18,000 ha with relatively high yields of 1.9 t/ha as reported by Rani et al. (2018). The mungbean area in these countries is expanding.

1.4 Mungbean Improvement Research

Mungbean improvement research is almost solely done by public sector organizations, including government research institutes and universities. The crop is of little interest to the private seed industry as seed yields are low and farmers tend to recycle their seed rather than buy certified seed. Unfortunately, most mungbean research programs are small and under-resourced as it not considered a strategic crop such as rice, wheat, or cotton.

Important improvement research has been done by the World Vegetable Center, which has had a mungbean improvement program since 1972. The Center's genebank holds 14,187 genebank accessions of mungbean (the genebank's second largest collection after soybean) and has shared accessions and breeding lines with mungbean breeders worldwide (Schreinemachers et al. 2014). The center has helped to establish and strengthen mungbean improvement programs in China, Pakistan, Myanmar, and Thailand. Countries with strong national mungbean research programs include China, India, and Australia. Other countries with significant mungbean research programs are Bangladesh, Pakistan, Thailand, and the Philippines. International collaboration has always been a key feature of various mungbean improvement programs in Asia, and the World Vegetable Center has played an important role to connect national programs, share germplasm, and test advanced breeding lines in different countries. In the following, we describe the World Vegetable Center program with reference to the national programs.

Traditional mungbean varieties reached maturity in 90–110 days, were indeterminate, shattered pods, were susceptible to disease, were low-yielding (about 400 kg/ha), and had small grains (Shanmugasundaram et al. 2009). From 1972 to 1981, the WorldVeg breeding program therefore focused on the development of varieties with resistance to abiotic and biotic stresses, high yield, and earliness to fit mungbean into the cereal cropping systems of Asia. Early success was achieved in the early 1980s when researchers crossed climate-resilient and disease-resistant (powdery mildew and *Cercospora* leaf spot) lines from India with high yielding, early maturing, and uniform maturing lines from the Philippines. These crosses were tested in international mungbean trials, and two lines were identified as particularly promising (VC1973A and VC2778A), which were released as KPS1 and KPS2 in Thailand and became dominant mungbean varieties in Thailand, China (released as Zhong Lu #1 and E Lu #2, respectively), and several other countries (Shanmugasundaram et al. 2009).

From 1981 to 1996, the focus of the program expanded to include mungbean yellow mosaic disease (MYMD) resistance, which had become the major constraint to mungbean production in large parts of South Asia and Myanmar. A shuttle breeding program between Thailand and Pakistan was financially supported by the UK government. The Nuclear Institute for Agriculture and Biology (NIAB) in Pakistan was the first to report MYMD resistance, which they had obtained through the irradiation (mutation breeding) of a local variety. NIAB and the World Vegetable Center collaborated to cross these resistant lines with KPS1, which after several generations led to two advanced MYMD-resistant lines: NM92 ("NIAB Mungbean 1992") and NM94 (Ali et al. 1997). These lines were then introduced to other countries in South Asia and promoted to farmers (Shanmugasundaram et al. 2009).

In 2010, the WorldVeg breeding program was relocated from Thailand to India to be able to screen more effectively for MYMD resistance. The establishment of the International Mungbean Improvement Network (IMIN) in 2016, supported by the Australian Centre for International Agricultural Research (ACIAR), strengthened partnerships between WorldVeg and the NARS of India, Bangladesh, Myanmar, and Australia. The program now is focused on exploiting the full potential of the available mungbean genetic resources by analyzing diversity and creating mungbean core and mini-core collections for breeding (Nair et al. 2012; Schafleitner et al. 2015). This IMIN has meanwhile expanded to include public and private sector partners in other mungbean-producing countries in Africa and Asia.

1.5 Conclusion

Mungbean is produced on at least 7.2 million ha globally. The main producing countries are India, Myanmar, China, Indonesia, Thailand, and Kenya. The mungbean market is segmented into a grain segment (South India, Myanmar, and Africa), a noodle segment (mostly China), a bean sprout segment (East and Southeast Asia), and a paste segment (East and Southeast Asia). There are different varieties for sprout production and grain production. The World Vegetable Center has played a key role in mungbean improvement by coordinating germplasm exchange between countries in Asia, conserving mungbean germplasm, and together with national partners developing improved mungbean varieties with high yield, short duration, and disease resistance.

References

Abd El-Salam MS, El-Metwally IM, Abd El-Ghany HM, Hozayn M (2013) Potentiality of using mung bean as a summer legume forage crop under Egyptian condition. J Appl Sci Res 9(2):1238–1243

Ali M, Malik IA, Sabir HM, Ahmad B (1997) The mungbean green revolution in Pakistan. Technical Bulletin No. 24, AVRDC. Shanhua, Taiwan, p 66

AMA (2014) Australian mungbean industry strategic plan 2015–2019. Queensland, Australian Mungbean Association

Cao D, Li H, Yi J, Zhang J, Che H, Cao J et al (2011) Antioxidant properties of the mungbean flavonoids on alleviating heat stress. PLOS One 6:e21071

Chen WL (2003) Republic of China. In: Shanmugasundaram S (ed) Processing and utilization of legumes. Report of the APO seminar on processing and utilization of legumes Japan, Chapter 2. Asian Productivity Organization, Tokyo, Japan, 9–14 Oct 2000

Dahiya PK, Linnemann AR, Van Boekel MAJS, Khetarpaul N, Grewal RB, Nout MJR (2015) Mung bean: technological and nutritional potential. Crit Rev Food Sci Nutr 55:670–688

Haque M, Miah MM, Ali A, Luna A (2014) Adoption of mungbean technologies and technical efficiency of mungbean (*Vigna radiata*) farmers in selected areas of Bangladesh. Bangladesh J Agric Res 39:113–125

Kim DK, Jeong SC, Gorinstein S, Chon SU (2012) Total polyphenols, antioxidant and antiproliferative activities of different extracts in mungbean seeds and sprouts. Plant Foods Hum Nutr 67:71

Lee SH (2003) Republic of Korea. In: Shanmugasundaram S (ed) Processing and utilization of legumes. Report of the APO seminar on processing and utilization of legumes Japan, Chapter 7. Asian Productivity Organization, Tokyo, Japan, 9–14 Oct 2000

Ministry of Agriculture, Livestock and Irrigation (2016) Myanmar agriculture at a glance 2016. Department of Planning, Ministry of Agriculture, Livestock and Irrigation, Nay Pyi Taw

Nair RM, Giri RR, Boddepalli VN, Prasad KVSV, Devulapalli R, Blümmel M (2019) Variation in grain yield, fodder quality and animal intake in two dual purpose legume crops: mungbean and vegetable soybean grown in semi-arid tropical India. Legume Res. https://doi.org/10.18805/LR-4078

Nair RM, Schafleitner R, Kenyon L, Srinivasan R, Easdown W, Ebert AW (2012) Genetic improvement of mungbean. SABRAO J Breed Genet 44:177–190

Pataczek L, Zahir Z, Ahmad M, Rani S, Nair R, Schafleitner R, Cadisch G, Hilger T (2018) Beans with benefits—the role of mungbean (*Vigna radiata*) in a changing environment. Am J Plant Sci 9:1577–1600. https://doi.org/10.4236/ajps.2018.97115

Rani S, Schreinemachers P, Kuziyev B (2018) Mungbean as a catch crop for dryland systems in Pakistan and Uzbekistan: a situational analysis. Cogent Food Agric 4:1499241

Schafleitner R, Nair RM, Rathore A, Wang YW, Lin CY, Chu SH et al (2015) The AVRDC—the world vegetable center mungbean (*Vigna radiata*) core and mini core collections. BMC Genom 16:344

Schreinemachers P, Ebert A, Wu MH (2014) Costing the ex situ conservation of plant genetic resources at AVRDC—the world vegetable center. Genetic Res Crop Evol 61:757–773

Schreinemachers P, Sequeros T, Rani S, Rashid MA, Gowdru NV, Rahman MS, Ahmed MR, Nair RM (2019) Counting the beans: quantifying the adoption of improved mungbean varieties in South Asia and Myanmar. Food Secur 11(3):623–634

Shanmugasundaram S (2003) Present situation and economic importance of legume in Asia and Pacific region. In: Processing and utilization of legumes. Report of the APO seminar on processing and utilization of legumes Japan. Asian Productivity Organization, Tokyo, Japan, pp 17–52, 9–14 Oct 2000

Shanmugasundaram S, Keatinge JDH, d'Arros Hughes J (2009) The mungbean transformation. Diversifying crops, defeating malnutrition. In: Spielman DJ, Pandya-Lorch R (eds) Millions fed: proven successes in agricultural development. International Food Policy Research Institute, Washington, DC, pp 103–108

USDA (2009) China—peoples republic of, edible bean annual. Global Agricultural Information Network (GAIN), GAIN Report Number: CH9088. USDA Foreign Agricultural Service, Washington, DC

USDA (2014) Thailand—grain and feed annual. Global Agricultural Information Network (GAIN), GAIN Report Number: TH4021. USDA Foreign Agricultural Service, Washington, DC

Mungbean Genetic Resources and Utilization

2

Gayacharan, Sunil Archak, Kavita Gupta, Veena Gupta, Vandana Tyagi and Kuldeep Singh

Abstract

Mungbean also known as green gram is an important food legume crop. It is the most widely grown crop among five cultivated Asiatic Vignas, viz. mungbean, urdbean, mothbean, adzukibean, and ricebean. The crop is expanding to non-traditional growing areas mainly due to its short duration of life cycle, high nutritional value, low input requirement, soil ameliorating properties, and high global demand. The crop fits well in cereal crop-based cropping systems in warm humid regions of South, East, and Southeast Asia. Mungbean cultivation covers over six million ha of land worldwide with annual production of around three million tons of grains. Narrow genetic base, disease pest proneness, and photo–thermo-sensitivity are the major problems resulting in the poor yield of the crop. The utilization of very few parental genotypes in mungbean breeding programs has led to the narrowed genetic base of the mungbean varieties. This has posed a serious threat to the mungbean cultivation amid newly emerging pests and pathogens of the crop. Worldwide over 43,000 mungbean germplasm are conserved ex situ which are playing crucial role in enhancing yield as well as resilience to the crop against biotic and abiotic stresses. Recent evaluation and characterization of ex situ collections of mungbean germplasm revealed substantial amount of useful variability in mungbean. Transboundary movement of germplasm has also helped in the development and release of varieties in several countries. Enhanced and efficient utilization of ex situ conserved mungbean and related wild species germplasm in breeding programs with the help of modern genomics tools would help in the development of desired genotypes with higher yield potential. In this chapter, we have discussed the collection, conservation, and utilization of mungbean and wild Vigna species germplasm.

Gayacharan · S. Archak · K. Gupta · V. Gupta · V. Tyagi · K. Singh (✉)
ICAR-National Bureau of Plant Genetic Resources, New Delhi, India
e-mail: kuldeep.singh4@icar.gov.in; director.nbpgr@icar.gov.in

2.1 Introduction

Mungbean is native of Indo-Burma region of Asia where it was first probably domesticated, believed to have originated in the Indian gene center (Jain and Mehra 1978). The wild progenitor species of munglocan has been designated as *V. radiata* var. *sublobata* and are of Indian origin (Chandel et al. 1984). The wild *V. radiata* var. *sublobata* occurs in Tarai region, sub-Himalayan tract, and sporadically in western and eastern peninsular tracts of India (Arora and Nayar 1984). India is also the primary center of diversity for mungbean (Arora 1988). Mungbean diversity is well dispersed throughout the continent from Himalayas in north

R. M. Nair et al. (eds.), *The Mungbean Genome*, Compendium of Plant Genomes,
https://doi.org/10.1007/978-3-030-20008-4_2

to southern peninsular and northeastern region. The Indo-gangetic plains are considered as secondary center of diversity for mungbean (Bisht et al. 1998a). In early days, mungbean seed was carried by emigrants and traders from Asia to the Middle East, East Africa, Latin America, parts of South America and Australia (Poehlman 1991). It is the most widely distributed among the Asiatic *Vigna* species, and its production is steadily increasing (Kim et al. 2015). Currently, the crop is cultivated throughout the South and Southeast Asia, including India, Pakistan, Bangladesh, Sri Lanka, Myanmar, Thailand, Philippines, Laos, Cambodia, Vietnam, Indonesia, Malaysia, South China, and Taiwan. In the USA, it was grown as early as 1835 as the Chickasaw pea. It is also grown to a lesser extent in many parts of Africa and USA (Oklahoma) and reintroduced in many parts of Australia. However, it did not become a major commercial crop in these countries. In India, the crop is mainly grown in states of Rajasthan, Maharashtra, Gujarat, Odisha, Bihar, Andhra Pradesh, and Madhya Pradesh.

Vigna species including wild relatives flourish in hot humid weather of subtropical to tropical regions. Mungbean is fast-growing crop and completes its life cycle from 50 to 90 days. It requires rainfall of 600–1000 mm/year. Optimal temperature required for vegetative growth ranges from 28 to 30 °C, however, some related wild *Vigna* species like *V. umbellata*, *V. angularis*, *V. trilobata*, etc. can sustain temperature few degrees cooler than the mungbean. The available breeders' varieties or landraces cannot sustain below 15 or above 45 °C of temperature. The crop is highly sensitive to waterlogging conditions and tolerant to saline soils to some extent. The crop can be grown in well-drained loamy to sandy loamy soils with a pH range of 5–8.

2.1.1 Taxonomic Classification

Mungbean belongs to the family Fabaceae, subgenus *Ceratotropis* in the genus *Vigna* Savi. Earlier mungbean was known as *Phaseolus aureus* before many species of the genus moved to a new genus *Vigna* (Lambrides and Godwin 2007). The genus *Vigna* consists of large group of cultivated crops and wild relatives distributed in Asia and Africa. It comprised of around seven subgenera and 19 sections with around hundred species (Maxted et al. 2004; Singh et al. 2006; Takahashi et al. 2016) out of which seven (She et al. 2015) species are most commonly cultivated around the world. The two species are African originated (*Vigna unguiculata* L. and *Vigna subterranea* L.) and other five species are from Asiatic group known to be originated in Indian subcontinent (*V. radiata* L., *V. mungo* L., and *V. aconitifolia* Jacq.) and in the Far East Asia (*V. angularis* Willd. and *V. umbellata* (Vavilov 1926; Smartt 1985). The species *V. radiata* has four direct subspecies, i.e., *V. radiata* subsp. *radiata, V. radiata* subsp. *sublobata* (L.) R. Wilczek, *V. radiata* var. *radiata* (L.) R. Wilczek, and *Vigna radiata* var. *setulosa* (Dalzell) Ohwi & H. Ohashi.

2.1.2 Mungbean Gene Pool

Cross compatibility among *Vigna* species is not so well defined, and so their gene pool, but in general there is no cross compatibility barrier between domesticated forms and their closest relatives (Tomooka et al. 2014). There are few studies on such wide hybridization for widening genetic base of *Vigna radiata* using *Vigna mungo* (Gosal and Bajaj 1983), *Vigna umbellata* (Pandian et al. 2008), *V. trilobata* (Pandiyan et al. 2012), and interspecific barriers could be easily overcome. Though interspecific crosses of *Vigna radiata* with other *Vigna* species like *V. mungo, V. radiata* var. *sublobata, V. radiata* var. *setulosa, V. trilobata, V. trinervia, V. hainiana, V. dalzelliana* are possible, cross-barrier problems particularly due to incompatibility in chromosomal pairing also have been observed in some cases (Pandian et al. 2008). Tomooka et al. (2011) have classified gene pool of mungbean as *V. radiata* as *V. radiata* var. *radiata* and its closest wild relative, i.e., *V. radiata* var. *sublobata* in gene pool 1 (GP-1). *V. mungo, V. subramaniana, V. grandiflora, V. stipulacea, V. tenuicaulis, V. trinervia,* and *V. umbellata* in GP-2, and other species in sections *V. aconitifolia* and *V.*

angularis in GP-3. However, some research groups have deviations from this gene pool classification like *V. radiata* var. *setulosa* which is a wild form placed in GP-1, *V. aconitifolia* in GP-2, and *V. umbellata* in GP-3 (Pratap et al. 2014a). From recent hybridization work, it is observed that *V. glabrescens* gives fertile progenies when crossed with *V. radiata* without any crossing barriers, though the species is grouped in GP-3. This indicates that there is a need to revisit *Vigna* gene pool classification and molecular tools along with conventional crossing-based studies will give robust grouping of *Vigna* species.

2.2 Mungbean Germplasm Collections and Introductions

2.2.1 Collections

Ex situ conservation of plant genetic resources (PGRs), harboring a wide range of diversity, is important to develop new varieties to face the challenges posed by adverse effects of climate change and to meet the food security aspects of increasing population worldwide. Grain legumes are among the topmost crops, which can help in addressing both these issues effectively. Importance of conservation of plant genetic resources for food and agriculture (PGRFA) was realized since the Neolithic era when human started practicing farming, seed selection, and seed storage. However, the systemic explorations, collections, and conservation of crop germplasm started in 1916 by renowned Russian geneticist N. I. Vavilov, and today over 7.4 million accessions are conserved ex situ worldwide (https://cdn.croptrust.org). The N. I. Vavilov Research Institute of Plant Industry (VIR), St. Petersburg, Russia, is one of the largest genebanks which holds over 230,000 accessions of various crops including 863 accessions of mungbean. These mungbean collections are the oldest collections, and most of them were collected between the years 1910 and 1927.

In India, mungbean germplasm collections started way back in 1925 to collect mungbean landraces from all over India and Burma (Bose 1932). During that period, mungbean germplasm collection efforts were made in several states of the country, but there was no coordination or exchange of the germplasm among them (Bisht et al. 1998b). Extensive and organized explorations were taken up throughout the Indian sub-continent only after the establishment of ICAR-National Bureau of Plant Genetic Resources (NBPGR) in 1976 (Rana et al. 2016). NBPGR is a nodal agency for germplasm introductions, collections, and conservation in India. The National Genebank at ICAR-NBPGR has documented more than 11,000 accessions of mungbean including 7,453 indigenous accessions and 3,588 exotic accessions (Figs. 2.1 and 2.2). Of these at present 3,927 accessions comprising of 3,392 indigenous collections from 28 states of India and 535 exotic collections from 12 countries across the world are conserved in long-term storage conditions (−20 °C). Maximum of the diversity collected and conserved is from Rajasthan and Gujarat, whereas diversity from northeastern states is least explored (Fig. 2.3). Fifty-six out of 108 mungbean breeders' varieties developed through various crop improvement programs on mungbean are also part of this collection (Fig. 2.4). To give due credit to breeders' germplasm other than the varietal genotypes, ICAR-NBPGR is also providing soft protection through Plant Germplasm Registration Committee (PGRC) under the aegis of ICAR. Till date, thirteen germplasm have been registered in PGRC (Fig. 2.4, Table 2.1) with novel unique traits.

The importance of mungbean as a crop was also realized worldwide and AVRDC-The World Vegetable Center, Taiwan, took the lead in collection, conservation, exchange, and varietal improvement programs at international level. The center maintains the second largest mungbean germplasm collections (6,700 accessions, Schafleitner et al. 2015). The Southern Regional Plant Introduction Station, University of Georgia, under USDA's National Plant Germplasm System conserves around 3,928 accessions of mungbean (https://www.ars-grin.gov/). 2,250 accessions are ex situ conserved by Field Crops Research Center, Department of Agriculture, Bang Khen, Thailand (www.fao.org/docrep/013/i1500e/Thailand.pdf). 1,076 acccssions are ex

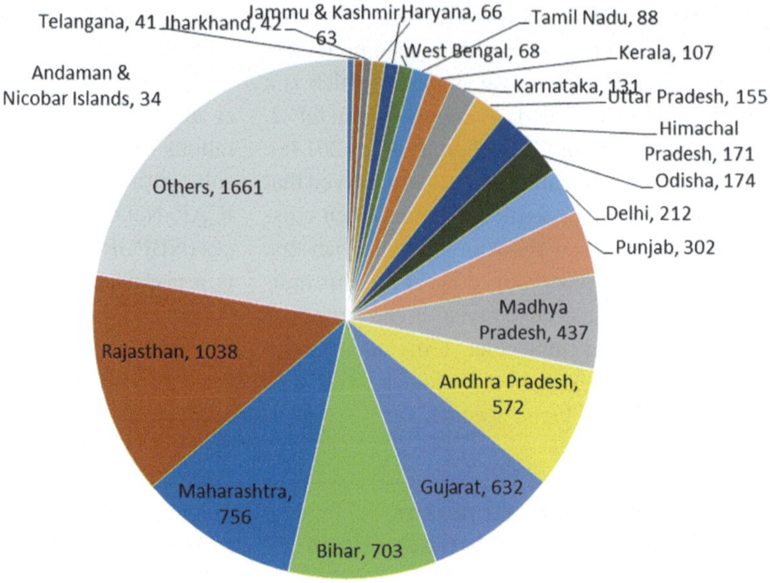

Fig. 2.1 State-wise distribution of indigenous collections documented at National Genebank, ICAR-NBPGR

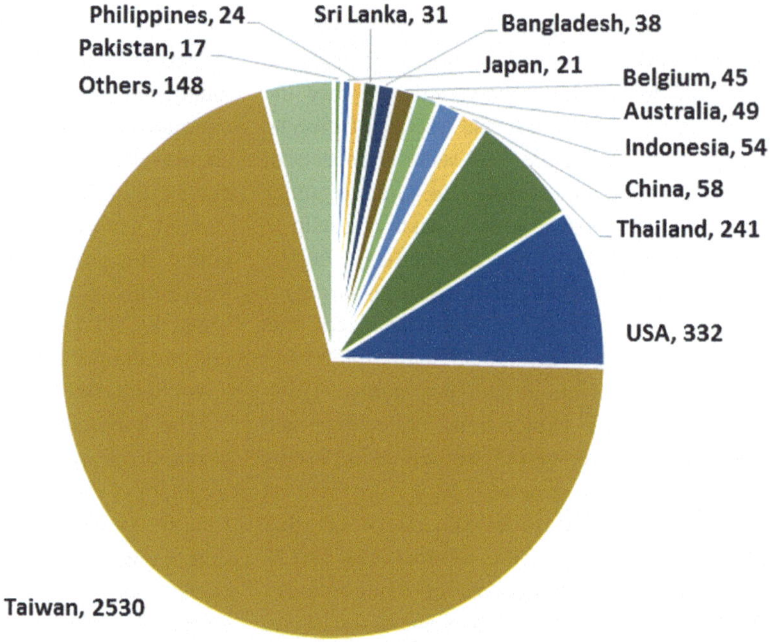

Fig. 2.2 Country-wise distribution of exotic collections documented at National Genebank, ICAR-NBPGR

situ conserved by The Genetic Resources Center, NARO, Tsukuba, Japan (https://www.gene.affrc.go.jp/databases-plant_search_en.php). Around 1,006 accessions of mungbean are conserved in the genebank of Department for Plant Genetic Resources, Austrian Agency for Health and Food Safety (AGES), Austria (https://www.genbank.at/en/ecpgr-vigna.html). The Australian Grains Genebank (AGG), Horsham, Victoria, conserves 1,385 accessions of mungbean germplasm. The national genebank managed by Institute of Crop Germplasm Resources of the Chinese Academy of Agricultural Sciences (CAAS) is another major custodian of mungbean germplasm.

Fig. 2.3 Highlights of phenotypic variation present in mungbean gene pool. **a**. Field view, **b**. seed coat color variability, **c**. variability for pod length and pod thickness (each pod contains 14–15 seeds), **d**. a lodging resistant and around 120 cm tall accession (IC553601), **e**. leaf size variation

Worldwide over 43,000 mungbean germplasm are conserved ex situ (Nair et al. 2013). Over 9,000 mungbean seed samples are kept under black box system in Svalbard Global Seed Vault as safety duplicates by thirteen different national and international genebanks (https://www.nordgen.org/sgsv/index.php?).

2.2.2 Introductions

In India, there is a single-window system for the exchange of small samples of plant germplasm (including genetically modified crops) meant for research, and ICAR-NBPGR is the authorized nodal institution. It regulates the import of seeds/planting material for research under the provisions of Plant Quarantine (Regulation of Import into India) Order, 2003, of the Destructive Insects and Pests (DIP) Act of 1914. The plant introductions include germplasm, elite strains, improved varieties, genetic stocks, and related species from various parts of the world.

Introductions are done time to time based on the breeders or researchers' requirement. Nearly four thousand diverse accessions of mungbean were introduced in India from various countries like Australia, Bangladesh, Bhutan, Brazil, Canada, China, Czech Republic, Ghana, Germany, Ethiopia, Indonesia, Italy, Japan, Madagascar, Malaysia, Nepal, Netherlands, Philippines, Russia, Sri Lanka, Surinam, Sweden, Taiwan, Thailand, Uruguay, USA, and Uzbekistan since 1976. Some of the important introductions having highly desirable traits for the crop improvement are mentioned here. EC118889, EC118894, EC118895, and EC118898 were procured for traits like high grain yield potential, wide adaptability, synchronous and early maturity, large seeded and having resistance to *Tobacco mosaic virus*, lodging. Mungbean germplasm lines, viz. EC318985 to 319057, EC390990–EC390993, were introduced from AVRDC, Taiwan, having useful traits like resistance to charcoal rot, leaf crinkle virus, tolerance to drought and flood and photo-insensitive. Four accessions EC393407–EC393410 were introduced from Bangladesh having important traits like long pods with large, shiny green seed, and synchronous maturity. There

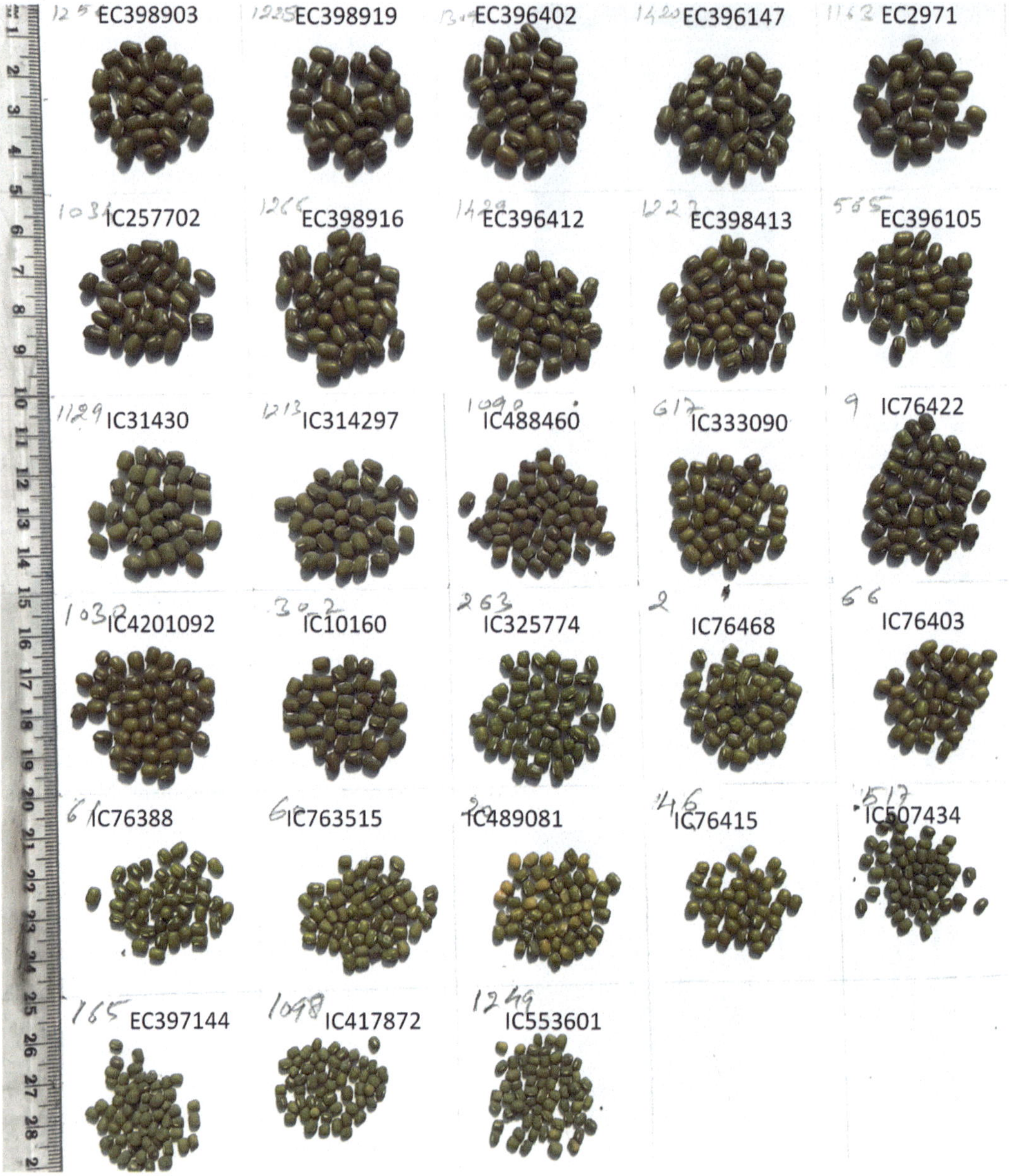

Fig. 2.4 Few representative accessions of mungbean showing seed size and seed coat luster variability

were several other mungbean introductions like six accessions viz. EC391170–EC391175 from Indonesia for high grain yield trait. Similarly, other notable introductions made are EC396424–EC396425, EC718740–EC718745 for important traits like early maturity, resistant to mungbean yellow mosaic disease (MYMD), *pea yellow mosaic virus* and bold seed. Preferred traits for introduction in mungbean are short duration, bold seed (>6 g), photo–thermo-insensitivity, bruchids resistance, pre-harvest sprouting tolerance, resistance against powdery mildew, and *Cercospora* leaf spot disease. Many of the introductions were further utilized in mungbean varietal development programs, few of which are mentioned in Table 2.2.

Table 2.1 Trait-specific germplasm identified in mungbean

Trait	Accessions ID
Cercospora leaf spot resistance	EC118895, EC124083A, EC124084A, EC124089A, EC245968
Powdery mildew resistance	EC118898, EC155745, EC155747, EC318985, V4718
Mottle virus resistance	EC124098, EC124104, EC124111, EC245968
Root rot disease resistance	EC124105
Charcoal rot resistance	EC245968, EC319008
Leaf crinkle virus resistance	EC319011
Mungbean yellow mosaic disease (MYMD) resistance	IC118998, K-1, EC397139, IC305286, IC305291, IC305284, IC472115, IC364130, EC319013, EC501570, EC565626, NM-92, NM-94, **IC573451, IC573453, IC573454, IC573455, IC573456**
Nematode resistance	IC212049
Bruchid (*Callosobruchus chinensis* and *C. maculatus*) resistance	EC0398897 (V2802), EC0398896 (V2709)
Extra early flowering (25–27 days)	EC398944, IC073537, IC507478, EC398883, EC398880, EC398890, IC076477, IC507476_2, IC314609_2, EC398953, IC073392, EC396129, EC396423, IC076422, IC119106, EC398890, IC39335, IC39332
Early maturity (48–50 days)	IC008822-3B, IC314302, IC076422, IC119106, IC314609_1, IC447908, EC398944, EC592177, IC314562, EC398901, EC398955, EC501566, **IC0589309, IC0589310**
Synchronous flowering and maturity (5–8 days of flowering period)	EC396115, IC076414, EC397140_2, IC488968, IC076422, IC507389, IC488582, IC118970, EC398923, IC076601_1, EC397138, IC314437, IC076378, EC398915, IC076417_1, EC397141, IC314691, IC076463, EC396143, IC076370, IC507524_1
High pod length (12–16 cm)	EC398937, EC398904, EC398887, EC398935, EC398902, EC398936, IC418452, IC148442, VI003370
Pentafoliate leaf small leaflets	**IC296679**
Trilobed leaflets	IC76451
Purple stem, petiole and leaf veins	IC19420, EC9129
High number of seeds per pod ($\geq$ 14.5–17.1)	IC507314, IC418452, IC507342_2, IC548274, IC488554_1, IC148415, IC252010_1, IC076388, IC076460, IC507337, VI004979, VI002090
100 seed weight (7.5–7.89 g)	EC398923, EC398903, EC398884, EC396413, EC396116, EC398919, EC396154, IC507459, EC398882, **IC0418452** (9.43), **IC296771** (6.2 g), VI005034, VI001244, VI000946
Lodging resistance and plant height	IC553601 ($\sim$ 120 cm), VI014178, VI005030
High protein content	EC251557, PLM-350, **IC296771** (27.8%), **IC573456** (25.8%)
High iron content (mg/100 g)	**IC573449** (9.18), **IC573450** (11.7), **IC573451** (11.8), **IC573454** (10.59), **IC573455** (11.35), **IC573456** (8.29)
High zinc content (mg/100 g)	**IC573453** (3.56), **IC573456** (4.19)
Photosensitive nature	**IC0546478**

Note Accessions in bold are registered germplasm at ICAR-NBPGR, New Delhi; prefix EC means exotic collection and IC means indigenous collection. Accessions starting with prefix 'V' or 'VI' belong to World Vegetable Center collections

Table 2.2 Utilization of exotic germplasm for pulse improvement

Accession	Country of origin/source	Varieties released in India
China moong	China	Shining Moong 1, Sunaiana, RMG62, Jalgaon 781, DGGV-2
NM9473	Pakistan/AVRDC, Taiwan	Pusa 9531
NM92	Pakistan/AVRDC, Taiwan	Pusa Vishal
NM94	Pakistan/AVRDC, Taiwan	SML688
V2164	AVRDC, Taiwan	SML134
V3484	AVRDC, Taiwan	Pusa101, WGG2
VC1137-213 (M 178)	AVRDC, Taiwan	Pusa105
CES44	Philippines	AAU34
MG50-10	Philippines	Co5, Co6
VC6368 (ML 26)	AVRDC, Taiwan	Pant mung 2, Pant mung 5
VC6367(44-55-2)	Thailand	IPM 410-3 (Shikha)
Iranian germplasm	Iran	PS 16
VC 6368 (46-40-4)	AVRDC, Taiwan	UPM 98
VC 6370 (30-65)	AVRDC, Taiwan	UPM 98-1

2.3 Process to Access Genetic Resources

Plant germplasm exchange process and related regulatory mechanisms are the same for each crop. Exchange of plant germplasm in India is regulated by Biological Diversity Act (BDA), 2002, in conjuncture with Plant Quarantine Order 2003 (PQO-2003). BDA defines that who and how any non-Indian can access the germplasm from India while Plant Quarantine Order ensures that the exchanged germplasm is free of pest, pathogen, and weed (of quarantine importance). The import/introduction of plant germplasm into India is governed by the PQO-2003 (Regulation of Import into India). Director, ICAR-NBPGR, has been authorized to issue import permit for import of germplasm, transgenic or genetically modified organisms for research purposes for further distribution to the researchers in the country. For commercial and bulk import, permission is granted based on the recommendations of EXIM Committee of Department of Agriculture, Cooperation and Farmers Welfare, Government of India.

BDA, 2002, was enacted in compliance to Convention on Biological Diversity (CBD). As per the provisions of the Act, a non-Indian can access any biological resource occurring in India with the prior approval of National Biodiversity Authority (NBA). Section 2.3 (2) of the BDA, 2002, defines the non-Indian entity. However, access under Collaborative Research Project which is compliant to the International Collaborative Guidelines issued by the Ministry of Environment, Forest and Climate Change (MoEFCC), is exempted (provisions in Sect. 2.5). For accessing germplasm resources which are not covered under any collaborative research project, the indentor can apply online at http://absefiling.nic.in/NBA/login/auth.

The germplasm exchange (import/export) is being carried out with about thirty countries and international agricultural research centers for augmenting diverse genetic resources and making them available to researchers/breeders/users for utilization in various research programs.

ICAR-NBPGR carries out this important activity under an established procedure.

2.4 Mungbean Germplasm Characterization and Evaluation

Characterization and evaluation are the first and foremost important activity for the utilization of germplasm in crop improvement programs. Characterization is the description and establishment of diagnostic characters of a plant germplasm. Generally, characterization traits are highly inheritable and qualitative in nature. Evaluation is the description of plant germplasm expression in optimum or adverse environmental conditions to reveal its potential useful variability available within the germplasm. Evaluation traits are quantitative in nature but are highly important for crop improvement. Major objective of the evaluation is to identify trait-specific germplasm of breeders/researchers' importance. In general, characterization and preliminary evaluation (for yield and yield-attributing traits) are done simultaneously while evaluation for biotic and abiotic stresses is done separately.

Like in any other genebank, characterization and evaluation are the main activity of NBPGR. Total 1,532 accessions were well characterized using 19 qualitative and 19 quantitative traits during 1993 (Kawalkar et al. 1996). During 2016–18, further another around 1,500 indigenous and exotic collections were characterized and evaluated for 27 agro-morphological traits and biotic and abiotic stresses like MYMD disease and heat stress. Based on these morphological observations and statistical analyses, it was found that cultivated mungbean gene pool harbors very good amount of variation for various economically important traits like seed weight, flowering period, pod length, number of seeds/pod, and seed size (Table 2.3, Fig. 2.4). For qualitative traits, very good level of phenotypic variations is observed. Range of variation for phenotypic traits reported in mungbean germplasm in national and international studies is highlighted in Table 2.3. Recently, World Vegetable Centre has characterized global mungbean collections of 5,234 accessions for eight agro-morphological traits; viz. primary leaf length, primary leaf width, plant height at flowering, plant height at maturity, days to 50% flowering, pod length, seeds per pod, and 1,000 seed weight were characterized (Schafleitner et al. 2015). It was observed that good amount of phenotypic variability is present for certain traits in global mungbean collections in terms of Shanon's diversity index which was 0.82 (average of all traits). Overall, variability for root nodulation, petiole length, number of primary branches, number of pod bearing peduncles, flowering period, plant height and seed weight, seed coat color, and flowering period was observed comparatively higher than the rest of the traits studied (Bisht et al. 1998a; Schafleitner et al. 2015).

Several other research groups also have characterized mungbean germplasm for understanding genetic variability, genetic divergence, and trait association (Abna et al. 2012; Bisht et al. 1998b; Chattopadhyay et al. 2008; Hakim 2016; Rahim et al. 2010; Sangiri et al. 2008; Singh et al. 2014; Tantasawat et al. 2010; Yimram et al. 2009). Simultaneously, mungbean germplasm has also been evaluated against important biotic and abiotic stresses like bruchids (Somta et al. 2007; Somta et al. 2008), nematodes (Mukhtar et al. 2017), MYMD (Akhtar et al. 2011; Ghafoor et al. 1992; Iqbal et al. 2011; Jalaluddin and Shaikh 1981; Shad et al. 2006; Sudha et al. 2013a, b), *Cercospora* leaf spot, and powdery mildew (Chankaew et al. 2011; Raje and Rao 2002; Reddy et al. 1987).

The characterization and evaluation studies done in mungbean have resulted in the identification of several trait-specific mungbean germplasm, some of which are listed in Table 2.1.

2.5 Development of Mungbean Core and Mini-Core

Core and mini-core are the output of characterization and evaluation of a large set of ex situ collections of any crop germplasm. The core

Table 2.3 Phenotypic variation available in cultivated mungbean gene pool

Trait	Range	Phenotypic CV (%)
Primary leaf length (cm) #	3.3–6.4	10.64
Primary leaf width (cm) #	1.1–2.8	13.38
Terminal leaf length (cm)	5.0–13.6	20.9
Terminal leaf width (cm)	1.5–12.9	21.06
Petiole length (cm)	1.0–25.0	38.34
No. of primary branches	1.0–9.0	34.37
Nodulation	0.0–98	76.39
First pod bearing node	2.0–8.0	7.62
No. of pod bearing peduncle	2.0–36.0	52.61
Plant height at flowering (cm) #	9.0–68.0	28.21
Plant height at maturity (cm) #	12.0–94.0	30.14
Days to initial flowering	24.0–80.0	15.01
Days to 50% flowering #	39.0–73.0	8.79
Days to 50% flowering	25.0–90.0	14.50
Flowering period	5.00–50.00	32.92
Days to initial maturity	37.0–90.0	11.13
Days to 80% maturity	48.0–110.0	11.16
Pod length (cm)	5.0–16.0	14.95
Pod length (cm) #	5.0–17.1	13.75
Number of seeds per pod	5.0–17.0	13.12
Number of seeds per pod #	5.8–15.4	11.29
100 seed weight (g)	1.58–7.89	30.18
100 seed weight (g) #	2.027–7.58	24.6

Source Catalogs of NBPGR and recently mungbean characterization unpublished data
#Mungbean characterization data from AVRDC-The World Vegetable Centre (Schafleitner et al. 2015)

concept was first given by Frankel (1984) with the objective of enhancing germplasm utilization efficiency particularly in crops with large number of ex situ collections where the selection of suitable germplasm for breeding and research becomes a difficult task. Till date, two cores have been developed in mungbean. First mungbean core was developed by NBPGR utilizing its 1,532 ex situ collections available by that time in National Genebank (Bisht et al. 1998a). A subset of 152 accessions was developed based on 38 agro-morphological descriptors. AVRDC-The World Vegetable Center, Taiwan, has developed another mungbean core of 1,481 accessions from 5,234 global mungbean collections conserved in its genebank (Schafleitmer et al. 2015). This core was developed by random selection of 20% of the accessions after geographical stratification and cluster analysis of eight phenotypic traits. Further to bring down this number to a more workable size, entire 1,481 core accessions were genotyped using 20 simple sequence repeats (SSRs) and a mini-core set of 296 accessions was developed.

2.6 Mungbean Varietal Development Programs

Germplasm evaluation and characterization have significantly contributed to the crop improvement and varietal release through direct selection or as donor in biparental crossing programs. At the inception of mungbean varietal development programs, mungbean varieties were mainly released by selection and purification of landraces (Singh et al. 2016). However, Indian National Agricultural Research System (NARS) lead by All India Coordinated Research Project (AICRP) took systematic efforts and several varieties were developed through extensive nationwide crossing programs. This resulted in significant yield enhancement despite the spread of several devastating diseases like MYMD, leaf crinkle, powdery mildew, etc. As a result, till 2018 total 108 mungbean varieties have been released in India among which 52 are released by Central Variety Release Committee (CVRC) and 56 are released by State Variety Release Committees (SVRCs) (Project Coordinator, AICRP on MULLaRP report, 2018). Parallel to Indian mungbean improvement programs, there are other countries like Australia, Bangladesh, Myanmar, Pakistan, Philippines, Taiwan, China, Thailand, etc., where significant work has been done in mungbean. AVRDC-The World Vegetable Centre, Taiwan, is heading international mungbean improvement programs. Several varieties have been released worldwide from national programs or through International Mungbean Improvement Network program of AVRDC (Shanmugasundaram et al. 2009).

Traditional mungbean cultivars have multiple drawbacks like seed shattering, indeterminate growth habit, small seed, highly prone to diseases like mungbean yellow mosaic virus disease (MYMD), and long maturity period of usually 90–110 days (Shanmugasundaram et al. 2009). Due to these constraints, mungbean yield in traditional farming systems could not reach beyond 400 kg/ha. However, diverse germplasm utilization in breeding programs has led to the development of several mungbean varieties having yield potential over 2.0 tones/ha with other desirable traits. AICRP on MULLaRP has been very instrumental in improving this Indian origin crop in the country. Mungbean varieties for several traits like early and synchronous maturity, pod length, MYMD resistance, bold seed, storage quality, summer season and regional environment specific as well as varieties of wider adaptability, etc., have been released and are in process of development. Recently, a mungbean variety Virat (IPM 205-7) developed, from a cross between IPM2-1 and EC398889, matures within 52–55 days and gives average yield of 1–1.2 tones/ha. Similarly, there are other varieties like Shikha (IPM410-3), Kanika (IPM 302-2), and Varsha (IPM 2K14-9) which yield 1.2–1.4 tons/ha and have wider adaptability.

2.7 Mungbean Crop Wild Relatives (CWRs) and Their Utilization

Mungbean crop wild relatives are distributed throughout the Asian countries (Fig. 2.5). Indian gene center harbors several *Vigna* wild species like *V. radiata* var. *sublobata, V. umbellata, V. vexillata, V. aconitifolia, V. trilobata, V. stipulacea, V. dalzelliana, V. trinervia* var. *bourneae, V. mungo var. sylvestris, V. khandalensis, V. hainiana, V. pilosa, V. grandis, V. marina, V. capensis, V. angularis var. nipponensis,* etc. Over 2,000 collections of wild *Vigna* species are conserved ex situ in Indian National Genebank at ICAR-NBPGR, New Delhi (Table 2.4). Few of the *Vigna* species like *V. aconitifolia* and *V. umbellata* are found both as wild and cultivated forms and are under active cultivation in many Indian states. Apart from being a genetic resource, *Vigna* CWRs play other important roles; they are used for various purposes like food grain (*V. marina, V. umbellata, V. stipulacea, V. trilobata*), green pods (*V. umbellata, V. minima, V. subterranea*), forage (*V. luteola*), green manure (*V. hosei, V. parkeri, V. stipulacea*), tubers (*V. adenatha, V. vexillata, V. ambacensis, V. angivensis, V. fischeri, V. monophylla, V. reticulata*), cover crop (*Vigna trinervia*), ornamental (*V. caracalla*), etc., and many times these have multiple uses (Tomooka et al. 2011).

Vigna wild relatives inhabit extreme and diverse environments like dry harsh (*V. aconitifolia* and *V. aridicola*), sandy saline seashore (*V. marina*), swampy marshland habitat (*V. luteola* and *V. adenantha*), limestone outcrops (*V. exilis*), high altitudes (*V. angularis* var. *nipponensis* and *Vigna nakashimae*) which shows that *Vigna* CWRs are goldmines for adaptive traits to

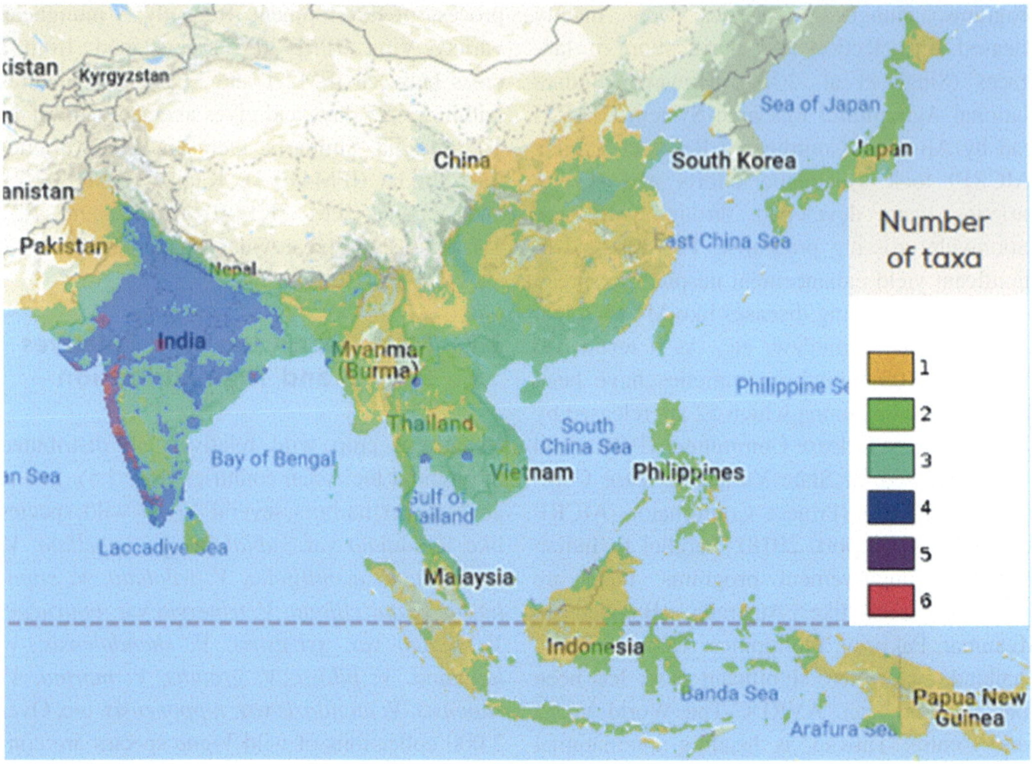

Fig. 2.5 Taxon richness of CWRs of *V. radiata*. *Source* https://www.cwrdiversity.org/

improve cultivated *Vigna* species against all major abiotic stresses. *Vigna* CWRs are also rich source of resistance for several biotic stresses like *V. umbellata,* and *V. radiata* var. *sublobata* used for breeding MYMD-resistant mungbean varieties (Pandiyan et al. 2008; Sudha et al. 2013a, b; Sudha et al. 2015; Pratap et al. 2014a). *V. vexillata* possess resistance for cowpea mottle carmovirus (Ogundiwin et al. 2002) and legume pod borer *Maruca vitrata* (Jackai et al. 1996). Powdery mildew resistance is available in *V. stipulacea* (Tomooka et al. 2006). Storage insect pest is a major factor for postharvest losses in pulses, and mungbean is highly susceptible to storage pests. However, mungbean CWRs like *V. umbellata, V. minima, V. vexillata, V. reticulata, V. oblongifolia, V. luteola, V. reflexo-pilosa, V. umbellata, V. minima,* and *V. radiata var. sublobata* are identified as resistant for major storage pests like cowpea weevil (*Callosobruchus maculatus*), adzuki bean weevil (*C. chinensis*), and storage weevil (*Callosobruchus analis*) (Fujii and Miyazaki 1987; Tomooka et al. 1992; Tomooka et al. 2000; Kashiwaba et al. 2003). *V. umbellata* and *V. glabrescens* are reported to possess photo-thermo–insensitivity (Pratap et al. 2014b) a very much important trait for expanding mungbean cultivation in non-traditional areas and crop rotations.

Recently, it was realized that mungbean varietal genetic base is very narrow as very limited variability is used in mungbean varietal development programs. The pedigree information of most of the breeders' varieties released indicate that varieties are being bred using breeders' varieties, and only few of them are released using new sources of germplasm. Evaluation of entire collections of mungbean germplasm at ICAR-NBPGR including global mungbean mini-core developed by World Vegetable Center (unpublished data) during *kharif* 2016 and 2018 indicated that only a few of the accessions were resistant to MYMD (the most devastating disease of mungbean) within cultivated gene pool of mungbean.

Table 2.4 *Vigna* species and their distribution in Indian subcontinent

Species	Indian collections	Distribution (Indian states)
V. aconitifolia	2,629	Rajasthan, Gujarat, Odisha, Haryana, Maharashtra
V. adenantha	1	Coastal areas along the banks of backwaters, Kerala
V. angularis var. nipponensis	9	Arunachal Pradesh, Mizoram, Nagaland
V. capensis	11	Eastern and Western Himalaya
V. dalzelliana	65	Karnataka, Kerala
V. hainiana	16	Madhya Pradesh, Himachal Pradesh, Maharashtra
V. indica	–	Gujarat, Karnataka, Madhya Pradesh, Maharashtra, and Rajasthan
V. khandalensis	12	Maharashtra
V. konkanensis	–	Maharashtra
V. marina	3	Andaman and Nicobar islands, Kerala
V. minima	6	Andhra Pradesh, Goa, Gujarat, Rajasthan
V. mungo var. silvestris	20	Goa, Karnataka, Kerala, Madhya Pradesh, Maharashtra, Odisha, Rajasthan, Tamil Nadu
V. nepalensis	3	Nagaland
V. pilosa	18	Kerala, Karnataka
V. radiata var. setulosa	11	Kerala, Maharashtra, West Bengal
V. radiata var. *sublobata*	282	Andhra Pradesh, Goa, Himachal Pradesh, Karnataka, Kerala, Madhya Pradesh, Maharashtra, Odisha, Rajasthan, Tamil Nadu
V. sahyadriana	–	Northern Western Ghats (Maharashtra)
V. stipulacea	6	Tamil Nadu, Kerala, Madhya Pradesh, Chhattisgarh, Maharashtra, Odisha, Rajasthan, Tamil Nadu
V. subramaniana	–	Himachal Pradesh, Kerala, Punjab
V. trilobata	182	Andhra Pradesh, Karnataka, Kerala, Rajasthan, Tamil Nadu
V. trinervia	5	Odisha
V. trinervia var. bourneae	30	Goa, Karnataka, Tamil Nadu
V. umbellata	2,933	Arunachal Pradesh, Karnataka, Assam, Meghalaya, Nagaland, Himachal Pradesh, Uttarakhand, Jharkhand, West Bengal
V. vexillata	115	Andhra Pradesh, Goa, Himachal Pradesh, Kerala, Maharashtra
V. wightii	4	Kerala

Information not available in National Genebank portal (http://pgrportal.nbpgr.ernet.in)

However, recent developments in mungbean genomics and use of wild *Vigna* CWRs in pre-breeding programs are good sign for the crop. Mungbean varieties are being bred utilizing cross-compatible *Vigna* CWRs. HUM 1, Pant Moong 4, IPM99-125, IPM 205–7, and IPM 4094 are few recently released varieties which were developed from *Vigna radiata* × *Vigna mungo* crosses (Pratap et al. 2014a). There are some other *Vigna* species like *V. umbellata*, *V. glabrescens*, *V. trilobata*, *V. radiata* var. *sublobata*, *V. mungo* var. *sylvestris*, etc., which are currently being used in mungbean varietal development programs, but till now only limited success could be achieved.

2.8 Mungbean Genomic Resources and Their Importance

Genomics tools and genomic resources have become very essential part of crop improvement and genetic resources management programs. Since the last decade, there has been tremendous development in genomics technologies and these tools are helping in identifications of genes/QTLs for all kind of traits, introgression of traits, shortening of breeding cycles, development of new and ideal plant types, development of new variations through utilizing alien species or through mutational approaches, etc. Similarly, the genomics is playing great role in plant germplasm management. For example, DNA fingerprinting and bar coding of varieties and other novel germplasm are using set of molecular markers which helps in avoiding misuse of breeders' and farmers' material. Genotyping of conserved genebank material helps in identifying duplicates and saves waste of resources in germplasm maintenance. Diversity analysis of on-farm germplasm across the regions or globes and comparison with the conserved genebank collections helps in making strategy for future collections. There are several such applications of genomics where it is playing complementary roles in crop improvement and germplasm management.

In addition to nutritional qualities, certain traits of mungbean like small genome size, short life cycle, self-pollinating, and close genetic relationship with many other legume species make it a suitable model organism for genetic studies (Kim et al. 2015). Complete de novo genome sequencing of cultivated mungbean (*V. radiata* var. *radiata, VC1973A*) (Kang et al. 2014) and its wild relative (*V. radiata* var. *sublobata* TC1966) and de novo assembly of RNA-seq data of 22 accessions of other 18 *Vigna* species which was completed during 2014 itself highlights the importance of the mungbean crop. Not only the genomic DNA information but cytoplasmic genomes of mungbean were also decoded. Sequencing of chloroplast genome was done by Tangphatsornruang et al. (2009) while mitochondrial genome was sequenced by Lin et al. (2016). Once the whole genome sequencing (WGS) and transcriptome sequence data are available, genetic and genomic studies become easier. The genomic sequence information becomes the raw material for several studies like genome annotations, genome-wide association mapping (GWAS), development of genome-wide DNA markers, development of saturated linkage map, gene tagging, identification of other genomic information like small RNAs, microRNAs, transposons, etc. WGS of other related *Vigna* species like *V. unguiculata* (Cowpea Modern Breeding Consortium) and *V. angularis* (Kang et al. 2015) can further fasten comparative genomics studies in mungbean. Annotation of transcriptome sequences for functional genes has been carried out in mungbean (Moe et al. 2011; Gupta et al. 2014; Chen et al. 2015), and DNA sequence-based markers like SNPs and EST-SSRs were developed. Earlier to advancement in genome sequencing technologies, several DNA-based molecular markers were also developed (Kaga et al. 2000; Barkley et al. 2008), and they are still being used particularly which are linked to a trait of interest (Schafleitner et al. 2016). These markers were also used in making linkage mapping, and markers were linked to loci governing important traits like seed weight, seed coat color, resistance for powdery mildew, YMV, bruchids, *Cercospora* leaf spot (Kim et al. 2015). Recent update on *Vigna radiata* in NCBI database indicates that there are three assemblies for three genotypes viz. VC1973A, RIL59, and Kamphaeng Saen 1; however, only one (VC1973A) assembly was converted to discrete linkage groups of the mungbean genome. There are total 30,060 annotated genes, 49,192 identified proteins, and 135,798 SNPs available in the database.

2.9 Future Prospects and Challenges

Mungbean is one of the highest priced pulse crops. The cultivation area is consistently increasing across the globe. However, potential yield of the crop could not be realized due to its highly prone nature toward several devastating diseases, insect pests, and abiotic stresses like waterlogging, low moisture, and terminal heat in summer mung. Thermo- and photo-sensitivity is another major concern. Several superior varieties have been bred, but the process of breeding has narrowed the genetic base of the cultivated mungbean gene pool. Present germplasm screening studies show that resistance sources for disease like MYMD are not available in entire mungbean germplasm. Now broadening the genetic base along with continuing traditional breeding programs has become a challenge for mungbean crop improvement programs. The rich genetic resources of mungbean crop wild relatives like *V. umbellata, V. trilobata, V. glabrascens, V. radiata* var. *sublobata, V. mungo* var. *sylvestris, V. aconitifolia, V. marina,* etc., can play a very important role in the crop improvement. Conserved diversity, genomic resources, and advanced genomic tools are of immense importance in achieving the potential yield. Focus is on targeted germplasm collection based on gap analysis, geo-referencing of all the indigenous germplasm/superimposing it with soil and climate maps and generation of robust cores based on high-throughput genomic resources and phenotypic traits.

References

Abna F, Golam F, Bhassu S (2012) Estimation of genetic diversity of mungbean (*Vigna radiata* L. Wilczek) in Malaysian tropical environment. Afr J Microbiol Res 6:1770–1775

Akhtar KP, Sarwar G, Abbas G, Asghar MJ, Sarwar N, Shah TM (2011) Screening of mungbean germplasm against *Mungbean yellow mosaic virus* and its vector *Bemisia tabaci*. Crop Prot 30:1202–1209

Arora RK (1988) The Indian gene centre. Priorities and prospects for collection. In: Paroda RS, Arora RK, Chandel KPS (eds) Plant genetic resources: Indian perspectives. NBPGR, New Delhi, p 545

Barkley NA, Wang ML, Gillaspie AG, Dean RE, Pederson GA, Jenkins TM (2008). Discovering and verifying DNA polymorphisms in a mung bean [*V. radiata* (L.) R. Wilczek] collection by EcoTILLING and sequencing. BMC Res Notes 1:28

Bisht IS, Mahajan RK, Patel DP (1998a) The use of characterization data to establish the Indian mungbean core collection and assessment of genetic diversity. Genet Res Crop Ev 127–133

Bisht IS, Mahajan RK, Kawalkar TG (1998b) Diversity in greengram (*Vigna radiata* (L.) Wilczek) germplasm collection and its potential use in crop improvement. Ann Appl Biol 132:301–312

Bose RD (1932) Studies in Indian pulses. IV Mung or green gram (*Phaseolus aureus* Roxb). Indian J Agric Sci 2:607–624

Chandel KPS, Lester RN, Starling RJ (1984) The wild ancestors of urid and mung beans (*Vigna mungo* (L.) Hepper and *V. radiata* (L.) Wilczek). J Linn Soc 89(1):85–96

Chankaew S, Somta P, Sorajjapinun W, Srinives P (2011) Quantitative trait loci mapping of Cercospora leaf spot resistance in mungbean, *Vigna radiata* (L.) Wilczek. Mol Breed 28(2):255–264

Chattopadhyay K, Bhattacharya S, Mandal N, Sarkar HK (2008) PCR-based characterization of mung bean (*Vigna radiata*) genotypes from Indian subcontinent at intra-and inter-specific level. J Plant Biochem Biot 17:141–148

Chen H, Wang L, Wang S, Liu C, Blair MW, Cheng X (2015) Transcriptome sequencing of mung bean (*Vigna radiata* L.) genes and the identification of EST-SSR markers. PLOS One 10:e0120273

Frankel OH (1984) Genetic perspectives of germplasm conservation. In: Arber WK, Llimensee K, Peacock WJ, Starlinger P (eds) Genetic manipulation: impact on man and society. Cambridge University Press, Cambridge, pp 161–170

Fujii K, Miyazaki S (1987) Infestation resistance of wild legumes (*Vigna sublobata*) to azuki bean weevil, *Callosobruchus chinensis* (L.) (Coleoptera: Bruchidae) and its relationship with cytogenetic classification. Appl Entomol Zool 22:229–230

Ghafoor A, Zubair M, Malik BA, Iqbal SM (1992) Evaluation of selected germplasm of mungbean (*Vigna radiata* (L.) Wilczek). Pak J Bot 24:112–118

Gosal SS, Bajaj YPS (1983) Interspecific hybridization between *Vigna mungo* and *Vigna radiata* through embryo culture. Euphytica 32:129–137

Gupta SK, Bansal R, Gopalakrishna T (2014) Development and characterization of genic SSR markers for mungbean (*Vigna radiata* (L.) Wilczek). Euphytica 195:245–258

Hakim L (2016) Variability and correlation of agronomic characters of mungbean germplasm and their utilization for variety improvement program lukman hakim. Indones J Agri Sci 9:24–28

Iqbal U, Iqbal SM, Afzal R, Jamal A, Farooq MA, Zahid A (2011) Screening of mungbean germplasm against *Mungbean yellow mosaic virus* (MYMV) under field conditions. Pak J Phytopathol 23:48–51

Jackai LEN, Padulosi S, Ng Q (1996) Resistance to the legume pod borer, *Maruca vitrata* Fabricius, and the probable modalities involved in wild *Vigna*. Crop Prot 15:753–761

Jalaluddin M, Shaikh MAQ (1981) Evaluation of mungbean (*Vigna radiata* (L.) Wilczek) germplasm for resistance to *Mungbean yellow mosaic virus*. SABRAO J 13:61–68

Kaga A, Ishii T, Tsukimoto K, Tokoro E, Kamijima O (2000) Comparative molecular mapping in *Ceratotropis* species using an interspecific cross between azuki bean (*Vigna angularis*) and rice bean (*V. umbellata*). Theor Appl Genet 100:207–213

Kang YJ, Kim SK, Kim MY, Lestari P, Kim KH, Ha BK, Shim S et al (2014) Genome sequence of mungbean and insights into evolution within *Vigna* species. Nature Commun 5:6443

Kang YJ, Satyawan D, Shim S et al (2015) Draft genome sequence of adzuki bean (*vigna angularis*). Sci Rep 5:8069

Kashiwaba K, Tomooka N, Kaga A, Han OK, Vaughan DA (2003) Characterization of resistance to three bruchid species (*Callosobruchus* spp., Coleoptera, Bruchidae) in cultivated rice bean (*Vigna umbellata*). J Econ Entomol 96:207–213

Kawalkar TG, Bisht IS, Mahajan RK, Patel DP, Gupta PN, Chandel KPS (1996) Catalogue on greengram (*Vigna radiata* L. Wilczek) germplasm. NBPGR, New Delhi, pp 130

Kim SK, Nair RM, Lee J, Lee SH (2015) Genomic resources in mungbean for future breeding programs. Front Plant Sci 6:626

Lambrides CJ, Godwin I (2007) "Mungbean." In: Kole C (ed) Genome mapping and molecular breeding in plants pulses, sugar and tuber crops, vol 3. Springer, Berlin, pp 69–90

Lin CP, Lo HF, Chen CY, Chen LFO (2016) The complete mitochondrial genome of mungbean (*Vigna radiata* var. *radiata*, NM92) and a phylogenetic analysis of crops in angiosperms. Mitochondrial DNA Part A 27:3731–3732

Maxted N, Mabuza-Dlamini P, Moss H, Padulosi S, Jarvis A, Guarino L (2004) An ecogeographic survey: African Vigna. In: Systematic and ecogeographic studies of crop genepools, vol 10. International Plant Genetic Resources Institute, Rome, Italy

Moe KT, Chung JW, Cho YI, Moon JK, Ku JH, Jung JK, Park YJ (2011) Sequence information on simple sequence repeats and single nucleotide polymorphisms through transcriptome analysis of mungbean. J Integr Plant Biol 53:63–73

Mukhtar T, Arooj M, Ashfaq M, Gulzar A (2017) Resistance evaluation and host status of selected green gram germplasm against *Meloidogyne incognita*. Crop Prot 92:198–202

Nair RM, Schafleitner R, Kenyon L, Srinivasan R, Easdown W, Ebert AW, Hanson P (2013) Genetic improvement of mungbean. SABRAO J Breed Genet 44:177–190

Ogundiwin EA, Thottappilly G, AkenOva ME, Ekpo EJA, Fatokun CA (2002) Resistance to cowpea mottle carmovirus in *Vigna vexillata*. Plant Breed 121:517–520

Pandiyan M, Ramamoorthi N, Ganesh SK, Jebaraj S, Pgarajan P, Balasubramanian P (2008) Broadening the genetic base and introgression of MYMV resistance and yield improvement through unexplored genes from wild relatives in mungbean. Plant Mut Rep 2: 33–43

Pandiyan M, Senthil N, Suersh R, Chakravarthy N, Packiaraj D, Jagadeesh S (2012) Interspecific hybridization of *Vigna radiata* × *Vigna trilobata*. Euphytica 35:1017–1022

Poehlman JM (1991) The Mungbean. Oxford and IBH publishers, New Delhi

Pratap A, Malviya N, Tomar R et al (2014) *Vigna*. In: Pratap A, Kumar J (eds) Alien gene transfer in crop plants, achievements and impacts, vol 2. Springer, New York, pp 163–189. https://doi.org/10.1007/978-1-4614-9572-7_8

Pratap A, Basu PS, Gupta S, Malviya N, Rajan N, Tomar R, Singh NP (2014b) Identification and characterization of sources for photo- and thermo-insensitivity in Vigna species. Plant Breed 133:756–764

Rahim MA, Mia AA, Mahmud F, Zeba N, Afrin KS (2010) Genetic variability, character ASSOCIATION and genetic divergence in mungbean (*Vigna radiate L.* Wilczek). Plant Omics 3:1

Raje RS, Rao SK (2002) Screening of mungbean (*Vigna radiata* L. wilczek) germplasm for yellow mosaic virus, Cercospora leaf spot and powdery mildew. Legume Res 25:99–104

Rana JC, Gautam NK, Gayacharan Singh M, Yadav R, Tripathi K, Yadav SK, Bhardwaj R (2016) Genetic resources of pulse crops in India: an overview. Indian J Genet Pl Br 76:420–436

Reddy KS, Pawar SE, Bhatia CR (1987) Screening for powdery mildew (*Erysiphae polygoni* DC.) resistance in mungbean (*Vigna radiata* (L.) Wilczek) using excised leaves. Proc Plant Sci 97:365–369

Sangiri C, Kaga A, Tomooka N, Vaughan D, Srinives P (2008) Genetic diversity of the mungbean (*Vigna radiata*, Leguminosae) genepool on the basis of microsatellite analysis. Aust J Bot 55:837–847

Schafleitner R, Nair R, Rathore A, Wang YW, Lin CY, Chu SH, Lin PY, Chang JC, Ebert AW (2015) The AVRDC—The World Vegetable Center mungbean (*Vigna radiata*) core and mini core collections. BMC Genom 16:344

Schafleitner R, Huang SM, Chu SH, Yen JY, Lin CY, Yan MR, Long-fang OC et al (2016) Identification of single nucleotide polymorphism markers associated with resistance to bruchids (*Callosobruchus* spp.) in wild mungbean (*Vigna radiata* var. *sublobata*) and cultivated *V. radiata* through genotyping by sequencing and quantitative trait locus analysis. BMC Plant Biol 16:159

Shad N, Mughal SM, Farooq K, Bashir M (2006) Evaluation of mungbean germplasm for resistance against *Mungbean yellow mosaic begomovirus*. Pak J Bot 38:449

Shanmugasundaram S, Keatinge JDH, Hughes J (2009). Counting on beans: mungbean improvement in Asia. In: Spielman DJ, Pandya-Lorch R (eds) Millions fed: proven successes in agricultural development.

International Food Policy Research Institute, Washington, DC, pp 103–108

She CW, Jiang XH, Ou LJ, Lui J, Long KL, Zhang LH, Duan WT, Zhao W, Hu JC (2015) Molecular cytogenetic characterization and phylogenetic analysis of the seven cultivated *Vigna* species (Fabaceae). Plant Biol 17:268–280

Singh CM, Mishra SB, Anil P, Madhuri A (2014) Morphological characterization and discriminant function analysis in mungbean [*Vigna radiata* (L.) Wilczek] germplasm. Electron J Plant Breed 5:87–96

Singh M, Bisht IS, Sardana S, Gautam NK, Hussain Z, Gupta S, Singh BB, Dwivedi NK (2006) Asiatic *Vigna*. In: Dhillon BS, Saxena S, Agrawal A, Tyagi RK (eds) Plant genetic resources: foodgrain crops. Narosa Publishing House, ISPGR, ICAR-NBPGR, New Delhi, pp 275–301

Singh DP, Singh BB, Pratap A (2016) Genetic improvement of mungbean and urdbean and their role in enhancing pulse production in India. Indian J Genet Plant Breed 76:550–567

Smartt J (1985) Evolution of grain legumes III. Pulses in the genus *Vigna*. Exp Agric 21:87–100

Somta C, Somta P, Tomooka N, Ooi PC, Vaughan DA, Srinives P (2008) Characterization of new sources of mungbean (*Vigna radiata* (L.) Wilczek) resistance to bruchids, *Callosobruchus* spp. (Coleoptera: Bruchidae). J Stored Prods Res 44:316–321

Somta P, Ammaranan C, Ooi PAC, Srinives P (2007) Inheritance of seed resistance to bruchids in cultivated mungbean (*Vigna radiata*, L. Wilczek). Euphytica 155:47–55

Sudha M, Karthikeyan A, Nagarajan P, P Raveendran M, Senthil N, Pandiyan M, Angappan K, Ramalingam J, Bharathi M, Rabindran R, Veluthambi K, Balasubramanian P (2013) Screening of mungbean (*Vigna radiata*) germplasm for resistance to *Mungbean yellow mosaic virus* using agroinoculation. Can J Plant Pathol 35:424–430

Sudha M, Anusuya P, Mahadev NG, Karthikeyan A, Nagarajan P, Raveendran M, Balasubramanian P (2013b) Molecular studies on mungbean (*Vigna radiata* (L.) Wilczek) and ricebean (*Vigna umbellata* (Thunb.)) interspecific hybridisation for *Mungbean yellow mosaic* virus resistance and development of species-specific SCAR marker for ricebean. Arch Phytopath Plant Prot 46:503–517

Sudha M, Karthikeyan A, Shobhana VG, Nagarajan P, Raveendran M, Senthil N, Bharathi M (2015) Search for Vigna species conferring resistance to *Mungbean yellow mosaic virus* in mungbean. Plant Genet Res 13:162–167

Takahashi Y, Somta P, Muto C, Iseki K, Naito K, Pandiyan M et al (2016) Novel genetic resources in the genus *Vigna* unveiled from gene bank accessions. PLOS One 11:e0147568. https://doi.org/10.1371/journal.pone.0147568

Tangphatsornruang S, Sangsrakru D, Chanprasert J, Uthaipaisanwong P, Yoocha T, Jomchai N, Tragoonrung S (2009) The chloroplast genome sequence of mungbean (*Vigna radiata*) determined by high-throughput pyrosequencing: structural organization and phylogenetic relationships. DNA Res 17:11–22

Tantasawat P, Trongchuen J, Prajongjai T, Thongpae T, Petkhum C, Seehalak W, Machikowa T (2010) Variety identification and genetic relationships of mungbean and blackgram in Thailand based on morphological characters and ISSR analysis. Afr J Biotech 9:4452–4464

Tomooka NT, Isemura et al (2014) *Vigna* species. Broadening the genetic base of grain legumes. Springer, pp 175–208

Tomooka N, Kaga A, Isemura T, Vaughan DA (2011) *Vigna*. In: Kole C (ed) Wild crop relatives: genomic and breeding resources. Legume crops and forages, chapter 15, pp 291–311

Tomooka N, Kaga A, Vaughan DA (2006) The Asian *Vigna* (*Vigna* subgenus *Ceratotropis*) biodiversity and evolution. In: Sharma AK, Sharma A (eds) Plant genome: biodiversity and evolution. Part C: phanerogams (angiosperms–dicotyledons), vol 1. Science Publisher, Enfield, New Jersey, pp 87–126

Tomooka N, Kashiwaba K, Vaughan DA, Ishimoto M, Egawa Y (2000) The effectiveness of evaluating wild species: searching for sources of resistance to bruchid beetles in the genus *Vigna* subgenus Ceratotropis. Euphytica 115:27–41

Tomooka N, Lairungreang C, Nakeeraks P, Egawa Y, Thavarasook C (1992) Development of bruchid-resistant mungbean line using wild mungbean germplasm in Thailand. Plant Breed 109:60–66

Vavilov NI (1926) Studies on the origin of cultivated plants. Bull Appl Botany Genet Plant Breed (Leningrad) 16:1–248 (In Russian English summary pp 139–248)

Yimram T, Somta P, Srinivas P (2009) Genetic variation in cultivated mungbean germplasm and its implication in breeding for high yield. Field Crops Res 112: 260–266

Taxonomy of Mungbean and Its Relatives

3

Yu Takahashi and Norihiko Tomooka

Abstract

This chapter reviews the taxonomic treatment of mungbean (*Vigna radiata*) and its relatives and presents photographs of key characters and a phylogenetic tree. After Tomooka et al. (2002a) listed 21 species in the subgenus *Ceratotropis,* six new species have been described. Phylogenetic analysis revealed the positions of two new species, *Vigna indica* and *Vigna sahyadriana*, but the remaining four new species, *Vigna konkanensis, Vigna pandeyana, Vigna sathishiana,* and *Vigna yadavii* still need to be analyzed. We transferred *Vigna subramaniana* from the section *Ceratotropis* to section *Aconitifoliae*. Two unidentified gene bank accessions, NI1135 and JP245506, have been referred for assessment with regard to the relationships with the new species. The gene pool for mungbean has been expanded from Tomooka et al. (2005), with special emphasis on *Vigna vexillata* and NI1135. Drought and salt stress studies are cited to show the importance of systematic screening for stress tolerances, which is the first step toward exploiting novel genes expected in wild relatives.

Y. Takahashi · N. Tomooka (✉)
Genetic Resources Center, National Institute of Agrobiological Sciences, Tsukuba, Ibaraki, Japan
e-mail: tomooka@affrc.go.jp

3.1 Introduction

This chapter outlines the taxonomic treatment of mungbean (*Vigna radiata* (L.) Wilczek) and its gene pool with some comments on newly described species after Tomooka et al. (2002a). Genus *Vigna* Savi includes nine crop species that are cultivated mainly for human food or as animal feed in tropical to temperate regions. More than 80 species have been described in *Vigna*, and they are expected to include wild species with tolerance (or resistance) to external stresses, such as drought, salinity, acidity, alkalinity, disease, and pests. Most wild *Vigna* species have the same number of chromosomes as the *Vigna* crop species and can be crossed both within and/or between species. Therefore, they may represent sources of genetic material that can be used to transfer stress tolerance to crops. In recent years, *Vigna* has been taxonomically revised and new species have been described. Therefore, it is necessary to reconsider taxonomic systems and gene pools, including for these newly described species.

3.2 Genus *Vigna* Savi

There has been confusion regarding the taxonomy of genera *Vigna* Savi and *Phaseolus* L., which are closely related to each other in the family *Fabaceae* and have been repeatedly revised. Savi (1824) erected genus *Vigna* for the type species *Vigna luteola* (Jacq.) Benth. Then De Candolle

R. M. Nair et al. (eds.), *The Mungbean Genome*, Compendium of Plant Genomes,
https://doi.org/10.1007/978-3-030-20008-4_3

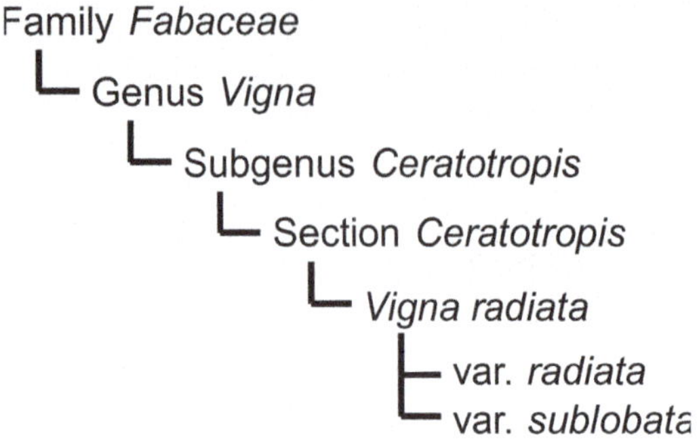

Fig. 3.1 Higher taxonomic ranks and intraspecific varieties of mungbean (*Vigna radiata*) referred in this chapter

(1825) listed mungbean as *Phaseolus* with the common bean (*Phaseolus vulgaris* L.). Wilczek (1954) then transferred mungbean to *Vigna* based on the morphology of its stipule and style beak. Verdcourt (1970) proposed a narrower definition for *Phaseolus*, which only retained the American species, and a broader definition for *Vigna* including eight new subgenera. He classified the cowpea *Vigna unguiculata* (L.) Walp. in the subgenus *Vigna* and mungbean in subgenus *Ceratotropis* (Fig. 3.1). Marechal et al. (1978) followed this proposal but classified *Vigna* into seven subgenera. Stipule, tubercle, keel, and style beak were provided as diagnostic traits for distinguishing between *Vigna* and *Phaseolus* (reviewed by Tomooka et al. 2002a, 2011). Subsequently, phylogenetic studies reduced the number of subgenera in *Vigna* to five; Thulin et al. (2004) transferred the subgenus *Macrorhynchus* from genus *Vigna* and established a new genus *Wajira*, and Delgado-Salinas et al. (2011) split subgenus *Sigmoidotropis* into several distinct genera inhabiting the Americas. Conversely, the number of species in *Phaseolus* has increased from 31 (Marechal et al. 1978) to 76 (Freytag and Debouck 2002). Therefore, it is necessary to revise the diagnostic characteristics of *Vigna* and other related genera.

At present, *Vigna* is divided into five subgenera: *Ceratotropis*, *Haydonia*, *Lasiospron*, *Plectrotropis*, and *Vigna*. Three of the subgenera include crop species: *Plectrotropis* and *Vigna*, which originated in Africa, include cowpea, Bambara nut (*Vigna subterranea* (L.) Verdc.), and tuber cowpea (*Vigna vexillata* (L.) A. Rich.), and *Ceratotropis*, which originated in Asia, includes mungbean and five other crops (Fig. 3.2).

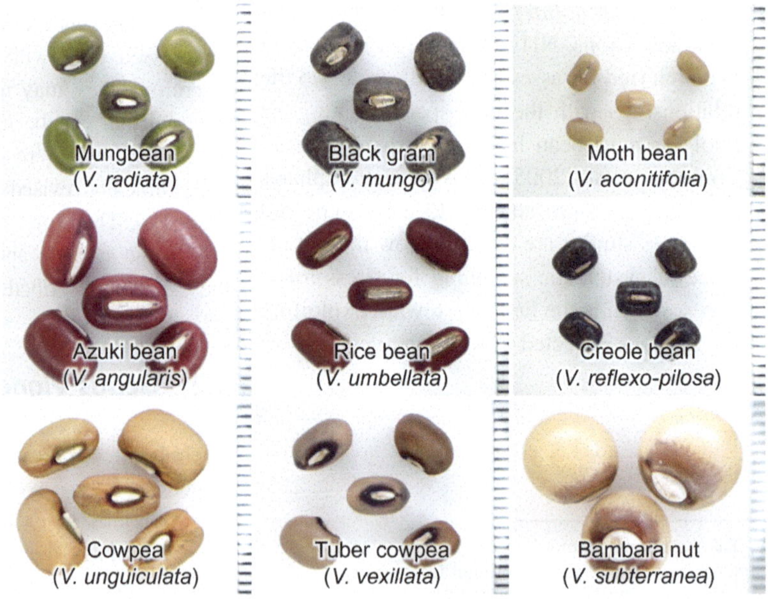

Fig. 3.2 Typical seeds of nine *Vigna* crop species

3.3 Subgenus *Ceratotropis* (Piper) Verdcourt

Relatively few specimens from Southeast Asia are housed in the European herbariums, which delayed the establishment of the taxonomic system for Asian *Vigna*, subgenus *Ceratotropis*. Piper and Morse (1914) originally erected *Ceratotropis* in *Phaseolus*, and then Verdcourt (1969) transferred it to genus *Vigna*. Marechal et al. (1978) described 17 species in *Ceratotropis* based on a survey of European herbariums; Tomooka et al. (2002a) listed 21 species and established three sections: *Aconitifoliae*, *Ceratotropis*, and *Angulares*. They accepted 17 of the species listed by Marechal et al. (1978) and added four new species described by Tateishi and Maxted (2002) and Tomooka et al. (2002b).

Species in subgenus *Ceratotropis* have yellow flowers, a keel curled to the left between 160° and 360°, and a peltate stipule (Fig. 3.3). The germination pattern and the presence or absence of appendage on standard and petiole of primary leaves were provided as diagnostic characters for identifying sections (Table 3.1, Fig. 3.3). Section *Aconitifoliae* includes the moth bean (*Vigna aconitifolia* (Jacq.) Marechal) and is found mainly in South Asia characterized by the absence of the appendage on standard and epigeal germination. Since *Vigna stipulacea* (Lam.) Kuntze in section *Aconitifoliae* has been recognized as showing hypogeal germination (Tomooka et al. 2002a; Javadi et al. 2011; Kang et al. 2014), section *Aconitifoliae* was considered to include species with epigeal and hypogeal germination. However, the elongation of the hypocotyl in *V. stipulacea* indicates epigeal germination, and the cotyledons usually appear at the ground surface level during germination (Fig. 3.3). Therefore, section *Aconitifoliae* contains only species with epigeal germination (Table 3.1). Section *Ceratotropis* includes mungbean and black gram (*Vigna mungo* (L.) Hepper) and is found mainly in South Asia to Southeast Asia characterized by the presence of the appendage on standard, epigeal germination, and sessile primary leaves. Section *Angulares* includes adzuki bean (*Vigna angularis* (Willd.) Ohwi et Ohashi), rice bean (*Vigna umbellata* (Thunb.) Ohwi et Ohashi), and creole bean (*Vigna reflexo-pilosa* Hayata) and is found mainly in Southeast Asia to East Asia characterized by the presence of the appendage on standard, hypogeal germination, and petiolate primary leaves.

Recently, six new species have been described in subgenus *Ceratotropis* from India (Dixit et al. 2011; Aitawade et al. 2012; Gaikwad et al. 2014; Latha et al. 2014; Gaikwad et al. 2015; Balan et al. 2017). Of the species described, Takahashi et al. (2016) accepted *Vigna indica* T. M. Dixit, K. V. Bhat *et* S. R. Yadav and *Vigna sahyadriana* Aitawade, K. V. Bhat *et* S. R. Yadav as distinct species based on DNA sequences and listed 23 species in subgenus *Ceratotropis* as 'Validated' (Table 3.2). Three of the remaining four recently described species, *Vigna sathishiana* A. P. Balan *et* S. V. Predeep, *Vigna konkanensis* Latha, K. V. Bhat, I. S. Bisht, Scariah, Joseph John *et* Krishnaraj, and *Vigna pandeyana* R. D. Gore, S. P. Gaikwad *et* S. D. Randive, are classified in section *Ceratotropis*, and the fourth, *Vigna yadavii* S. P. Gaikwad, R. D. Gore, S. D. Randive *et* K. U. Garad, is classified in section *Angulares* based on morphology. However, their genetic distinctness has not been clarified and thus is listed in Table 3.2 as 'Unknown.'

3.3.1 Section *Ceratotropis* N. Tomooka *et* Maxted

Tomooka et al. (2002a) classified *Vigna grandiflora* (Prain) Tateishi *et* Maxted, *V. mungo*, *V. radiata*, and *Vigna subramaniana* (Babu *ex* Raizada) Raizada in section *Ceratotropis*. Aitawade et al. (2012) described *V. sahyadriana* in section *Ceratotropis*, and this classification was confirmed by Takahashi et al. (2018a) with molecular phylogenetics, who also transferred *V. subramaniana* from section *Ceratotropis* to section *Aconitifoliae*. Therefore, the authors recognize four species in section *Ceratotropis*: *V. grandiflora*, *V. mungo*, *V. radiata*, and *V. sahyadriana*. *Vigna trinervia* (Heyne *ex* Wight *et* Arn.) Tateishi *et* Maxted was classified in section

Fig. 3.3 Photographs of diagnostic characters for distinguishing three sections. Ac: section *Aconitifoliae*. Ce: section *Ceratotropis*. An: section *Angulares*. Black arrows indicate the presence of an appendage on standard. White arrows indicate the position of cotyledon showing epigeal germination. Scale bar for 'Flower' and 'Standard' photographs: 2 mm. Scale in 'Germination' photographs indicates 1 cm order. Scale in 'Style' photographs: 1 mm

Table 3.1 Diagnostic characters for distinguishing three sections in the subgenus *Ceratotropis*

Characters	Section		
	Aconitifoliae	*Ceratotropis*	*Angulares*
Appendage on standard	Absent	Present	Present
Germination	Epigeal	Epigeal	Hypogeal
Primary leaves	Both	Sessile	Petiolate

Angulares based on its hypogeal germination and petiolate primary leaves (Fig. 3.3) but was clustered with the above four *Ceratotropis* species in a molecular phylogenetic tree (Fig. 3.4). Since *V. trinervia* also exhibits the characteristics of section *Ceratotropis*, such as pubescent pods and a dull seed coat, it might be necessary to revise the diagnostic characters used for section classification in the future.

3.3.1.1 *Vigna radiata* (L.) Wilczek

Linnaeus (1753), the founder of binominal nomenclature, originally named the domesticated mungbean *Phaseolus radiatus* L., and Roxburgh (1832) named the wild mungbean *Phaseolus sublobatus* Roxb. This was reviewed by Hara (1955), who stated: 'Thus I conclude that the name *P. radiatus* Linnaeus should be applied to the Mungbean, in the same manner as Prain and others have already interpreted it, and that *P. sublobatus* Roxb. is its wild variety.' Nonetheless, Hara used the name *P. radiatus* var. *setulosus* (Dalz.) Hara *comb. nov.* for the wild variety of mungbean in the same publication and listed *P. sublobatus* Roxb. *nom. nud.* as a synonym. Hara (1955) probably considered *P. sublobatus* Roxb. to be *nom. nud.* because he could not find the type specimen of *P. sublobatus* Roxb. neither in the Royal Botanic Gardens, Kew, nor the Natural History Museum, London. Ohwi and Ohashi (1969) described *V. radiata* (L.) Wilczek var. *setulosa* (Dalz.) Ohwi *et* Ohashi *comb. nov.* by citing *P. sublobatus* in Roxburgh (1832) and *P. setulosus* Roxb. as its synonyms. However, a Kew botanist described *V. radiata* (L.) Wilczek var. *sublobata* (Roxb.) Verdc. *comb. & stat. nov.* based on *P. sublobatus* Roxb. (Verdcourt 1970). Takahashi et al. (2018a) considered *V. radiata* var. *sublobata* as the correct name for wild mungbean because *P. sublobatus* Roxb. has priority over *P. setulosus* Dalz., accepting the treatment of *P. sublobatus* Roxb. as the valid name by Dr. Verdcourt. Until recently, most taxonomists were unable to distinguish wild mungbean from *V. grandiflora* and/or *V. trinervia* (reviewed by Takahashi et al. 2018a).

Domesticated mungbean and wild mungbean were proposed to be different species by Bairiganjan et al. (1985). However, they are cross-compatible with each other and are clustered in the phylogenetic tree (Takahashi et al. 2016). Therefore, we considered it appropriate to distinguish them as varieties (Takahashi et al. 2018a).

For those reasons, we considered that the correct name for domesticated mungbean is *V. radiata* (L.) Wilczek var. *radiata* and that the correct name for wild mungbean is *V. radiata* (L.) Wilczek var. *sublobata* (Roxb.) Verdc.

3.3.2 New Species of the Subgenus *Ceratotropis*

3.3.2.1 *Vigna sathishiana* A. P. Balan *et* S. V. Predeep

The holotype of *V. sathishiana* was found in Kerala in southwestern India, and it has smaller organs and different pollen grain morphologies compared with wild mungbean (Balan et al. 2017). We found that *V. sathishiana* is similar to NI1135 collected at an altitude of 1650 m in Himachal Pradesh in northern India (Fig. 3.5). Tomooka et al. (2002a) misidentified NI1135 as *V. subramaniana*, and thus, *V. subramaniana* was assigned to section *Ceratotropis*. Takahashi et al. (2018a) reported NI1135 as a candidate new species that is most closely related to wild mungbean but has its own distinct *rDNA-ITS* sequences (Fig. 3.4). As a result of artificial

Table 3.2 Described species in the subgenus *Ceratotropis*

Section	Scientific name and original description	Status[a]
Aconitifoliae	*Vigna aconitifolia* (Jacq.) Marechal in Bulletin du Jardin Botanique National de Belgique 39:160 (1969)	Validated
	Vigna aridicola N. Tomooka *et* Maxted in Kew Bulletin 57:613 (2002)	Validated
	Vigna indica T. M. Dixit, K. V. Bhat *et* S. R. Yadav in Rheedea 21:1 (2012)	Validated
	Vigna khandalensis (Santapau) Raghavan *et* Wadhwa in Current Science 41:429 (1972)	Validated
	Vigna stipulacea (Lam.) Kuntze in Revisio generum plantarum 1:212 (1891)	Validated
	Vigna subramaniana (Babu *ex* Raizada) Raizada in Indian Journal of Forestry 3:133 (1980)	Validated
	Vigna trilobata (L.) Verdc. in Taxon 17:172 (1968)	Validated
Angulares	*Vigna angularis* (Willd.) Ohwi *et* Ohashi in Journal of Japanese Botany 44:29 (1969)	Validated
	Vigna dalzelliana (Kuntze) Verdc. in Kew Bulletin 24:558 (1970)	Validated
	Vigna exilis Tateishi *et* Maxted in Kew Bulletin 57:625 (2002)	Validated
	Vigna hirtella Ridley in Journal of the Federated Malay States Museums 10:132 (1920)	Validated
	Vigna minima (Roxb.) Ohwi *et* Ohashi in Journal of Japanese Botany 44:30 (1969)	Validated
	Vigna nakashimae (Ohwi) Ohwi *et* H. Ohashi in Journal of Japanese Botany 44:30 (1969)	Validated
	Vigna nepalensis Tateishi *et* Maxted in Kew Bulletin 57:629 (2002)	Validated
	Vigna reflexo-pilosa Hayata in Journal of college of science, Imperial University of Tokyo 30:82 (1911)	Validated
	Vigna riukiuensis (Ohwi) Ohwi *et* Ohashi in Journal of Japanese Botany 44:31 (1969)	Validated
	Vigna tenuicaulis N. Tomooka *et* Maxted in Kew Bulletin 57:617 (2002)	Validated
	Vigna trinervia (Heyne *ex* Wight et Arn.) Tateishi *et* Maxted in Kew Bulletin 57:633 (2002)	Validated
	Vigna umbellata (Thunb.) Ohwi *et* Ohashi in Journal of Japanese Botany 44:31 (1969)	Validated
	Vigna yadavii S. P. Gaikwad, R. D. Gore, S. D. Randive *et* K. U. Garad in Biodiversity Data Journal 2:e4281 (2014)	Unknown
Ceratotropis	*Vigna grandiflora* (Prain) Tateishi *et* Maxted in Kew Bulletin 57:632 (2002)	Validated
	Vigna konkanensis Latha, K. V. Bhat, I. S. Bisht, Scariah, Joseph John *et* Krishnaraj in Journal of Plant Taxonomy and Geography 69:49 (2014)	Unknown
	Vigna mungo (L.) Hepper in Kew Bulletin 11:128 (1956)	Validated

(continued)

Table 3.2 (continued)

Section	Scientific name and original description	Status[a]
	Vigna pandeyana R. D. Gore, S. P. Gaikwad *et* S. D. Randive in Biodiversity Data Journal 3:e4606 (2015)	Unknown
	Vigna radiata (L.) Wilczek in Flore du Congo Belge et du Ruanda-Urundi 6:386 (1954)	Validated
	Vigna sahyadriana Aitawade, K. V. Bhat *et* S. R. Yadav in Rheedea 22:1 (2012)	Validated
	Vigna sathishiana A. P. Balan *et* S. V. Predeep in Journal of Japanese Botany 92:194 (2017)	Unknown

[a]*Validated* Distinctness as an independent species had been validated using DNA sequences and morphology by the authors of this chapter

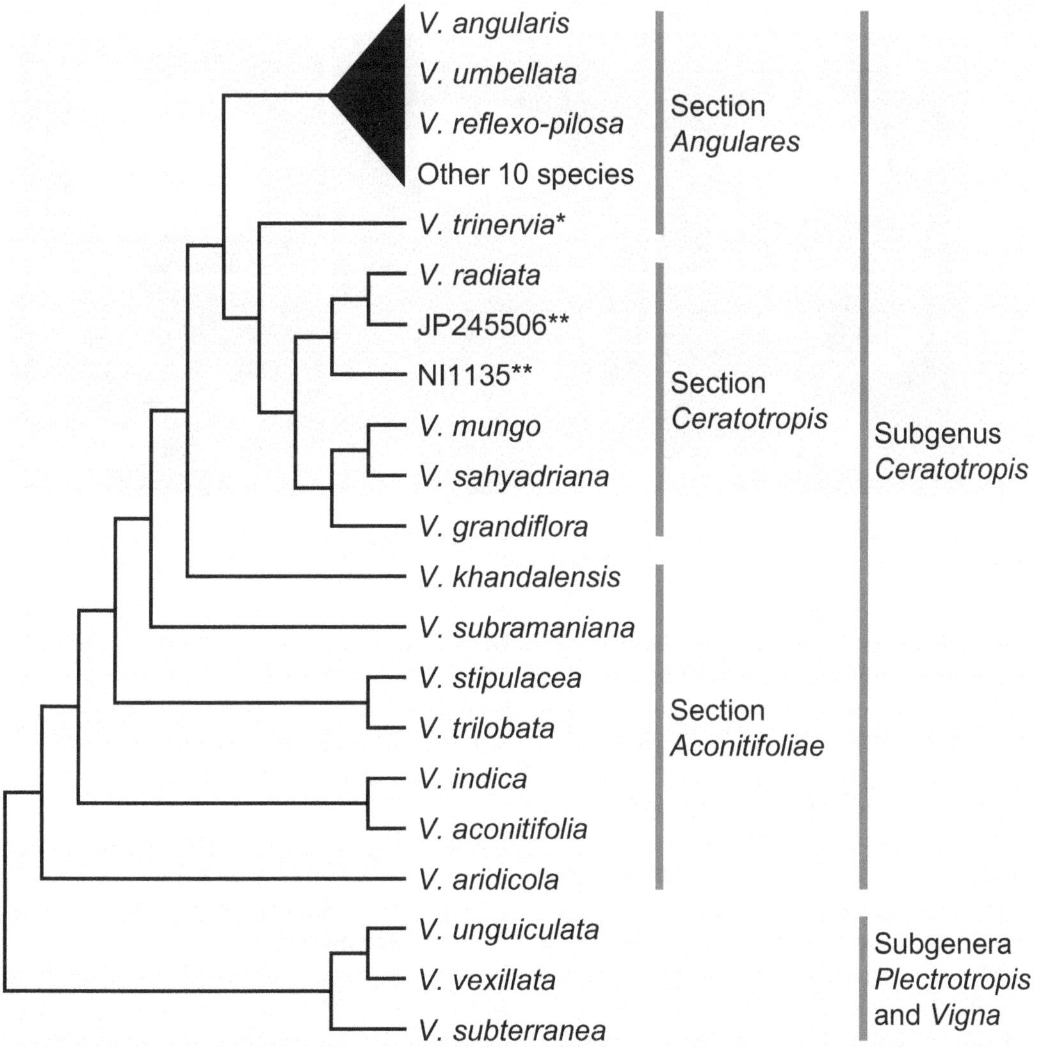

Fig. 3.4 Phylogenetic tree of the species in the genus *Vigna* inferred from *rDNA-ITS* sequences. **Vigna trinervia* was classified to section *Angulares* based on its morphology. **JP245506 and NI1135 are the unidentified accessions closely related to mungbean

Fig. 3.5 Photographs of *Vigna radiata* var. *sublobata* and the unidentified accessions. Scale in 'Seed' photographs indicated 1 mm order. Scale bar for 'Pod' and 'Flower' photographs: 2 mm

crossing between *V. radiata* and NI1135, we were unable to obtain F_1 seeds from crossing some accessions of domesticated mungbean and NI1135 but was able to obtain F_1 seeds using wild mungbean instead (Table 3.3). The F_1 hybrid plants grew well but exhibited high levels of sterility. However, BC_1 seeds could be obtained by backcrossing using domesticated mungbean accessions. The observation of partial sterility among cross-combinations revealed a partial reproductive barrier between *V. radiata* and NI1135. Thus, they are distinguishable in molecularly phylogeny and have developed under partial reproductive isolation from one another. Since a partial DNA sequence of NI1135 was deposited in DDBJ (*rDNA-ITS*: LC064353, *atpB-rbcL*: LC064363), genetic analysis of the holotype of *V. sathishiana* is expected to reveal the relationship between *V. sathishiana* and NI1135.

3.3.2.2 *Vigna konkanensis* Latha, K. V. Bhat, I. S. Bisht, Scariah, Joseph John *et* Krishnaraj

Vigna konkanensis, which is distributed in Maharashtra in Midwest India, is similar to wild mungbean but has glabrous leaves, stems, and inflorescences, and sparsely setose mature pods (Latha et al. 2014). We found that morphological

Table 3.3 Hybridization between *Vigna radiata* and closely related accession NI1135

Female	Male	No. of pod setting/crossing	Pod length (mm)	Seeds collected	Year
Direct cross using domesticated mungbean (Cultigen)					
Cultigen (JP78923)	NI1135	9/56	14–25	0	1998
Cultigen (JP31331)	NI1135	6/16	20–34	0	2003
NI1135	Cultigen (JP78923)	0/11	–	–	1998
Direct cross using wild mungbean (Wild)					
Wild (JP107877)	NI1135	3/8	27–50	24	1998
Wild (JP107877)	NI1135	14/32	10–49	38	2015
Wild (JP107875)	NI1135	0/7	–	–	2003
Wild (JP107875)	NI1135	0/73	–	–	2015
Wild (JP226874)	NI1135	14/26	10–55	11	2015
NI1135	Wild (JP107877)	0/3	–	–	1998
NI1135	Wild (JP107877)	9/11	22–41	37 (sterile)	2015
NI1135	Wild (JP107875)	7/37	15–37	28 (sterile)	2015
NI1135	Wild (JP226874)	7/12	22–52	40	2015
Backcross using domesticated mungbean (Cultigen)					
Cultigen (JP78923)	F1 (JP107877 × NI1135)	0/3	–	–	2003
Cultigen (JP31331)	F1 (JP107877 × NI1135)	4/10	50–60	24	2003
F1 (JP107877 × NI1135)	Cultigen (JP31331)	1/4	17	2	2003
BC1 [(JP107877 × NI1135) × JP31331]	Cultigen (JP31331)	5/28	31–58	21	2004

characters of *V. konkanensis* were similar to those of the Indian wild *Vigna* accession JP245506 (2008TN32 in Tomooka et al. 2008), which has glabrous inflorescences and mature pods, and sparsely setose leaves and stems (Fig. 3.5). Takahashi et al. (2016, 2018a) estimated the interspecific relationship based on *rDNA-ITS* sequences and found that JP245506 is most closely related to *V. radiata* (Fig. 3.4). Since partial DNA sequences of JP245506 have been deposited in DDBJ (*rDNA-ITS*: LC0643 54, *atpB-rbcL*: LC064364), genetic analysis of the holotype of *V. konkanensis* is expected to reveal the relationship among *V. konkanensis, V. radiata,* JP245506, and NI1135.

3.3.2.3 *Vigna pandeyana* R. D. Gore, S. P. Gaikwad *et* S. D. Randive

Vigna pandeyana, which is distributed in Maharashtra, Midwest India, is characterized by subterranean cleistogamous flowers (Gaikwad et al. 2015). We classify *V. pandeyana* in section *Ceratotropis*, because we identified the presence of the appendage on standard, epigeal germination, and sessile primary leaves in photographs and line drawings (Gaikwad et al. 2015). We consider *V. pandeyana* to be most closely related to *V. mungo* (L.) Hepper var. *silvestris* Lukoki, Marechal *et* Otoul, because it has a prominent hilum, lanceolate stipule, densely pubescent short pods, and golden yellow flowers. It seems also to be closely related to *V. sahyadriana*. However, Gaikwad et al. (2015) compared this species only to *Vigna dalzelliana* (Kuntze) Verdc. and *V. yadavii* in section *Angulares*.

We now consider that *V. pandiana* could be an intraspecific variation in *V. mungo*. In the mountainous areas of Western Nepal, Takahashi et al. (2017) came across plants of the wild adzuki bean ("N43", *V. angularis* (Willd.) Ohwi & Ohashi var. *nipponensis* (Ohwi) Ohwi & Ohashi) in Surkhet and plants of *Vigna hirtella* Ridley ("N68") in Doti, which seemed to have cleistogamous flowers and shorter pods close to or under the ground surface, while the same plants produced longer pods on the aerial shoots. We considered this dimorphic flower/pods formation near or under the soil surface to be an ecological adaptation to the sloped environment. Therefore, it is possible that some populations of *V. mungo* var. *silvestris* evolved these traits. Since partial sequences of *V. mungo* var. *silvestris* collected in India have been deposited in DDBJ (*rDNA-ITS*: LC064347, *atpB-rbcL*: LC064357), genetic analysis of the holotype of *V. pandeyana* is expected to reveal the relationship among *V. pandeyana, V. sahyadriana*, and *V. mungo* var. *silvestris*.

3.3.2.4 *Vigna yadavii* S. P. Gaikwad, R. D. Gore, S. D. Randive *et* K. U. Garad

Vigna yadavii, which is distributed in Maharashtra and Karnataka in Midwest India, is similar to *V. dalzelliana*, but differs in having subterranean cleistogamous flowers and a linear style beak (Gaikwad et al. 2014). We classify *V. yadavii* in section *Angulares*, because it has an appendage on standard, hypogeal germination, and petiolate primary leaves, based on photographs and line drawings in Gaikwad et al. (2014). The intraspecific variation of flowering positions has not been sufficiently observed in *V. dalzelliana*; therefore, we cannot deny the possibility that populations of *V. dalzelliana* with subterranean cleistogamous flowers exist. Tomooka et al. (2002a) listed a flat style beak as a useful diagnostic character of *V. dalzelliana*. However, we recently noticed that the style beaks of Myanmar accessions (JP210811, JP210820) are flat but that of an Indian accession (JP 235419) is linear in *V. dalzelliana* (Fig. 3.3). Therefore, it is necessary to verify the distinctness of *V. yadavii* and *V. dalzelliana* as different species. Since partial DNA sequences of *V. dalzelliana* collected in India have been deposited in DDBJ (*rDNA-ITS*: LC081997, *atpB-rbcL*: LC082249), genetic analysis of the holotype of *V. yadavii* is expected in future studies.

3.4 Gene Pool Concept for Mungbean

The gene pool concept is based on cross-compatibility between a target crop and its relatives and is useful for incorporating useful traits into target crops (Harlan and De Wet 1971). Tomooka et al. (2005) proposed the gene pool concept of mungbean. Pandiyan et al. (2010) succeeded in producing F_1 hybrids of mungbean crossed with 13 *Vigna* species showing different levels of fertility. Surprisingly, they obtained F_2 segregating generations in crosses between mungbean and *V. vexillata* (subgenus *Plectrotropis*). For this reason, *V. vexillata* is newly included in Gene pool 2 (Fig. 3.6).

The wild mungbean is placed in Gene pool 1. We consider that most species in subgenus *Ceratotropis* could be placed in Gene pool 2 because F_2 or BC_1 progenies could be produced with some species in the three sections listed in Fig. 3.6 (Egawa et al. 1996; Pandiyan et al. 2010). In section *Ceratotropis*, the accession 'NI1135' is placed in Gene pool 2. Although we could not yet obtain fertile F_1 hybrids between 'NI1135' and domesticated mungbean, F_1 hybrids with different levels of fertility were obtained when 'NI1135' was crossed with wild mungbean accessions (Table 3.3).

In section *Angulares*, *V. trinervia*, *Vigna tenuicaulis* N. Tomooka *et* Maxted, and *V. umbellata* are placed in Gene pool 2 because F_2 generation plants were obtained when they were crossed with mungbean (Egawa et al. 1996; Pandiyan et al. 2010). In the hybrid of mungbean and

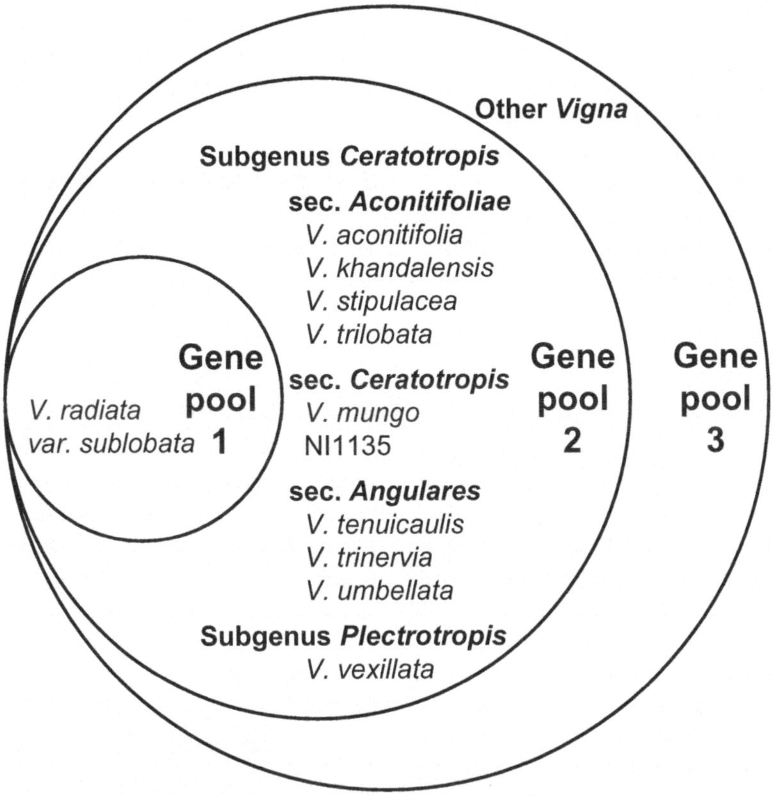

Fig. 3.6 Gene pool concept for mungbean. Modified from Tomooka et al. (2005). Gene pool 1 constitutes the biological species. Gene pool 2 includes species that can cross and result in some level of fertility. Gene pool 3 includes species for which radical techniques are required to transfer genes

V. tenuicaulis, a relatively high pod set (50%) was observed; however, hybridized pods became shriveled at around 20 days after pollination. Embryo culture using immature seeds could successfully produce nine F_1 plants out of 13 rescued embryos. However, the nine F_1 plants were almost completely sterile (average pollen stainability of 10.6%). Two F1 plants could bear one pod containing one seed each. From two F_2 seeds that germinated, one F_2 plant grew well, and average pollen stainability in the first growing year (from November 2004) was slightly increased (39.4%) compared with that of F_1 plants. Although this F_2 plant could not produce F_3 seeds in 2004, it started bearing pods each containing a single F_3 seed in 2005. The pollen stainability of the F_2 plant in February 2005 further increased to 89.9%.

In section *Aconitifoliae*, *V. aconitifolia, Vigna khandalensis* (Santapau) Raghavan *et* Wadhwa, *V. stipulacea*, and *Vigna trilobata* (L.) Verdc. are placed in Gene pool 2 because F_2 generations were obtained when they were crossed with mungbean (Pandiyan et al. 2010). Furthermore, it is considered that the next generation can be obtained by backcrossing, even in a combination in which the F_2 generation cannot be obtained from the F_1 hybrid.

Although no previous studies were found, species in subgenera *Haydonia*, *Lasiospron*, and *Vigna* are tentatively placed in Gene pool 3 (Other *Vigna* in Fig. 3.6). Most species, including cowpea, have $2n = 22$ chromosomes. However, some species with $2n = 20$ chromosomes have been reported in subgenera *Lasiospron* and *Vigna* (Sen and Bhowal 1960; Parida et al. 1990). The progenies of hybrids between parents with different numbers of chromosomes frequently become chromosomally unstable. Therefore, the priority for crossing trials is higher for $2n = 22$ species. However, the chromosomes of all *Vigna* species have not yet been observed, and it is necessary to determine the number of chromosomes for all species, including any newly described species.

3.5 Stress Tolerance of Mungbean and Its Relatives

Previous reports are few on the beneficial traits of species in genus *Vigna*, but there have been some reports on their tolerance (or resistance) against drought, salinity, acidity, alkalinity, disease, and pests (reviewed by Tomooka et al. 2011, 2014). Recently, Iseki et al. (2016, 2018) conducted systematic screenings for drought and salt tolerance in 23 species in subgenus *Ceratotropis* and some species in subgenera *Plectrotropis* and *Vigna*. The accession with the highest relative shoot dry weight under strong drought conditions was NI1135, and a trade-off between drought tolerance and growth rate (shoot dry weight) was observed (Fig. 3.7a). Most of the segregated progenies between mungbean and NI1135 are presumed to be distributed on the curve shown in Fig. 3.7a. We should try to select genotypes distribute along the straight line in Fig. 3.7a.

An accession of the coastal species *Vigna riukiuensis* (Ohwi) Ohwi *et* Ohashi in section *Angulares* had the highest relative shoot dry weight under 200 mM NaCl salt stress (Fig. 3.7 b; Iseki et al. 2016). However, they did not include coastal species/accessions from section *Ceratotropis* in their study, which might explain the failure to detect salt-tolerant accessions (Fig. 3.7b). Finding coastal species or ecotypes belonging to section *Ceratotropis* is of high priority for future field surveys. Although, to our knowledge, there are no reports on the successful hybridization of mungbean with *V. riukiuensis*, successful hybridization is likely based on the mungbean gene pool concept. Recently, an ecotype of *V. vexillata* was collected on Okinawa Island, Japan, that had adapted to saline coastal cliffs environment (Takahashi et al. 2018b).

3.6 Future Perspectives

As reviewed in this chapter, six new species in subgenus *Ceratotropis* of genus *Vigna* have been described in the last 10 years. This demonstrates that systematic field surveys to discover novel genetic resources are an urgent task with high priority. Some of the wild habitat that these species inhabit will probably be lost in the near future. Careful observation and description in the field to reveal the ecological adaptations of wild plants to their specific habitats might greatly increase the value of germplasm resources. Discovery of new *Vigna* species could expand gene pool of related crop species. Taxonomic description of a new species should be accompanied with genetic sequence data, which will provide clarification of the distinctness and phylogenetic relationship among closely related taxa at the molecular level. Interspecific hybridization research is of high priority to further understand the gene pool concept of mungbean. The discovery of a crossable combination expands the gene pool of mungbean, as was demonstrated by the successful production of fertile hybrids of mungbean and *V. vexillata*. Systematic screening studies to detect useful traits for breeding are recommended as the first step toward exploiting novel genes in wild relatives. A reasonable strategy might be to start screening using germplasm including species level diversity. Studies conducted by Iseki et al. (2016, 2018) are good examples of interspecific variations in Vigna. Exploitation of inter- and intraspecific variation in useful traits of *Vigna* could lead to the discovery of untapped valuable gene(s) for improving closely related crops such as mungbean. When useful wild germplasm is crossed with crops, linebreeding (continuous

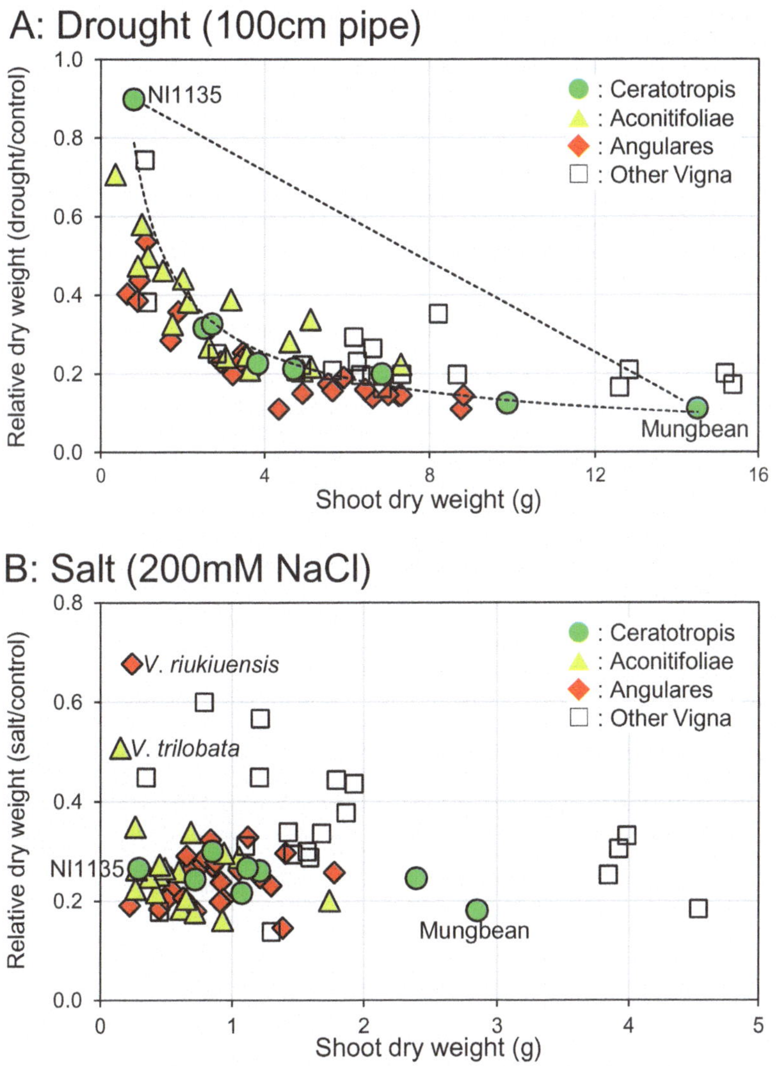

Fig. 3.7 Drought and salt tolerance of *Vigna* accessions belonging to different subgenera. Tolerance was evaluated based on the relative dry weight of accessions under stressful conditions in Iseki et al. (2016, 2018)

backcrossing), which can be an onerous task, is required to eliminate undesired traits and select useful lines. We believe that it is worth the effort.

References

Aitawade MM, Sutar SP, Rao SR, Malik SK, Yadav SR, Bhat KV (2012) Section*Ceratotropis* of subgenus *Ceratotropis* of *Vigna* (Leguminosae–Papilionoideae) in India with a new species from Northern Western Ghats. Rheedea 22:20–27

Bairiganjan GC, Panda PC, Choudhury BP, Patnaik SN (1985) Fabaceae in Orissa. J Econ Taxon Bot 7:249–276

Balan AP, Predeep AV, Udayan AS (2017) *Vigna sathishiana* (Fabaceae): a new species from Southern Western Ghats, India. J Jpn Bot 92(4):193–198

De Candolle AP (1825) Leguminosae. In: De Candolle AP (ed) Prodromus systematis naturalis regni vegetabilis 2, pp 95–524

Delgado-Salinas A, Thulin M, Pasquet R, Weeden N, Lavin M (2011) *Vigna* (Leguminosae) sensu lato: the names and identities of the American segregate genera. Am J Bot 98:1694–1715

Dixit TM, Sutar SP, Yadav SR, Bhat KV, Rao SR (2011) *Vigna indica*, a new name for *Vigna trilobata* var. *pusilla* and a note on section *Aconitifoliae* in India. Rheedea Rheedea 21:1–7

Egawa Y, Chotechuen S, Tomooka N, Ngampongsai S, Kitbamroong C (1996) Cross-compatibility among the subgenus *Ceratotropis* species of the genus *Vigna*. In: Egawa Y and Chotechuen S (eds) Phylogenetic differentiation of mungbean germplasm (subgenus *Ceratotropis* of the genus *Vigna*) and evaluation for breeding program. Under the collaborative research program between JIRCAS (Japan) and DOA (Thailand), pp 19–30

Freytag GF and Debouck DG (2002) Taxonomy, distribution, and ecology of the genus Phaseolus (Leguminosae–Papilionoideae) in North America, Mexico and Central America. Botanical Research Institute of Texas (BRIT), Fort Worth, TX, USA, 298 p

Gaikwad SP, Gore RD, Randive SD, Garad KU (2014) *Vigna yadavii* (Leguminosae: Papilionoideae), a new species from Western Ghats, India. Biodivers Data J 2: e4281

Gaikwad SP, Gore RD, Randive SD (2015) *Vigna pandeyana* (Fabaceae), a new species from Northern Western Ghats, India. Biodivers Data J 3:e4606

Hara H (1955) Critical notes on some type specimens of East-Asiatic plants in foreign herbaria 3. J Jap Bot 30:138–142

Harlan JR, De Wet JMJ (1971) Toward a rational classification of cultivated plants. Taxon 20(4):509–517

Iseki K, Takahashi Y, Muto C, Naito K, Tomooka N (2016) Diversity and evolution of salt tolerance in the genus *Vigna*. PLoS ONE 11:e0164711

Iseki K, Takahashi Y, Muto C, Naito K, Tomooka N (2018) Diversity of drought tolerance in the genus *Vigna*. Front Plant Sci 9:729

Javadi F, Tun YT, Kawase M, Guan K, Yamaguchi H (2011) Molecular phylogeny of the subgenus Ceratotropis (genus Vigna, Leguminosae) reveals three eco-geographical groups and Late Pliocene-Pleistocene diversification: evidence from four plastid DNA region sequences. Ann Bot 108:367–380

Kang YJ, Kim SK, Kim MY, Lestari P, Kim KH, Ha BK et al (2014) Genome sequence of mungbean and insights into evolution within *Vigna* species. Nat Commun 11:5443

Latha M, Scariah Sheen, Krishnaraj MV, Presannakumari KT, Bhat KV, Bisht IS, Joseph John K (2014) *Vigna konkanensis* (Fabaceae: Papilionoideae) a new species from the west coast of India. Webbia 69:49–52

Linnaeus C (1753) Species Plantarum, Stockholm: Impensis G. C. Nauk

Marechal R, Mascherpa JM, Stainer F (1978) Etude taxonomique d'un groupe complexe d'speces des genres Phaseolus et *Vigna* (Papilionaceae) sur la base de donnees morphologiques et polliniques, traitees par l'analyse informatique. Boissiera 28:1273

Ohwi J, Ohashi H (1969) Adzuki beans of Asia. J Jap Bot 44:29–31

Pandiyan M, Senthil N, Ramamoorthi N, Muthiah AR, Tomooka N, Duncan Jayaraj T (2010) Interspecific hybridization of *Vigna radiata* x 13 wild *Vigna* species for developing MYMV donar. Electron J Plant Breed 1:600–610

Parida A, Raina SN, Narayan RKJ (1990) Quantitative DNA variation between and within chromosome complements of *Vigna* species (Fabaceae). Genetica 82:25–133

Piper CV, Morse WJ (1914) Five oriental species of beans. Bull USDA 119:1–32

Roxburgh W (1832) Flora indica; or, descriptions of Indian plants, vol III. W. Thacker and Co., Calcutta and Parbury, Allen and Co. London, Serampore

Savi G (1824) Phaseoli. Nuovo Giornale dei Letterati 8. Dalla Tipografia della societa letteraria, Pisa, Italy, pp 106–123

Sen NK, Bhowal JG (1960) Cytotaxonomic studies on *Vigna*. Cytologia 25:195–207

Takahashi Y, Somta P, Muto C, Iseki K, Naito K, Pandiyan M, Natesan S, Tomooka N (2016) Novel genetic resources in the genus *Vigna* unveiled from gene bank accessions. PLoS ONE 11:e0164711

Takahashi Y, Nemoto K, Sharma S, Dongol DMS, Shrestha DS, Joshi BK, Ghimire KH, Joshi BK, Paudel MN, Tomooka N (2017) Collection and conservation of leguminous crops and their wild relatives in Western Nepal from October 29 to November 10, 2016. AREIPGR 33:295–329

Takahashi Y, Muto C, Iseki K, Naito K, Somta P, Pandiyan M, Natesan S, Tomooka N (2018a) A new taxonomic treatment for some wild relatives of mungbean (*Vigna radiata* (L.) Wilczek) based on their molecular phylogenetic relationships and morphological variations. Genet Res Crop Ev 65:1109–1121

Takahashi Y, Baba-Kasai A, Abe A, Akai K, Akiba M, Hirashima S, Kaga A, Tomooka N (2018b) Collection and conservation of legume genetic resources on Okinawa Island, Yagaji Island, Kouri Island and Hamahiga Island in 2017. AREIPGR 34:1–16

Tateishi Y, Maxted N (2002) New species and combinations in *Vigna* subgenus Ceratotropis (Piper) Verdc. (Leguminosae, Phaseoleae). Kew Bull 57:625–633

Thulin M, Lavin M, Pasquet R, Delgado-Salinas A (2004) Phylogeny and biogeography of *Wajira* (Leguminosae): A monophyletic segregate of *Vigna* centered in the horn of Africa region. Syst Bot 29:903–920

Tomooka N, Vaughan DA, Moss H, Maxted N (2002a) The Asian *Vigna*: Genus *Vigna* subgenus *Ceratotropis* genetic resources. Kluwer Academic Publishers, Dordrecht, the Netherlands

Tomooka N, Maxted N, Thavarasook C, Jayasuriya AHM (2002b) Two new species, new species combinations and sectional designations in *Vigna* subgenus *Ceratotropis* (Piper) Verdcourt (Leguminosae, Phaseoleae). Kew Bull 57(3):613–624

Tomooka N, Vaughan DA, Kaga A (2005) Mungbean (*Vigna* radiata (L.) Wilczek). In: Singh RJ, Jauhar PP (eds) Genetic resources, chromosome engineering and crop improvement. Grain legumes. CRC, vol I, Boca Rogue, FL, pp 325–345

Tomooka N, Senthil N, Pandiyan M, Ramamoorthi N, Kaga A, Vaughan DA (2008) Collection and conservation of leguminous crops and their wild relatives in Tamil Nadu, India, 2008. AREIPGR 24:113–125

Tomooka N, Kaga A, Isemura T, Vaughan DA (2011) *Vigna*. In: Kole C (ed) Wild crop relatives: genomic and breeding resources legume crops and forages. Springer, New York, pp 291–311

Tomooka N, Naito K, Kaga A, Sakai H, Isemura T, Ogiso-Tanaka E, Iseki K, Takahashi Y (2014) Evolution, domestication and neo-domestication of the genus *Vigna*. Plant Genet Res 12(S1):S168–S171

Verdcourt B (1969) New combinations of *Vigna* Savi (Leguminosae–Papilionoideae). Kew Bull 23:464

Verdcourt B (1970) Studies in the Leguminosae-Papilionoideae for the flora of tropical east Africa IV. Kew Bull 24:507–569

Wilczek R (1954) Phaseolinae. In: Robyns W et al (ed) Flore du Congo Belge et du Ruanda-Urundi, Spermatophytes, VI, L'Institut National pour l'Étude Agronomique du Congo Belge, Bruxelles, pp 260–409

Classical Genetics and Traditional Breeding in Mungbean

4

Harsh Kumar Dikshit, Gyan Prakash Mishra, Prakit Somta, Tun Shwe, A. K. M. M. Alam, Tejinderjit S. Bains and Ramakrishnan M. Nair

Abstract

Mungbean is a highly self-pollinating diploid species, with chromosome number of $2n = 22$. Varietal improvement in mungbean has benefited greatly from exemplary work done in the past on determination of genetics of key traits. Varietal introductions, pure line selection, recombination breeding and mutagenesis have been employed successfully in developing varieties. Development of high-yielding varieties with synchronous and early maturity (about 60 days), determinate growth habit, large-seed size, resistance to diseases (powdery mildew, Cercospora leaf spot and mungbean yellow mosaic disease) through hybridization and mutagenesis was a game changer in mungbean development. This enabled the expansion of the crop through good adoption by the farming community, particularly in Asia.

4.1 Introduction

This chapter covers information about the cytogenetics and nuclear DNA content in mungbean. The floral biology and the crossing techniques commonly followed are briefly described. The genetics of important agronomic traits and resistance to stresses and the breeding methods employed for varietal development have been discussed.

4.2 Cytogenetics

Mungbean (*Vigna radiata* (L.) Wilczek var. *radiata*) is a diploid with somatic chromosome number of $2n = 22$. Numerous karyotyping studies were conducted to estimate the dimensions of the mungbean chromosome (Table 4.1). Bhatnagar (1974) put forwarded $4L^{sm} + 4M^{sm} + 3M^{m}$ [L = long (2.7–3.5 μm), M = medium (1.9–2.6 μm, sm = sub median centromere and m = median centromere)] as karyotype formula for mungbean.

H. K. Dikshit · G. P. Mishra
Division of Genetics, ICAR – Indian Agricultural Research Institute, New Delhi, India

P. Somta
Department of Agronomy, Faculty of Agriculture, Kasetsart University, Kamphaeng Saen Campus, Kamphaeng Saen, Thailand

T. Shwe
Department of Agricultural Research, Yezin, Nay Pyi Taw, Myanmar

A. K. M. M.Alam
Pulses Research Centre, Bangladesh Agricultural Research Institute (BARI), Gazipur, Bangladesh

T. S. Bains
Punjab Agricultural University, Ludhiana, India

R. M. Nair (✉)
World Vegetable Center, South Asia, ICRISAT Campus, Patancheru, Hyderabad, Telangana, India
e-mail: ramakrishnan.nair@worldveg.org

R. M. Nair et al. (eds.), *The Mungbean Genome*, Compendium of Plant Genomes,
https://doi.org/10.1007/978-3-030-20008-4_4

Table 4.1 Morphology and dimensions of mungbean chromosome

Chromosome length (μm)		Number of chromosomes with			Reference
Total	Range	Median	Submedian		
		Centromere	Centromere	Satellite	
23.00	1.4–3.30	6	5		Krishnan and De (1965)
28.81	1.9–3.50	3	8		Bhatnagar et al. (1974)
11.97	0.8–1.39				Joseph and Bouwkamp (1978)
		5	6	2	Bhattacharya (1979)
24.80	1.37–3.35	6	5		Sahai and Rana (1980)
37.40	1.0–2.8	6	5		Sarbhoy (1980)
15.95	1.24–2.04	3	8	3	Sharma and Gupta (1982)
24.90	1.6–3.5	5	6	2	Lavania and Lavania 1983

Source Kumar et al. (2006)

4.3 Nuclear DNA Content

The nuclear DNA (2C DNA) content of the mungbean chromosome was estimated at 0.96 pg and 1.42 pg by Murray et al. (1979) and Johnston et al. (1999) respectively. It was assessed that the haploid genome has approximately 470 Mbps by Murray et al. (1979) and 579 Mbps by Arumuganathan and Earle (1991). Kang et al. 2014 constructed 421 Mb (80%) of the total estimated *V. radiata* var. *radiata* genome and identified 22,427 high-confidence protein-coding genes and 160 *Vigna* gene clusters.

4.4 Floral Biology

Mungbean is a highly self-pollinated crop (cleistogamous) with 4–5% outcrossing (Van Rheenen 1964). It produces green to dark yellowish papilionaceous flowers in terminal racemes or axillary clusters on long peduncles. The flower is a hermaphrodite having a gynoecium covered with a diadelphous androecium (9 + 1), two keels, two wings and one standard petal. Flower opening occurs between 6:00 and 8:00 AM and continues up to 11:00 AM. The standard crossing technique involves emasculation in the evening and pollination in the following morning (Boling et al. 1961; Singh and Malhotra 1975; Park and Yang 1978; Cupka and Edwards 1986; Khattak et al. 1998). However, crossing technique by emasculation and pollination at the same time in the morning can also be done (Tomooka et al. 2002).

4.5 Genetics of Traits

Classical genetic studies of mungbean commenced in the 1930s (Bose 1939). Fery (1980) published a set of rules proposed for the genus *Vigna* and a detailed literature review of the genetics of mungbean. The most comprehensive review on classical genetics of mungbean was made nearly 20 years ago by Poehlman (1991).

4.5.1 Plant Type and Growth Habit

Mungbean genotypes can be erect, semi erect, semispreading or twining. Sen and Ghosh (1959) and Khattak et al. (1999) reported twining habit under the control of a single dominant gene. Pathak and Singh (1963) reported single recessive gene for twining habit, while single dominant gene for semi-spreading habit. Further, single dominant gene for indeterminate growth habit was reported by Talukdar and Talukdar (2003), Isemura et al. (2012) and Li et al. (2018). The gene controlling determinacy is named *st-det5.9.1* by Isemura et al. (2012) and *VrDet1* by

Li et al. (2018). The causal mutation(s) in the promoter region of the *VrDet1* gene cause the difference between the indeterminacy and determinacy (Li et al. 2018). For days to maturity, epistatic gene action is reported (Rao et al. 1984; Malik and Singh 1983; Wilson et al. 1985).

4.5.2 Pigmentation

Pigmentation in mungbean can be observed on hypocotyls, epicotyls, stem, leaf rachis, petiole and peduncle. Purple hypocotyls are dominant over purple spotted and green hypocotyls, and purple spotted over green hypocotyls. Single dominant gene '*A*' for purple hypocotyl colour was reported (Bose 1939). Sen and Ghosh (1958) designated '*P*' gene for purple hypocotyls colour. Anthocyanin pigmentation in peduncle, petiole, stem, hypocotyls and epicotyls was reported to be governed by single dominant gene (Pathak and Singh 1963; Van Rheenen 1965; Verma and Krishi 1969; Misra et al. 1970; Virk and Verma 1977). Also, a single dominant gene '*Ppp1*' with pleiotropic effect controlling this trait was reported by Dwivedi and Singh (1986). Van Rheenen (1965) proposed symbol '*R*' for red colour of top of leaflet stalk, stem and hypocotyls. Appa Rao and Jana (1973) reported single recessive gene for peduncle, petiole, stem, hypocotyls and epicotyls anthocyanin pigmentation. Anthocyanin in hypocotyls was reported to be controlled by two supplementary genes viz. '*Sh*' and '*Ph*' with recessive epistatic interaction (Mukherjee and Pradhan 2002).

4.5.3 Stem Traits

The inheritance of stem fasciation in mungbean was reported to be controlled by single recessive gene '*fs1*' having a pleiotropic effect on the number of floral organs (Dwivedi and Singh 1990). Malik and Singh (1983) reported epistatic effects for higher branches per plant. Both additive and non-additive effects with preponderance of additive effects have been reported for plant height (Bhargava et al. 1966; Yohe and Poehlman 1975; Lal et al. 1982; Malik and Singh 1983; Reddy and Sreeramulu 1982; Rao et al. 1984; Wilson et al. 1985).

4.5.4 Leaf Traits

Large, ovate, entire and lobed trifoliate leaves have been observed in mungbean. Pentafoliate leaf shape is known to be governed by a single gene (Chhabra 1990). Soehendi et al. (2006) reported that hybridization between large-heptafoliate leaves and small-pentafoliate leaves mungbean mutants resulting in F_1 plants with normal trifoliate leaves and F_2 plants segregated in a 9 : 3 : 3 : 1 ratio of large-trifoliate : large-heptafoliate : small-pentafoliate : small-heptafoliate leaves. They proposed symbols *N1*, *n1* and *N2*, *n2* for genes controlling leaflet number. Genotypes *N1_N2_*, *n1n1N2_*, *N1_n2n2* and *n1n1n2n2* were suggested for the above leaf traits, respectively (Soehendi et al. 2006). Gene symbol *A_ B_*, *A_bb*, *aaB_* and *aabb* were suggested for dark green, green, greenish yellow and pale green seed coats, respectively. The results suggested that the genes controlling leaflet size and leaflet number are unlinked. Recently, mungbean mutant showing unifoliate leaf has been identified (Jiao et al. 2019). Unifoliate leaf is recessive to the normal trifoliate leaf and is controlled by a single gene, designated *un* (Jiao et al. 2019). Several researchers (Singh and Mehta 1953; Pohle 1972; Ramamoorthi et al. 1994; Talukdar and Talukdar 2003; Singh 1980; Chhabra 1990) have reported that lobed trifoliate leaf is dominant over the entire leaf. However, there are also reports of two dominant genes, '*Tlb1*' and '*Tlb2*' with duplicate gene action for trilobed leaves (Sareen 1982). Narrow lanceolate leaf is reported to be controlled by two recessive genes, '*nl1*' and '*nl2*' (Dwivedi and Singh 1986).

4.5.5 Inflorescence/Flower Traits

Simple and compound inflorescence have been reported in mungbean. Sen and Ghosh (1959) reported simple inflorescence under control of two dominant genes (*Í1'* and *Í2'*), while double recessive homozygous genotypes result in compound inflorescence. Singh and Singh (1971) reported that single cluster per node is governed by dominant gene '*C*' and its recessive counterpart '*c*' governs three clusters per node. Monogenic recessive inheritance of induced sterility was reported by Saini et al. (1974). Four colours of standard petal have been recorded in mungbean. Light yellowish olive colour was dominant to olive yellow colour and was designated by gene '*O*' (Bose 1939). Murty and Patel (1973) proposed gene symbols *Pg*, *Pb* and *Pn* for allelic series conditioning pale green yellow, bright yellow and naphthalene yellow flower colour. Additive gene action (Yohe and Poehlman 1975; Luthra et al. 1979; Reddy and Sreeramulu 1982; Rao et al. 1984; Malik and Singh 1983; Wilson et al. 1985) and partial dominance (Luthra et al. 1979) have been reported for days to 50% flowering.

4.5.6 Pod Traits

Verma and Krishi (1969) reported that pod shattering is dominant to non-shattering and is governed by a single gene. Singh et al. (1975) reported that resistance to pod shattering in interspecific cross as quantitative trait. Quantitative genetic analysis revealed that pod shattering is conditioned by two loci (Isemura et al. 2012). Swollen pod-tip was reported dominant (gene '*Tp*') over tapering pod-tip (Sen and Ghosh 1958). Khadilkar (1963) reported that pod-pubescence is dominant which is also governed by independent duplicate genes. However, a single dominant gene is reported controlling plant pubescence trait (Murty and Patel 1973). Additive gene effects (Singh and Jain 1971; Singh and Singh 1971; Lal et al. 1982; Luthra et al. 1979; Reddy and Sreeramulu 1982; Wilson et al. 1985) and partial dominance (Malhotra et al. 1980) have been reported for inheritance of pod length. Mungbean mutant possessing twin podded was generated by inducted mutation using gamma ray (Dheeranupattana 1985). The twin podded mutant also showed larger seed size than the wild type. The twin podded is controlled by two genes with dominant and recessive epistasis (Dheeranupattana 1985).

4.5.7 Seed Traits

Two types of seed coat lustre are present in mungbean, dull versus shiny. Dull seed coat is dominant to shiny seed coat and is conditioned by a single gene, designated *D* (Rheenen 1965). Diversity for seed-coat colour (yellow, green, amber, brown, yellow mosaic green mosaic, tan and black) has been recorded in mungbean. Black, black-spotted and dull-green seed-coat colours were found to be dominant over green, non-spotted and shiny green colour, respectively. Khattak et al. (1999) and Lambrides et al. (2004) reported monogenic inheritance for seed-coat colour. Bose (1939) reported that two independent dominant genes control the seed-coat colour where gene '*A*' conditions the green colour, while gene '*B*' conditions the dark green colour. Gene symbol *A_ B_*, *A_bb*, *aaB_* and *aabb* were suggested for dark green, green, greenish yellow and pale green seed coats, respectively. Chen et al. (2001) suggested that the inheritance of black and green seed colours was controlled by a single gene (*B*), black being dominant over green. Sen and Ghosh (1958) suggested that three gene pairs '*BL*', '*bf*' and '*G*' condition blue sap colour, buff sap colour and green chloroplast which together define seed-coat colour. Rheenan (1965) reported dominant allele 'A' and 'Sp' for green and spotted seed coat. Four gene models ('*B*', '*M*', '*Br*' and '*G*') were proposed by Murty and Patel (1973) for defining genotypes for different seed-coat colour. Further, five major genes with non-allelic interactions were reported by Chhabra et al. (1990). Thakare et al. (1980) identified a green cotyledon mutant in mungbean from cultivar S8 and found that green cotyledon is governed by single recessive gene '*gc*'.

Seed hardness in mungbean is reported to be a dominant character which is governed by single gene pair (Singh et al. 1983, 2005; Lawn et al. 1988), designated Hd_1 Hd_1 (Singh et al. 1983).

QTL analysis revealed that four loci control hardseededness (Humphry et al. 2005; Isemura et al. 2012). The major locus on LG1 explained up to 34% of the trait variation (Isemura et al. 2012). Additive gene effects (Bhargava et al. 1966; Singh and Jain 1971; Singh and Singh 1971; Yohe and Poehlman 1975; Reddy and Sreeramulu 1982; Malik and Singh 1983; Rao et al. 1984; Imrie et al. 1985; Wilson et al. 1985) and dominance gene action (Singh and Singh 1971; Singh and Jain 1971 and Rao et al. 1984) for seed weight have been reported. Apart from these, four QTLs for seed weight were identified by Alam et al. (2014b).

4.5.8 Photoperiod Response

Verma (1971) reported that photo insensitiveness was dominant over the photo sensitiveness and was governed by single gene. Tiwari and Ramanujam (1976) observed dominance of earliness and photo insensitivity in F_1 generation. However, in F_2 generation, digenetic control was observed. Swindell and Poehlman (1978) reported dominant or partially dominant gene for sensitivity to photoperiod. Islam et al. (1998) reported two recessive genes for photoperiod sensitivity.

4.5.9 Yield Traits

Seed yield is important trait with complex inheritance. Seed yield is associated with many component traits that make direct and indirect contribution to the ultimate response. Both additive (Yohe and Poehlman 1975; Luthra et al. 1979; Reddy and Seeramulu 1982; Malik and Singh 1983; Rao et al. 1984; Payasi 2015) and non-additive genes (Singh and Jain 1971; Singh and Singh 1971 Lal et al. 1982) and epistatic interaction (Murty and Patel 1973; Singh and Singh 1971) have been reported for seed yield in mungbean. Further, partial dominance (Singh and Jain 1971; Singh and Singh 1971; Rao et al. 1984) and overdominance (Malik and Singh 1983; Luthra et al. 1979) have also been reported for seed yield. From the studies reported, it is quite evident that seed yield is complex trait and mode of inheritance varies with parent genotype.

Seed weight is a key trait contributing to seed yield and affects consumer preference and processing. Seed weight is principally under the control of several genes with additive effects (Yohe and Poehlman 1975; Imrie et al. 1985; Wilson et al. 1985; Malik et al. 1986; Alam et al. 2014), although genes with dominant or overdominant effect are also reported for the trait (Malik and Singh 1983; Rao et al. 1984; Luthra et al. 1979).

Pods per plant is a key agronomic trait which is found determined by additive gene action (Bhargava et al. 1966; Singh and Jain 1971; Yohe and Poehlman 1975; Malhotra et al. 1980, Reddy and Sreeramulu 1982; Rao et al. 1984), partial dominance to overdominance (Singh and Jain 1971; Singh and Singh 1971 and Luthra et al. 1979). Further, non-additive effects were more pronounced than the additive effects for the expression of this trait (Singh and Singh 1971; Alam et al. 2014).

Seeds per pod is an important yield component and additive gene action regulating this trait has been reported by several researchers (Yohe and Poehlman 1975; Luthra et al. 1979; Lal et al. 1982; Reddy and Sreeramulu 1982; Malhotra 1983; Alam et al. 2014). Also partial dominance to overdominance (Singh and Jain 1971; Singh and Singh 1971; Luthra et al. 1979) and epistasis (Malik and Singh 1983) regulation for this trait have been reported.

Pods per cluster is important determinant for yield per plant. Additive effects (Bhargava et al. 1966; Malhotra et al. 1980; Reddy and Sreeramulu 1982; Malik and Singh 1983; Wilson et al. 1985) and partial to overdominance (Singh and Singh 1971) are mainly reported for this trait.

4.5.9.1 Biotic and Abiotic Stress Resistance

Although several insects and diseases attack and cause yield reduction in mungbean, common insects and diseases that have been extensively studied in mungbean, viz. bruchids (*Callosobruchus* spp.), powdery mildew disease caused by fungus *Erysiphe polygoni*, Cercospora leaf spot disease caused by fungus *Cercospora canescens* and MYMD caused by MYMV and MYMIV. Resistance to these insect pests and diseases are each controlled by single or a few genes.

Bruchid resistance in mungbean is controlled by a single dominant locus, *Br*, with few modifying genes (Kitamura et al. 1988; Somta et al. 2007). The *Br* locus also confers resistance to pod sucking bug (***Riptortus pedestris*** Fab.) (Ishimoto and Kitamura 1993). Powdery mildew resistance is controlled by either single or two dominant genes (AVRDC 1979; Reddy et al. 1994; Khajudparn et al. 2007). Three dominant genes for the resistance were named *Pm*1, *Pm*2 and *Pm*3. Combination of dominant alleles at *Pm*1 and *Pm*2 resulting in immune resistance (Reddy et al. 1994; Reddy 2009), while dominant allele at *Pm*3 alone conferring immune resistance (Reddy 2009). Cercospora leaf spot is governed by a single dominant gene (AVRDC 1974; Thakur et al. 1977; Chankaew et al. 2011). The gene symbol *F* was proposed for the Cercospora leaf spot resistance (Thakur et al. 1977). MYMD resistance is controlled by single dominant gene (Lekhi et al. 2018) or single recessive gene (Malik et al. 1986; Khattak et al. 2000; Thakur et al. 1977). Bacterial leaf spot caused by *Xanthomonas phaseoli* is conditioned by a single dominant gene (Thakur et al. 1977). The gene symbol *Bls* was proposed for the resistance (Fery 1980). The gene is inherited independently of the genes for resistance to Cercospora leaf spot and MYMD.

In case of the abiotic stress, genetics of the resistance has been studied for calcareous soil (iron deficiency chrolosis) only. Nopparat et al. (1997) reported that the resistance is controlled by the two genes with inhibitory action, although a single dominant gene action is also possible. However, Srinives et al. (2010) reported that resistance is conditioned by a single dominant gene, designated *IR*, with a few modifying genes.

4.6 Breeding Methods

Ranali and Cubero (1997) discussed the basis of genetic improvement in legumes and the application of breeding methods, including introduction, hybridization, early generation selection and mutation, along with molecular markers that offer opportunity to enhance precision.

4.6.1 Introduction

Introduction is a primary approach in crop improvement, in which introduced germplasm is directly released as variety. In last two decades, lot of mungbean germplasm was introduced into several countries of Asia and Africa from World Vegetable Centre. The introduced germplasm/breeding line possessed earliness, bold seed size and long pod with up to 18 seeds per pod. In India, the germplasm was utilized for broadening the genetic base of mungbean. Introductions have been successful in the development of Pusa 105, Pusa 9531, Pant Moong 5, Pusa Vishal and SML 668. In Thailand and China, the breeding lines from the World Vegetable Centre, such as VC1973A and VC2778A, were selected and released as mega varieties. Those breeding lines/varieties have contributed to the great expansion of mungbean cultivation in Thailand and China (Srinives 1996). At present, in China, the breeding lines are used to cross with local germplasm to develop superior varieties. In Pakistan, the breeding lines were crossed with local germplasm with resistance to yellow mosaic disease, resulting in popular varieties: NIAB Mung 92 (NM92) and NIAB Mung 94 (NM94) that revolutionized mungbean cultivation in this country (Ali et al. 1997). In semi-arid areas of eastern Kenya, mungbean introductions from the World Vegetable Centre were tested for adaptability and seed yield. AVMU 0801, AVMU 1003 and AVMU 8501 were identified as farmer preferred lines through farmer participatory evaluation and finally released as varieties (Karimi et al. 2019).

4.6.2 Pure Line Breeding

Pure line selection is the step preceding introduction of a line, in which the selection of better plant types is made from an already existing genetically heterogeneous population or landrace. These superior plant types are identified as the result of natural selection pressure, which helps to evolve new plant types with strong genetic potential. These variants are fixed by breeders through a continuous cycle of selfing and selection (Gupta and Kumar 2006; Tickoo et al. 2006). In India, many mungbean varieties are developed from pure line selections (Srinives 1996).

4.6.3 Recombination Breeding

Hybridization is the most common method used by plant breeders for combining desirable traits. Better recombinants can be obtained through intraspecific or interspecific hybridization. Landraces, exotic collections and primitive forms are key sources of rare alleles for useful traits in plant breeding. At the World Vegetable Centre, superior breeding lines were developed by hybridizing between high yielding, large seeded, synchronous maturing and photo-insensitive germplasm from the Philippines and disease-resistant germplasm from India (Fernandez and Shanmugasundaram 1988). Several of such the breeding lines are used as parents in crossing to generate new recombinants with better-desired traits. The prior knowledge of parental performance, combining ability and genetics of trait to be incorporated is essential for the breeding of high-yielding genotypes. The knowledge of yield contributing traits helps plant breeder in selection of appropriate breeding material in segregating generations. After hybridization in mungbean breeding material can be advanced following pedigree, bulk, recurrent, backcross or single-seed methods of selection. Interspecific hybridization often results in pre-breeding material which is subsequently utilized by the breeders for varietal improvement. Dahiya and Singh (1986) compared efficiency of selection methods including single-seed descent (SSD), mass selection and selective intermating in mungbean in which progeny developing after two cycles of selection were evaluated for yield and seven agronomic and yield-related traits. Number of high-yielding lines, mean yield of top 10% lines and mean of the highest yielding line were used to determine the relative efficacy of each selection method. They found that selection after two cycles of selective intermating was the best method for developing high-yielding lines and that SSD was the least efficient method. Later, Gill et al. (1995) compared efficiency of four selection methods, viz., honeycomb (HC), pedigree selection (PS), SSD and bulk selection (BS) in mungbean based on the basis of the mean of the lines, the range, the number of superior lines over the best check and the proportion of the top 10% lines in all the crosses and generations. They found that (i) HC is the most efficient method for yield per plant and yield-related traits, (ii) PS, SSD and BS were not different and (iii) HC and SSD methods were suitable for generating superior lines with high seed yield and pods per plant.

For breaking undesirable linkage and accumulating desirable traits, recurrent selection and population improvement have been suggested. Burton (1997) suggested use of early generation testing. This method helps in discarding inferior progenies and reducing population load. In this method, F_2, F_3 and even F_4 families are subjected to early generation selection depending upon the target trait and inferior families are rejected. Interspecific hybridization involving mungbean and black gram (*Vigna mungo* (L.) Hepper) has led to the development of four mungbean varieties (Pant M 4, HUM 1, Meha, PM 6) with improved plant types. Important traits like sympodial bearing, non-shattering, stable MYMD resistance, etc., can be transferred to mungbean from urdbean. Interspecific hybridization between mungbean and rice bean (*Vigna umbellata* (Thunb.) Ohwi and Ohashi) have resulting in mungbean lines with resistance to bruchids (seed weevils) (Mariyammal et al. 2019) and MYMD (Mathivathana et al. 2019).

4.6.4 Mutagenesis

Induced mutations are useful for traits lacking variability in primary gene pool. Mutations may occur spontaneously or can be induced artificially. Effectiveness and efficiency of mutagen are important. Effectiveness of mutagen is associated with mutation per unit dose of mutagens and efficiency is related to changes like sterility, injury and lethality (Goud 1967). Mutation can be induced in seed as well as in the vegetative portion of the plant. The effect and efficiency of mutagen are measured by its effect on genotype which varies with dosage and nature of mutagen. Khan et al. (2006) reported that ethlyl methane sulphonate (EMS) exhibits high mutagenic efficiency in comparison to other chemical mutagens. Gunasekaran et al. (1998) compared efficiency of gamma rays and ethidium bromide in generating variation for different agronomic traits. They reported that gamma rays were more effective in causing genetic changes breaking linkages. Variation for high protein content and yield in mungbean was induced by Chakraborty et al. (1998) using gamma irradiation. Variation for yield and related agronomic traits was induced in mungbean by different researchers (Tah and Saxena 2009; Ahmed et al. 2015; Dewanjee and Sakar 2017; Wani et al. 2017; Das and Baisakhi 2018). Srinives et al. (2000) and Tah (2006) obtained leaf mutants, pod mutants and semi-dwarf plants utilizing gamma irradiation.

Singh and Kole (2006) used EMS and studied genetic variability for agronomic traits. They obtained branchless and multifoliate mutants. Improvement in resistance to powdery mildew, Cercospora leaf spot and cowpea weevil through gamma radiation-induced mutation was reported by Wongpiyasatid et al. (1999). Mutation breeding has been used to develop improved cultivars in mungbean either through mutation breeding directly or by involving mutants as a parent in crossing programmes (Ahloowalia et al. 2004; Gopalakrishna and Reddy 2009). Till date, 38 varieties have been developed using mutation breeding. In India, 16 varieties have been developed through mutation breeding. Most cultivars are early maturing, high yielding and tolerant/resistant to YMV (Ahloowalia et al. 2004). In addition to these popular varieties, Pusa Vishal and SML 668 have been developed through selection in mutant lines NM92 and NM94, respectively. These varieties are early maturing, bold seeded, high yielding and tolerant to MYMD. Mutant varieties NM92 and NM98 are popular in Pakistan and in other countries like Bangladesh and Myanmar. In Thailand, mutation breeding is a main method in developing high-yielding varieties. Mutant variety Chai Nat 72 is popular in the country for its higher yield, larger seed size and better resistance to alkaline soil than the wild type variety. All the currently popular varieties in the country were developed by mutation breeding.

4.7 Conclusion

Traditional breeding has contributed in a great way in developing varieties which have been well adopted by farmers. The chapters on breeding for biotic stresses and abiotic stresses in this book covers more detailed information, with examples. Greater use of mungbean germplasm held in genebanks and also of related species would help in broadening the genetic base of mungbean varieties. This will be critical as the area under the crop expands and new pests and diseases emerge.

References

Ahloowalia BS, Maluszynski M, Nichterlein K (2004) Global impact of mutation-derived varieties. Euphytica 135:187–204

Ahmad HM, Ahsan M, Ali Q, Javed I (2015) Path coefficient analysis for various quantitative traits of mungbean (*Vigna radiata* L.) at different radiation levels. IJAVMS Online First 25 Dec 2015. https://doi.org/10.5455/ijavms.20130111041610

Alam AKMM, Somta P, Srinives P (2014a) Generation mean and path analyses of reaction to mungbean yellow mosaic virus (MYMV) and yield-related traits in mungbean (*Vigna radiata* (L.) Wilczek). SABRAO J Breed Genet 46:150–159

Alam AKMM, Somta P, Muktadir MA, Srinives P (2014b) Quantative trait loci associated with seed weight in mungbean (*Vigna radiata* (L) Wilczek). Kasetsert J (Nat Sci)48:197–204

Ali M, Malik IA, Sabir HM, Ahmad B (1997) The mungbean green revolution in Pakistan. Asian Vegetable Research and Development Center, Shanhua, Taiwan

Appa Rao S, Jana MK (1973) Inheritance of anthocyanin coloration in *Phaseolus* mutants. Ind Sci Cong Ass Pro 60:302

Arumuganathan K, Earle ED (1991) Nuclear DNA content of some important plant species. Plant Mol Bio Rep 9:211–215

AVRDC (1974) Progress report 1974. Asian Vegetable Research and Development Center, Shanhua, Taiwan, Republic of China

Bhargava PD, Johri JN, Sharma SK, Bhatt BN (1966) Morphological and genetic variability in green gram. Indian J Gen Plant Breed 26:370–373

Bhattacharya S (1979) Study of karyotype, report of B chromosomes and polytene chromosomes in *Phaseolus aureus*Roxb. Curr Sci 48:635–637

Bhatnagar CP (1974) Cytotaxonomic studies in the genus *Phaseolus*. Indian J Gene 34A:800

Boling M, Snader DA, Matlock RS (1961) Mungbean hybridization technique. Agron J 53:54–55

Bose RD (1939) Studies in Indian pulses, IX. Contributions to the genetics of mung (*Phaseolus radiates* Linn. Syn. *Ph. Aureus* Roxb.). Indian J Agri Sci 9:575–594

Burton JW (1997) Soybean (*Glycine max* (L.) Merr.). Field Crops Res 53:171–186

Chankaew S, Somta P, Sorajjapinun W, Srinives P (2011) Quantitative trait loci mapping of *Cercospora* leaf spot resistance in mungbean, *Vigna radiata* (L.) Wilczek. Mol Breed 28:255–264

Chakraborty RK, Bhowmik A, Hossain T, Mian MAK (1998) Induction of mutation in mungbean (*Vigna radiata*) through gamma-irradiation. Ann Bangladesh Agri 8(129):136

Chen HM, Liu XH, Chen HM, Liu XH (2001) Inheritance of seed color and luster in mungbean (*Vigna radiata*). Hunan Agri Sci Tec Newslett 2(1):8–12

Chhabra AK (1990) Inheritance of lobed and pentafoliate leaves in mungbean. Indian J Pulses Res 3:69–72

Chhabra AK, Singh VP, Kharb RPS (1990) Multifactor inheritance of seed coat colour in green gram (*Vigna radiata* (L.). Wilczek). Euphytica 47:153–158

Cupka TB, Edwards LH (1986) A new technique for crossing mungbeans 1. Crop Sci 26:830–831

Dahiya N, Singh VP (1986) Comparison of single seed descent, selective intermating and mass selection for seed size in greengram (*Vigna radiata* (L.) Wilczek). Theor Appl Genet 72:678–681

Das TR, Baisakhi B (2018) Mutation induced variability and early prediction of high yielding mutants in green gram [*Vigna radiata* (L.) Wilczek]. Int J Curr Microbiol App Sci 3228–3236

Dewanjee S, Sakar KK (2017) Evaluation of performance of induced mutants in mungbean [*Vigna radiata* (L.) Wilczek]. Leg Res Inter J 41(2):213–217

Dheeranupattana S (1985) Inheritance of twin podded mungbean mutant. Master thesis, Kasetsart University, Thailand

Dwivedi S, Singh DP (1986) Inheritance of people pigmentation and seed colour in green gram. Crop Impro 13:63–66

Dwivedi S, Singh DP (1990) Inheritance of fascination in mungbean (*Vigna radiata* L. Wilzeck). Indian J Genet Plant Breed 50:81–82

Fernandez GCJ, Shanmugasundaram S (1988) The AVRDC mungbean improvement program: the past, present and future. In: Shanmugasundaram S, McLean BT (eds) Mungbean: proceeding of the second international symposium. AVRDC, Shanhua, Taiwan, pp 58–70

Fery RL (1980) Genetics of *Vigna*. In: Janick J (ed) Horticultural reviews, vol 2. AVI Publishing Co., Inc, Wesport, CT, pp 311–394

Gill JS, Verma MM, Gumber RK, Brar JS (1995) Comparative efficiency of four selection methods for deriving high-yielding lines in mungbean [*Vigna radiata* (L.) Wilczek]. Theor Appl Genet 90:554–560

Gopalakrishna T, Reddy KS (2009) Mutation breeding of food legumes. In: Ali M, Kumar S (eds) Milestones in food legumes research. Indian Institute of Pulses Research, Kanpur, India, pp 206–224

Goud JV (1967) Induced mutations in bread wheat. Indian J Genet 27:40–55

Gupta S, Kumar S (2006) Urd bean breeding. In: Ali M, Kumar S (eds) Advances in mung bean and urd bean. Indian Institute of Pulses Research, Kanpur, India, pp 149–168

Humphry ME, Lambrides CJ, Chapman SC, Aitken EAB, Imrie BC, Lawn RL, Mcintyre CL, Liu CJ (2005) Relationships between hard seededness and seed weight in mungbean (*Vigna radiata*) assessed by QTL analysis. Plant Breed 124:292–298

Imrie BC, Ahmed ZU, Eefens JPJ (1985) Heritability of seed weight in mungbean. SABRAO J 17:173–175

Isemura T, Kaga A, Tabata S, Somta P, Srinives P, Shimizu T, Jo U, Vaughan DA, Tomooka N (2012) Construction of a genetic linkage map and genetic analysis of domestication related traits in mungbean (***Vigna radiata***). PLoS ONE 7(8):e41304

Ishimoto I, Kitamura K (1993) Inhibitory effects of adzuki bean weevil—resistant mungbean seeds on growth of bean bug. Japan J Breed 43:75–80

Islam MT, Mian MAK, Hoque ME, Afzal MA (1998) Inheritance of photoperiod sensitivity in mungbean (*Vigna radiata* (L.) Wilzeck. Thai J Agric Sci 31:217–223

Jiao K, Li X, Su S, Guo W, Guo Y, Guan Y, Hu Z, Shen Z, Luo D (2019) Genetic control of compound leaf development in the mungbean (*Vigna radiata* L.). Horticul Res 6:23

Johnston JS, Bennett MD, Rayburn AL, Galbraith DW, Price HJ (1999) Reference standards for determination of DNA content of plant nuclei. Am J Bot 86:609–613

Joseph LS, Bouwkamp JC (1978) Karyomorphology of several species of *Phaseolus* and *Vigna*. Cytologia 43:595–599

Kang YJ, Kim SK, Kim MY, Lestari P, Kim KH, Ha B-K, Jun TH, Hwang WJ, Lee T, Lee J et al (2014) Genome sequence of mungbean and insights into evolution within Vigna species. Nat Commun 5:5443

Karimi R, Nair RM, Ledesma D, Mutisya DL, Muthoni L (2019) Performance and participatory evaluation of green gram genotypes in the semi-arid environments of Eastern Kenya. East African Agric Forest J 83:119–136

Khadilkar BT (1963) Inheritance of two characters in green gram (*Phaseolus aureus* Roxb. Piper). Akola Agric College Mag 3:29–32

Khajudparn P, Wongkaew S, Thipyapong P (2007) Mungbean powdery mildew resistance identification of genes for powdery mildew resistance in mungbean. Afr Crop Sci Conf Proc 8:743–745

Khan S, Wani MR, Parveen K (2006) Sodium azide induced high yielding early mutant in lentil. Agric Sci Digest 26(1):65–66

Khattak GSS, Haq MA, Rana SA, Elahi T, Srinives P (1998) An efficient technique for crossing mungbean (*Vigna radiata* (L.) Wilczek). Thai J Agric Sci 31 (4):577–582

Khattak GSS, Haq MA, Rana SA, Srinives P, Ashraf M (1999) Inheritance of resistance to mungbean yellow mosaic virus (MYMV) in mungbean (*Vigna radiata* (L) Wilczek). Thai J Agric Sci 32(1):49–54

Khattak GSS, Haq MA, Ashraf M, Elahi T (2000) Genetics of mungbean yellow mosaic virus (MYMV) in mungbean (*Vigna radiata* (L.) Wilczek). J Genet Breed 54:237–243

Kitamura K, Ishimoto M, Sawa M (1988) Inheritance of resistance to infestation with azuki bean weevil in Vigna sublobata and successful incorporation to *V. radiata*. Jap J Breed 38:459–464

Krishnan R, De DN (1965) Studies on pachytene and somatic chromosomes of *Phaseolus aureus* Roxb. Nucleus 8:7–16

Kumar S, Singh BB, Singh DP (2006) Genetics and cytogenetics of mungbean. In: Ali M, Kumar S (eds) Advances in mungbean and urdbean. Indian Institute of Pulses Research, Kanpur, India, pp 40–68

Lal S, Gupta SN, Tomer YS (1982) Line × tester analysis in mungbean. Haryana Agri Uni J Res 12:438–443

Lambrides CJ, Lawn RJ, Godwin ID, Manners J, Imrie BC (2004) Segregation distortion for seed testa color in mungbean (*Vigna radiata* (L.) Wilczek). J Heredity 95(6):532–535

Lavania UC, Lavania S (1983) Karotype studies in Indian pulses. Genetica Agraria 37:299–308

Lawn RJ, Williams RW, Imrie BC (1988) Potential of wild germplasm as a source of tolerance to environmental stresses in mungbean. In: Shanmugasundaram S, McLean BT (eds) Mungbean: proceeding of the second international symposium. AVRDC, Shanhua, Taiwan, pp 136–145

Lekhi P, Gill RK, Kunal Kaur S (2018) Identification of molecular marker linked to mungbean yellow mosaic virus (MYMV) resistance in *Vigna radiata* (L.) Wilczek. Electr J Plant Breed 9:839–845

Li S, Ding Y, Zhang D, Wang X, Tang X, Dai D, Jin H, Lee SH, Cai C, Ma J (2018) Parallel domestication with a broad mutational spectrum of determinate stem growth habit in leguminous crops. Plant J 96(4):761–771+

Luthra OP, Arora ND, Singh RK, Chaudhary BD (1979) Genetic analysis for metric traits in mungbean (*Vigna radiata* Wilczek). Haryana AgriUni J Res 9:19–24

Malhotra RS, Gupta PK, Arora ND (1980) Diallel analysis over environments in mungbean. Indian J Gene Plant Breed 40:64–65

Malik BPS, Singh VP (1983) Genetics of some metric traits in green gram. Indian J Agri Sci 53:1002–1005

Malik LA, Sarwar G, Ali Y (1986) Genetic studies in mungbean (*Vigna radiata* (L.) Wilczek). I: inheritance of tolerance to mungbean yellow mosaic virus and some morphological characters. Pak J Bot 18:189–198

Mariyammal I, Seram D, Samyuktha SM, Karthikeyan A, Dhasarathan M, Murukarthick J, Kennedy JS, Malarvizhi D, Yang TJ, Pandiyan M, Senthil N (2019) QTL mapping in *Vigna radiata* × *Vigna umbellata* population uncovers major genomic regions associated with bruchid resistance. Mol Breed 39:110

Mathivathana MK, Murukarthick J, Karthikeyan A, Jang W, Dhasarathan M, Jagadeeshselvam N, Sudha M, Vanniarajan C, Karthikeyan G, Yang TJ, Raveendran M, Pandiyan M, Senthil N (2019) Detection of QTLs associated with mungbean yellow mosaic virus (MYMV) resistance using the interspecific cross of *Vigna radiata* × *Vigna umbellata*. J Appl Genet. https://doi.org/10.1007/s13353-019-00506-x

Misra RC, Sahu RC, Tripathy D (1970) A note on heterosis in green gram (*Phaseolus aureus* Roxb.). Current Sci 39:190–191

Mukherjee A, Pradhan K (2002) Genetics of lobation in trifoliate leaf of mungbean. Abstr Perspect Cytol Genet 11:88

Murray MG, Palmer JD, Cuellar RE, Thompson WF (1979) DNA sequence organization in the mungbean genome. Biochemistry 18:5259–5266

Murty BK, Patel GJ (1973) Inheritance of some morphological characters in mungbean. Basilal Amrital College Agric Mag 25:1–9

Nopparat S, Srinives P, Kaveta R, Jintakanon S (1997) An inheritance of mungbean tolerance to microessential element denciency in Takhli soil series. In: Romkaew J, Kaveeta R, Srinives P, Thanomsub W, Ngampongsai S, Phoomthaisong J (eds) Proceedings of national mungbean research conference VII. Phitsanulok, Thailand, pp 78–82 (in Thai with English abstract)

Park HG, Yang CY (1978) The mungbean breeding program. In: Proceedings of 1st-International

Mungbean Symposium. Asian Vegetable Research and Development Centre, R.O.C., Shanhua, Taiwan, pp 214–216
Pathak JN, Singh B (1963) Inheritance studies in green gram. Indian J Genet Plant Breed 23:215–218
Payasi DK (2015) Genetic variability analysis for seed yield and its components in mungbean (*Vigna radiata* L. Wilczek). Inter J Plant Breed Genet 9:177–188
Poehlman JM (1991) The mungbean. Oxford and IBH Publishing Co., New Delhi, p 375
Pohle YS (1972) Spontaneous mutation in mung (*Phaseolus aureus* Roxb.). Sci Cult 38:142
Ramamoorthi N, Veerabadhiran P, Ramasamy A, Jehangir KS (1994) Inheritance of lobed leaf in mungbean. Madras Agric J 81:685
Ranali P, Cubero JI (1997) Bases for genetic improvement of grain legumes. Field Crops Res 53(1–3):69–82
Rao SS, Singh SP, Singh SK (1984) Estimation of additive, dominance, digenic epistatic interaction effects for yield and its components in mungbean (*Vigna radiata* (L.) Wilczek). Leg Res 7:6–12
Reddy KS (2009) Identification and inheritance of a new gene for powdery mildew resistance in mungbean (*Vigna radiata* L. Wilczek). Plant Breed. 128:521–523
Reddy KS, Pawar SE, Bhatia CR (1994) Inheritance of powdery mildew (Erysiphe polygoni D.C.) resistance in mungbean [*Vigna radiata* (L.) Wilczek]. Theor Appl Genet 88:945–948
Reddy PRR, Sreeramulu C (1982) Heterosis and combining ability for yield and its components in green gram (*Vigna radiata* (L.) Wilczek). GeneticaAgraria 36:297–308
Rheenen HA (1965) The inheritance of some characters in the mungbean, *Phaseolus aureus* Roxb. Genetica 36:412–419
Sahai S, Rana RS (1980) Homology and differentiation in *Phaseolus*. Indian J Genet Plant Breed 40:311–315
Saini RG, Minocha JL, Singh A (1974) Sterile mutants in *Phaseolus aureus*. Sci Cult 40:37–38
Sarbhoy RK (1980) Karyological studies in the genus *Phaseolus* Linn. Cytologia 45:363–373
Sareen PK (1982) Genetic analysis of the trilobite leaf mutant in mungbean (*Vigna radiata* var. *aureus*). Current Sci 51:1028–1029
Sen NK, Ghosh AK (1959) Genetics studies in green gram. Indian J Genet Plant Breed 19:210–227
Sharma PC, Gupta PK (1982) Karyotypes in some pulse crops. Nucleus 25:181–185
Singh DP (1980) Note on the inheritance of lobed leaf in green gram. Indian J Genet Plant Breed 40:608–609
Singh TP, Malhotra RS (1975) Crossing technique in mungbean (*Phaseolus aureus*Roxb.). Curr Sci 44:64–65
Singh D, Mehta TP (1953) Inheritance of lobed leaf margin in mung (*Phaseolus aureus*L.) Wilczek). Curr Sci 2:348
Singh DP, Sharma BL, Dwivedi S (1983) Inheritance of hard seeds in interspecific crosses of mungbean. Indian J Gnut 43:378–379
Singh I, Gill MS, Bains TS, Singh (2005) Genetic analysis of hardseededness in mungbean (*Vigna radiata* L. Wilzeck). Crop Improv 32(2):170–172
Singh P, Singh IB, Singh U, Singh HG (1975) Interspecific hybridization between mung (*Phaseolus aureus*Roxb.) and urd (*Phaseolus mungo* L.). Sci Cult 41:233–234
Singh R, Kole CR (2006) Delineation of EMS induced genetic variability in some agronomic traits in mungbean (*Vigna radiata* L. Wilczek). Crop Res 32:94–96
Singh KB, Jain RP (1971) Analysis of diallel cross in *Phaseolus aureus* Roxb. Theor Appl Genet 41:279–281
Singh TP, Singh KB (1971) Mode of inheritance and gene action for yield and its components in *Phaseolus aureus*. Canadian J Genet Cytol 14:517–525
Soehendi R, Chanprame S, Toojinda T, Srinives P (2006) Inheritance and AFLP tagging of leaflet mutants in mungbean (*Vigna radiata* (L.) Wilczek). Kasetsart J (Nat Sci) 40:566–572
Somta P, Ammaranan C, Ooi PAC, Srinives P (2007) Inheritance of seed resistance to bruchids in cultivated mungbean (*Vigna radiata*, L. Wilczek). Euphytica 155:47–55
Srinives P (1996) Mungbean breeding: past, present and future. In: Srinives P, Kitbamroong C, Miyazaki S (eds) Mungbean germplasm: collection, evaluation and utilization for breeding program, JIRCAS working report number 2. Japan International Research Center for Agricultural Sciences, Ministry of Agriculture, Forestry and Fisheries, pp 73–82
Srinives P, Hual AN, Saengchot S, Ngampongsai S, Srinives P (2000) The use of wild relatives and gamma radiation in mungbean and blackgram breeding. In: The seventh international workshop on genetic resources. Part 1: wild legumes, 13–15 Oct 1999. Ministry of Agriculture, Forestry and Fisheries (MAFF), Ibaraki, Japan, pp 205–218
Swindell RE, Poehlman JM (1978) Inheritance of photoperiod response in mungbean (*Vigna radiata* (L.) Wilczek). Euphytica 27:325–333
Tah PR (2006) Studies on gamma ray induced mutations mungbean [*Vigna radiata* (L.) Wilczek]. Asian J Plant Sci 5(1):61–70
Tah PR, Saxena S (2009) Induced synchrony in pod maturity in mungbean (*Vigna radiata*). Int J Agric Bio 11:321–324
Talukdar T, Talukdar D (2003) Inheritance of growth habit and leaf shape in mungbean (*Vigna radiata* (L.). Wilczek). Indian J Genet Plant Breed 63:165–166
Thakare RG, Pawar SE, Joshua DC (1980) Green cotyledon mutant in mungbean (*Vigna radiata* (L.). Wilczek). Curr Sci 49:945–946
Thakur RP, Patel PN, Verma JP (1977) Genetical relationships between reactions to bacterial leaf spot, yellow mosaic and *Cercopsora* leaf spot diseases in mungbean (*Vigna radiata*). Euphytica 26:765–774
Tickoo JL, Lal SK, Chandra N, Dikshit HK (2006) Mungbean breeding. In: Ali M, Kumar S

(eds) Advances in mungbean and urdbean Indian Institute of Pulses Research. Kanpur, India, pp 110–148

Tiwari AS, Ramanujam S (1976) Genetics of flowering response in mungbean. Indian J Genet Plant Breed 36:418–419

Tomooka N, Vaughan DA, Moss H, Maxted N (2002) The Asian Vigna: subgenus ceratotropis genetic resources. Kluwer Academic Publisher, Dordrecht, The Netherlands

Van Rheenen HA (1964) Preliminary study of natural cross-fertilization in mungbean. Neth J Agric Sci 12 (4):260–262

Verma SNP, Krishi JN (1969) Inheritance of some qualitative characters in green gram (*Phaseolus aureus* Roxb.). Indian J Hered 1:105–106

Verma SNP (1971) Inheritance of photosensitivity in mungbean (*Phaseolus aureus* Roxb.). Mysore J AgriSci 5:477–480

Virk DS, Verma MM (1977) A dominant mutation in *Vigna radiata* var. *radiata*. Crop Impro 4:115–116

Wani MR, Dar AR, Tak A, Amin I, Shah NH et al (2017) Chemo-induced pod and seed mutants in mungbean. SAARC J Agri 15(2):57–67

Wilson D, Mercy ST, Nayar NK (1985) Combining ability in green gram. Indian J Agric Sci 55:665–670

Wongpiyasatid A, Chotechuen S, Hormchan P, Srihuttagum M (1999) Evaluation of yield and resistance to powdery mildew, Cercospora leaf spot and cowpea weevil in mungbean mutant lines. Kasetsart J (Nat Sci) 33:204–15

Yohe JM, Poehlman JM (1975) Regressions, correlations and combining ability in mungbeans (*Vigna radiata* (L.) Wilczek). Trop Agric 52:343–352

Breeding Progress and Future Challenges: Biotic Stresses

5

Aditya Pratap, Col Douglas, Umashanker Prajapati, Gita Kumari, Abdul Rasheed War, Rakhi Tomar, Abhay K. Pandey and Sonali Dubey

Abstract

Mungbean is a short-duration legume crop cultivated in South Asia, Southeast Asia and Australasia. Its cultivation is rapidly spreading to other parts of the world. Insect pests and diseases are the major constraints in increasing the productivity of mungbean crop. The important diseases in mungbean include mungbean yellow mosaic, anthracnose, powdery mildew, Cercospora leaf spot, dry root rot, halo blight, bacterial leaf spot and tan spot. The major insect-pests of mungbean are stem fly, thrips, aphids, whitefly, pod borers and bruchids. Development of host plant resistance to insect pests and diseases in mungbean by breeding for resistance is an alternative, economical and environment-friendly approach. Though breeding for resistance to insect pests and diseases has been extensively studied in mungbean, the success rate in stabilizing the resistance has been less due to the development of insect biotypes, new strains in pathogens and the environmental interactions. This chapter covers the insect and disease resistance sources in mungbean, resistant traits, the genetic basis of resistance and different breeding methods involved in breeding for insect and disease resistance.

5.1 Introduction

The Asiatic *Vigna* species belong to the subgenus *Ceratotropis* of which 4 species are globally recognized for their high agronomic importance. These include *V. radiata* (L.) Wilczek (green gram or mungbean); *V. mungo* (L.) Hepper (black gram or urdbean); *V. angularis* (W) Ohwi and Ohashi (adzuki bean); and *V. aconitifolia* (Jacq.) (moth bean) (Pratap et al. 2015); besides, *V. umbellata* (Thumb.) Ohwi and Ohashi (rice-bean) and *V. glabrescens* Marechal, Mascherpa and Stainier (tua pea) are of little importance. Among these, mungbean, also known as green gram, is economically the most important as indicated by its area, production and consumption at the global level (Kumar et al. 2006; Tomooka et al. 2007; Nair et al. 2013). Mungbean is a warm season, short-day plant that has been grown in India since ancient times. Besides India, it is widely grown in South Asia and Southeast Asia and also in Africa, South America and Australia and serves as a major source of

A. Pratap (✉) · U. Prajapati · G. Kumari · R. Tomar · S. Dubey
ICAR-Indian Institute of Pulses Research, Kalyanpur, Kanpur 208024, Uttar Pradesh, India
e-mail: Aditya.Pratap@icar.gov.in

C. Douglas
Agri-Science Queensland, Department of Agriculture and Fisheries, Hermitage Research Facility, 604 Yangan Road, Warwick, QLD 4370, Australia

A. R. War · A. K. Pandey
World Vegetable Center, South Asia, ICRISAT Campus, Patancheru, Hyderabad, Telangana, India

R. M. Nair et al. (eds.), *The Mungbean Genome*, Compendium of Plant Genomes,
https://doi.org/10.1007/978-3-030-20008-4_5

dietary protein for the vast majority of vegetarian people (AVRDC 2012; Clarry 2016). As of now, mungbean is grown over an area of 6.0 million ha globally with the production of 3.5 million tonnes. However, despite an average yield potential of >1.2 t/ha for most of the released mungbean varieties, the average productivity is still <0.7 t/ha in India and <1.0 t/ha in several other mungbean-growing countries (Pratap et al. 2019). Several factors such as biotic and abiotic stresses, environmental fluctuations and high genotype × environment interaction affect the yield of mungbean. Among the biotic stresses, diseases alone can lead to a yield reduction of 10–100%, while weeds may cause 50–90 and insect pests up to 20–55% yield loss (Rana et al. 2016), depending upon the stage of the crop, the severity of the stress and prevailing environmental factors. To stabilize the mungbean production and improve its productivity, it is important to develop the cultivars that are either resistant to and/or withstand the insect and disease pressure. Advanced technologies such as genomics, proteomics and metabolomics have paved way for the in-depth studies in the genetic basis of insect–plant and pathogen–plant interactions, which in turn can be applied to design effective crop improvement strategies.

5.2 Major Biotic Stresses

Among the biological constraints, diseases impart the most serious constraints, which may limit its productivity besides affecting the physical quality of seeds, leading them unusable. Mungbean is prone to several viral, fungal and bacterial diseases (Khattak et al. 2000; Pandey et al. 2018; Noble et al. 2019) (Table 5.1). Among the viral diseases, mungbean yellow mosaic disease (MYMD) is the most important disease of mungbean (Singh and De 2006; Kitsanachandee et al. 2013), caused by Begomovirus and transmitted by whitefly. The MYMD (Fig. 5.1a) on *Vigna* species was first time reported by Mclean (1941) from western India in the late 1940s in lima bean and later in mungbean from the Indian Agricultural Research Institute, New Delhi (Naraini 1960), followed by Pakistan (Ahmad and Harwood 1973). From India, 32–78% yield reduction in mungbean grains has been reported (Khattak et al. 2000). However, yield reduction was higher (100%) at early growth stages (Kitsanachandee et al. 2013). Urdbean leaf crinkle disease caused by urdbean leaf crinkle virus (ULCV) is an emerging viral disease of mungbean in South Asia and Southeast Asia (Singh et al. 1988).

Although 35 fungal diseases are reported to affect mungbean globally, only very few of them are widespread and economically important. Fungal diseases (Fig. 5.1b–e) of common occurrence are Cercospora leaf spot (CLS) [*Cercospora* spp.], powdery mildew (*Erysiphe polygoni* DC, *Podosphaera fusca* (Fr.) U. Braun and Shishkoff), dry root rot and Macrophomina blight [*Macrophomina phaseolina* (Tassi) Goid], anthracnose (*Colletotrichum* sp.), Rhizoctonia root rot and web blight (*Rhizoctonia solani* Kuhn) (Singh et al. 2011; Pandey et al. 2018). Bacterial diseases such as halo blight (*Pseudomonas savastanoi* pv. *phaseolicola*) and tan spot (*Curtobacterium flaccumfaciens* pv. *flaccumfaciens*) (Fig. 5.1f, g) are economically the most significant diseases of mungbean in Australia (Ryley and Tatnell 2011). Bacterial leaf spot caused by *Xanthomonas campestris* pv. *Vigna radiata* is prevalent in India (Thakur et al. 1977). All the diseases together can cause significant yield losses of up to 10–100% (Rana et al. 2016). All bacterial pathogens are seed-borne and can persist in stubbles, and varietal resistance is recognized as the cornerstone of integrated disease management (Noble et al. 2019). Previously thought to be restricted to Australia, recent reports indicate that their distribution may be more widespread (Sun et al. 2017).

Insect pests are one of the major constraints in mungbean production and take a heavy toll on the yield. Insect pests attack different parts of the mungbean plant including roots, shoots, flowers and pods at different growth stages. The damage by insect pests is either direct, where damage is caused by direct feeding by the insects such as pod borers, thrips, aphids and/or indirect, where insects act as vectors of some serious diseases

Table 5.1 Characteristic symptoms and distribution of common fungal and bacterial diseases of mungbean

Disease	Causal pathogen(s)	Characteristic symptoms	Distribution
Fungal diseases			
Cercospora leaf spot	*Cercospora cruenta* Sacc. (*Mycosphaerella cruenta* Latham), *C. canescens* Ell. and Martin, C. kikuchii Matsumato and Tomoyasum *(M. phaseoli)* *C. dolichi* Ell. and Evr *C. caracallae* (Speg.) Greene	Small leaf spots (1–5 mm) with brown to greyish centres and reddish border	Taiwan (Hartman et al. 1993), Thailand (Wongpiyasatid et al. 1999), India (Zhimo et al. 2013), Pakistan (Iqbal et al. 2004)
Powdery mildew	*Erysiphe polygoni* DC., *Podosphaera fusca*	White powdery coating on leaves, stems and pods	Australia (Ryley and Tatnell 2011), Taiwan (Hartman et al. 1993), Thailand (Kasettranan et al. 2010), India (Mandhare and Suryawanshi 2008)
Anthracnose	*Colletotrichum lindemuthianum* (Sacc. and Magn.) Bri. and Cav. (*Glomerella lindemuthianum* (Sacc. and Magn.) Shear) *C. Capsici* (Syd.) Butler and Bisby, *C. dematinum* (Pers. ex Fr.) *Grov., C. truncatum* (schw.) Andrus and Moore, *C. graminicola (*Ces.) Wilson	Circular, brown, sunken spots with dark centres and bright red orange margin leaves	India (Kaur et al. 2011), Pakistan (Bashir et al. 1985)
Dry root rot and macrophomina blight	*Macrophomina phaseolina* (Tassi) Goid (*Rhizoctonia bataticola* (Taub) Butler)	Dark brown patch on stem with black dot-like sclerotia and brown pycnidia	India (Choudhary et al. 2011), Pakistan (Khan and Shuaib 2007)
Rhizoctonia root rot and web blight	*Thanatephorus cucumeris* (Frank) *Donk.* (*Rhizoctonia solani* Kuhn.)	Necrotic small circular brown spots, fungal hyphae are seen spreading like spider web on the affected leaves with sclerotia	India (Reddy et al. 1992; Jhamaria and Sharma 2002)
Alternaria leaf spot	*Alternaria alternata* (Fr.) Keissler	Leaf spots with concentric rings leading to 'shot holes'	India (Maheshwari and Krishna 2013)
Rust	*Uromyces appendiculatus* (Pers.) *Unger. Phakopsora pachyrhizi* Syd	Reddish brown pin head uredo pustules surrounded by yellow	India (Satyagopal et al. 2014)
Bacterial diseases			
Bacterial leaf spot	*Xanthomonas campestris* pv. *Vigna radiata* Dye, mungbean strain	Brown raised spots on both surfaces which later become necrotic, water-soaked or with translucent border	India (Thakur et al. 1977)
Halo blight	*Pseudomonas syringae pv. phaseolicola* (Burk.) Young Dye and Wilkie	Water-soaked spots surrounded by a greenish yellow halo	Australia (Noble et al. 2019), China (Sun et al. 2017), India (Patel and Jindal 1972)
Tan spot	*Curtobacterium flaccumfaciens* pv. *flaccumfaciens*	Papery brown lesions originating on the leaf margins and spreading inwardly	Australia (Diatloff and Imrie 2000)

Fig. 5.1 **a–g** Symptoms of major diseases in mungbean. **a** mungbean yellow mosaic disease, **b** cercospora leaf spot, **c** powdery mildew, **d** dry root rot, **e** anthracnose, **f** halo blight and **g** tan spot. (from Pandey et al. 2018)

such as *mungbean yellow mosaic virus* (MYMV), *bean common mosaic virus* (BCMV) and *peanut bud necrosis virus* (PBNV). The common insect pests of mungbean along with the country of distribution are given in Table 5.2; Kooner and Chhabra (1980) enlisted 12 potential species of defoliators, pod borers, leafhoppers, aphids and stem borers which commonly affect mungbean in India. However, the major pests are stem fly, thrips, aphids, pod borers, whitefly and bruchids (Chiang and Talekar 1980; Kooner et al. 2006; Gentry 2010; Mbeyagala et al. 2017; Fig. 5.2a–e).

5.3 Plant Genetic Resources

Genetic resources in crop plants have evolved over thousands of years surviving all odds against nature and therefore provide a reservoir of useful genes for various survival traits. The wild and weedy relatives of crop plants grow in harsh environments and therefore provide an important source of adaptation-related traits and resistance to biotic and abiotic stresses. Therefore, their collection, evaluation, characterization, documentation and utilization in crop improvement are of

Table 5.2 Key insect pests of mungbean

Scientific name	Common name	Distribution and references
Madurasia obscurella Jacoby	Galerucid beetle	Australia (Gentry 2010), India (Kooner et al. 2006), Uganda (Mbeyagala et al. 2017)
Ophiomyia phaseoli (Tryon) *Melanagromyza sojae* (Zehntner)	Bean fly/Stem fly	Australia (Gentry 2010), Bangladesh (Islam et al. 1984; Rahman et al. 1981), India (Sahoo and Patnaik 1994; Kooner et al. 2006), Indonesia (Indiati et al. 2017), Pakistan (Khattak et al. 2004), South Africa (DAFF 2010), Taiwan (Chiang and Talekar 1980), Thailand (Srinives 1991), Uganda (Mbeyagala et al. 2017) India (Singh 1982), South Africa (DAFF 2010), Taiwan (Chiang and Talekar 1980), Thailand (Srinives 1991)
Bemisia tabaci (Gennadius)	Whitefly	Australia (Gentry 2010), Bangladesh (Rahman et al. 1981), India (Ganapathy and Durairaj 1995; Yadav and Singh 2006; Kooner et al. 2006), Indonesia (Indiati et al. 2017), Nigeria (Asawalam and Constance 2018), Pakistan (Khattak et al. 2004), Taiwan (AVRDC 1998), Thailand (Srinives 1991), Uganda (Mbeyagala et al. 2017)
Empoasca spp.	Green jassid	Australia (Gentry 2010), Bangladesh (Hossain et al. 2004), India (Sahoo and Patnaik 1994; Yadav and Singh 2006), Nigeria (Asawalam and Constance 2018), Taiwan (AVRDC 1998), Pakistan (Khattak et al. 2004), Thailand (Srinives 1991), Uganda (Mbeyagala et al. 2017), Indonesia (Indiati et al. 2017)
Polyphagotarsonemus latus (Banks)	Bean mite	India (Kooner et al. 2006; Duraimurugan and Tyagi 2014)
Aphis craccivora Koch	Black aphid	India (Sahoo and Patnaik 1994; Yadav and Singh 2006; Kooner et al. 2006; Swaminathan et al. 2012), Australia (Gentry 2010), Uganda (Mbeyagala et al. 2017), Thailand (Srinives 1991), Bangladesh (Hossain et al. 2004), South Africa (DAFF 2010), Ethiopia (Abate et al. 1982)
Acherontia styx (Westwood)	Til hawk moth	India (Das 1999; Kooner et al. 2006)
Spilosoma obliqua Walker	Bihar hairy caterpillar	India (Kooner et al. 2006), Bangladesh (Islam et al. 1984), Pakistan (Khattak et al. 2004), Uganda (Mbeyagala et al. 2017), Ethiopia (Abate et al. 1982)
Spodoptera litura (Fabricius)	Tobacco caterpillar	India (Kooner et al. 2006; Swaminathan et al. 2012), Bangladesh (Islam et al. 1984), Pakistan (Khattak et al. 2004), Indonesia (Marwoto 2008; Indiati et al. 2017)
Maruca testulalis (Geyer)	Spotted caterpillar	India (Kooner et al. 2006; Swaminathan et al. 2012), Australia (Gentry 2010), Uganda (Mbeyagala et al. 2017), Bangladesh (Rahman et al. 1981; Hossain et al. 2004), Thailand (Srinives 1991), Indonesia (Indiati et al. 2017), Ethiopia (Abate et al. 1982)
Moorei (Butler)	Red hairy caterpillar	India (Chhabra and Kooner 1998; Kooner et al. 2006; Swaminathan et al. 2012)
Anomis flava (Fab.)	Green semilooper	India (Swaminathan et al. 2012), Australia (Gentry 2010)
Ootheca bennigsen Weise *Ootheca mutabilis* (Sch.)	Bean foliage beetles	Uganda (Mbeyagala et al. 2017), Ethiopia (Abate et al. 1982)
Helicoverpa armigera (Hubner)	Gram pod borer	India (Kooner et al. 2006), Australia (Gentry 2010), Uganda (Mbeyagala et al. 2017), Bangladesh (Rahman et al. 1981; Hossain et al. 2004), Thailand (Srinives 1991), Ethiopia (Abate et al. 1982)

(continued)

Table 5.2 (continued)

Scientific name	Common name	Distribution and references
Megalurothrips spp. *Caliothrips indicus* (Bagnall)	Thrips	India (Duraimurugan and Tyagi 2014), Australia (Gentry 2010), Uganda (Mbeyagala et al. 2017), Pakistan (Khattak et al. 2004), Bangladesh (Rahman et al. 1981; Hossain et al. 2004), Taiwan (Chiang and Talekar 1980), Thailand (Srinives 1991), Indonesia (Indiati 2015; Indiati et al. 2017), India (Yadav and Singh 2006), Pakistan (Afzal et al. 2002)
Clavigralla gibbosa Spinola	Pod bug	India (Kooner et al. 2006; Swaminathan et al. 2012), Uganda (Mbeyagala et al. 2017), Thailand (Srinives 1991), Ethiopia (Abate et al. 1982)
Mylabris pustulata (Thunberg)	Blister beetle	India (Kooner et al. 2006; Swaminathan et al. 2012; Duraimurugan and Tyagi 2014), Uganda (Mbeyagala et al. 2017), Ethiopia (Abate et al. 1982)
Lampides boeticus Linnaeus	Blue butterfly	India (Swaminathan et al. 2012)
Nezara viridula (Lin.)	Green vegetable stink bug	India (Swaminathan et al. 2012), Australia, Uganda (Mbeyagala et al. 2017), Indonesia (Indiati et al. 2017), South Africa (DAFF 2010), Ethiopia (Abate et al. 1982)
Anoplocnemis spp.	Giant coreid bug	Australia (Gentry 2010), Uganda (Mbeyagala et al. 2017), Ethiopia (Abate et al. 1982)
Riptortus spp.	Brown bean bug	India (Swaminathan et al. 2012), Australia (Gentry 2010), Uganda (Mbeyagala et al. 2017), Indonesia (Indiati et al. 2017)
Creontiades pacificus (Stål)	Brown mirid	Uganda (Mbeyagala et al. 2017), Australia (Gentry 2010),
Megacopta cribraria (Fab.)	Kudzu bug	Indonesia (Indiati et al. 2017)
Callosobruchus maculatus (Fab.) *Callosobruchus chinensis* L.	Stored pests—bruchids	India (Swaminathan et al. 2012), Uganda (Mbeyagala et al. 2017), Australia (Gentry 2010), South Africa (DAFF 2010), Ethiopia (Abate et al. 1982), Taiwan (Fernandez and Talekar 1990)

utmost importance. Globally, the mungbean germplasm collections are maintained at different places including Indian Council of Agricultural Research (ICAR)-NBPGR; the University of the Philippines; The World Vegetable Center (erstwhile Asian Vegetable Research and Development Center, AVRDC), Taiwan; the Institute of Crop Germplasm Resources of the Chinese Academy of Agricultural Sciences; and the Plant Genetic Resources Conservation Unit of the University of Georgia, USA (Ebert et al. 2013). The current global holdings of mungbean include 24918 accessions among which 4104 accessions are maintained at ICAR-National Bureau of Plant Genetic Resources (ICAR-NBPGR) (Singh et al. 2017). Simultaneously, >1800 accessions including 193 wild accessions are also maintained in medium-term storage facility of ICAR-Indian Institute of Pulses Research, Kanpur. Several collections of other *Vigna* species are also maintained at ICAR-NBPGR which can be useful genetic resources for mungbean improvement programme through distant hybridization. Table 5.3 provides the details of such germplasm resources available at ICAR-NBPGR genebank.

5.4 Sources of Resistance to Diseases

A number of reports are available which have identified sources of resistance against MYMD of mungbean (Manivannan et al. 2001; Pathak and Jhamaria 2004; Kumar et al. 2006; Kaur et al.

Fig. 5.2 **a–e** Damage by major insect pests in mungbean. **a** stem fly, **b** cowpea aphid, **c** seedling thrips, **d** pod borer and **e** bruchids

2007; Yadav and Brar 2010; Iqbal et al. 2011; Zhimo et al. 2013; Suman 2015). Mungbean lines ML 109, ML 111, ML 161, LM 214 were initially reported as resistant to MYMD (Sandhu 1978). Later, Singh (1982) reported that out of 777 lines screened, 22 were highly resistant to this disease. Rajarathinam et al. (1990) also reported Vamban 1 as highly resistant to this disease. Under the All India Coordinated Research Project on MULLaRP (AICRP on MULLaRP) crops, a large number of lines were screened against MYMD and several lines including Pant Mung 3, Pant Mung 2, MN 303, DPU 88-31, PDM 54 and DU 3 were reported as resistant (Singh et al. 2002). Sharma and Dubey (1984) screened a large number of mungbean lines against ULCV and reported D 215, HPM 1, Madana 1, M 58, ML 12 and T44 as immune to this disease. Singh

Table 5.3 Plant genetic resources maintained in the genebank of ICAR-NBPGR

S. no.	Common name	Species	Indigenous collection (IC)	Exotic collection (EC)
1	Adzuki bean	*Vigna angularis*	89	98
2	Beach pea	*Vigna marina*	2	0
3	Black gram	*Vigna mungo*	2091	7
4	Black gram	*Vigna mungo* var. *mungo*	4	0
5	Cowpea	*Vigna catjang*	0	2
6	Cowpea	*Vigna sesquipedalis*	0	1
7	Cowpea	*Vigna sinensis*	0	19
8	Cowpea	*Vigna unguiculata*	2556	1062
9	Cowpea	*Vigna unguiculata* subsp. *cylindrical*	0	4
10	Cowpea	*Vigna unguiculata* subsp. *unguiculata*	4	0
11	Green gram	*Vigna radiata* var. *setulosa*	2	0
12	Moth bean	*Vigna aconitifolia*	1474	37
13	Mungbean	*Vigna radiata*	3387	534
14	Ricebean	*Vigna umbellata*	1883	144
15	Vigna	*Vigna* sp.	12	0
16	Wild black gram	*Vigna mungo* var. *silvestris*	16	0
17	Wild Vigna	*Vigna angularis* var. *nipponensis*	9	0
18	Wild Vigna	*Vigna bourneae*	4	0
19	Wild Vigna	*Vigna dalzelliana*	28	0
20	Wild Vigna	*Vigna hainiana*	6	0
21	Wild Vigna	*Vigna khandalensis*	1	0
22	Wild Vigna	*Vigna minima*	1	0
23	Wild Vigna	*Vigna nepalensis*	3	0
24	Wild Vigna	*Vigna pilosa*	3	0
25	Wild Vigna	*Vigna radiata* var. *sublobata*	227	0
26	Wild Vigna	*Vigna stipulacea*	6	0
27	Wild Vigna	*Vigna sylvestris*	1	0
28	Wild Vigna	*Vigna trilobata*	141	0
29	Wild Vigna	*Vigna trinervia*	2	0
30	Wild Vigna	*Vigna trinervia* var. *bourneae*	11	0
31	Wild Vigna	*Vigna vexillata*	107	1
32	Yard long bean	*Vigna unguiculata* subsp. *sesquipedalis*	88	1

Source www.genebank.nbpgr.ernet.in (accessed on 31.01.2019)

Table 5.4 Sources of resistance to major diseases in mungbean

Genotype/accession number	Country	References
Yellow mosaic disease		
IW 3390, EC 398897, TM-11-07, TM-11-34, PDM-139, IPM-02-03, IPM-02-14, Pusa-0672, Pusa-0871, CO-7 and MH-521	India	Reddy and Singh (1995); Mohan et al. (2014)
Powdery mildew		
LGG-460, Vaibhav, BPMR-145, TARM-18, Phule M-2003-3, Phule M- 2002-13, Phule M-2002-17, Phule M- 2001-3 and Phule M-2001-5	India	Mandhare and Suryawanshi (2008)
V2159, V4189, V4207, V4574, V4668, V4990 (resistant), R/HR: V3912, V4186 (resistant/highly resistant) V1104, V4631, V4658, V4662, V4717, V4883 (highly resistant)	Taiwan	Hartman et al. (1993)
R:M5-10 and M5-25	Thailand	Wongpiyasatid et al. (1999)
Cercospora leaf spot		
LGG-460, GM-02-08, GM-02-13 and GM-03-03, NM-98, 98-cmg-003, C2/94-4-42, NM-1, NM-2, 98cmg-018, BRM-188, CO-3, Basanti, PDM-11, BARI Mung-2 and VC3960-88	India	Haque et al. (1997)
V1471, V2757, V2773, V4718, V5036	Taiwan	Hartman et al. (1993)
M5-22 and M5-25	Thailand	Wongpiyasatid et al. (1999)
NCM 255-2, NCM 257-6, ML-267, NCM 251-1, NCM 259-2, NCM 251-13, NCM 257-2, NM-92, NCM 251-12, VC-3960-A88, NCM 257-10, NCM-209, Mung-6 C1/94-4-19, VC 3960-A89 (resistant) BRM-188, NM-98, C2/94-4-42, 98-cmg-003, NM-2, NM-1, 98cmg-018, Basanti, CO-3, PDM-11, VC3960-88, BARI Mung-2 (highly resistant)	Pakistan	Iqbal et al. (2004)
Urdbean leaf crinkle virus		
D 215, HPM 1, Madana 1, M 58, ML 12, T44, V 2182 and V 2294	India	Singh and Dubey (1982)
Dry root rot		
40504, NCM 257-5, 40457, NCM 251-4, 6368-64-72 (resistant) HR: NCM 252-10 and 40536 (highly resistant)	Pakistan	Khan and Shuaib (2007)

et al. (1988) reported AVRDC lines V 2182 and V 2294 as resistant to ULCV. Mungbean lines BPMR 1 and BPMR 115 were reported as resistant to powdery mildew by Singh and Gurha (2005). Mandhare and Suryawanshi (2008) identified resistance sources for powdery mildew, viz. Vaibhav, BPMR-145, TARM-18, Phule M-2003-3, Phule M-2002-13, Phule M-2002-17, Phule M-2001-3 and Phule M-2001-5. Hegde (1999) reported that four genotypes, viz. DHMC 9601, DHMC 9602, DHMC 9603 and DHMC 9604, were highly resistant to powdery mildew under natural epiphytotic conditions.

For other diseases, Yadav et al. (2014) reported mungbean genotype LGG-460 as highly resistant against CLS under disease epiphytotic conditions while GM-02-08, GM-02-13 and GM-03-03 were categorized as resistant. Haque et al. (1997) identified 12 genotypes, viz. NM-98, 98-cmg-003, C2/94-4-42, NM-1, NM-2, 98cmg-018, BRM-188, CO-3, Basanti, PDM-11, BARI Mung-2 and VC3960-88, that were highly resistant to Cercospora leaf spot. Table 5.4 describes the promising resistant lines of mungbean which can be used for transferring disease resistance.

5.5 Sources of Resistance to Insect Pests

Whitefly is the vector of MYMV in mungbean and causes indirect losses of 80–100% (Kitsanachandee et al. 2013; Nair et al. 2017). In addition to transmitting the MYMD, the direct yield losses by whitefly in mungbean range between 17 and 71% (Chhabra and Kooner 1998; Mansoor-Ul-Hassan et al. 1998). In mungbean, various insect-resistant sources have been identified through a series of screening methods. Khattak et al. (2004) reported NM 92 as resistant to whiteflies. In another study, Yadav and Dahiya (2000) reported ML 803, ML 839, PDM 91-249 and PBM 5 as resistant sources against whitefly. Likewise, Kooner and Cheema (2007) identified ML 1265 and ML 1229 as highly resistant to this pest, and these lines have been extensively used as resistant sources in mungbean breeding programmes in India. The other genotypes reported as resistant to whitefly are TMB 36 and RMG 1004 (Singh and Singh 2014) and ML 1774 and ML 1779 (Cheema et al. 2015). These could contribute as important sources for mungbean breeding to whitefly resistance. Nymphs and adults of bean blossom thrips or flower thrips [*Megalurothrips distalis* (Karny)] cause heavy yield reduction of the crop by feeding on the pedicles and stigma of flowers (Chhabra and Kooner 1985a, b). Malik (1990) observed that summer mungbean genotypes SML 77, UPM 82-4 and Pusa 107 were resistant to *M. distalis* under natural as well as artificial screen house conditions. The cultivars Co 3, Co 4 and Co 5 were also reported to be less susceptible to thrips (Lal 1987). Chhabra (2001) reported mungbean genotypes PIMS 2, PIMS 3, CO 3, ML 5 and ML 337 as resistant to thrips. NM-92 has also been reported to be resistant to thrips in India and Pakistan (Khattak et al. 2004; Kooner et al. 2005). MH 3153 recorded the lowest number of thrips per leaf among eight advanced mungbean genotypes/cultivars in Pakistan (Nadeem et al. 2014).

For stem fly, *Ophiomyia* sp., Talekar (1990) reported 3 mungbean cultivars, viz. V2396, V3495 and V4281, as resistant. CIAT accessions such as G05253, G05776, G02005 and G02472 are highly resistant to bean fly and are recommended as the potential sources resistant to this pest (Abate 1990). For sweet potato whitefly, *Bemisia tabaci* (Gennadius), 43 accessions of mungbean were identified as resistant out of >2000 genotypes (Chhabra et al. 1980, 1988; Kooner 1998; Kooner and Cheema 2007; Kooner et al. 1977). In other studies, Kooner et al. (1997) reported the mungbean lines, viz. ML 1, ML 6, ML 7, P 290, P 292, P 131, P 293, P 325, P 364 and 11,148, as least susceptible to *B. tabaci* and MYMV.

Chhabra et al. (1988) reported ML 337, ML 423 and ML 438 cultivars as least susceptible to Jassids. For cowpea aphid, *Aphis craccivora* Koch JRUM 1, JRUM 11, JRUM 33, DPI 703, LAM 14-2, UPM 83-6 and UPM 83-10, Pusa 115, PDM 116 and ML 353 have been reported as resistant (Sahoo and Hota 1991). The cultivars J1, LM 11, P526 and Co3 are less susceptible to the pod borer, *Helicoverpa armigera* (Hub.) (Lal 1987). Further, Chhabra et al. (1988) reported that genotypes ML 337, ML 423 and ML 428 are resistant to this pest. Swarnalatha (2007) reported that LGG 505, ML 267, LGG 502, LGG 407, LGG 460 and LGG 485 were resistant to the legume pod borer, *M. testulalis*. Genotypes MGG 364, MGG 365 and MGG 363 have been reported as tolerant to Maruca pod borer damage (Choragudi et al. 2012).

Among the stored grain pests, bruchids, *Callosobruchus maculatus* (Fab.) and *C. chinensis* (L.) cause extensive damage to mungbean if not properly stored (Cheema et al. 2017). Initially, a wild mungbean accession TC1966 (*V. radiata* var. *sublobata* (Roxb.) Verdc.) was identified as a potential source of resistance to *C. maculatus* and *C. chinensis* (Fujii and Miyazaki 1987; Fujii et al. 1989; Lambrides and Imrie 2000; Kashiwaba et al. 2003). TC1966 was extensively used in breeding programmes for developing bruchid-resistant mungbean. More recently, two of the accessions (V2802 and V2709) were confirmed to possess complete resistance to *C. chinensis* and *C. maculatus* (Somta et al. 2007). Reduced survival and prolonged developmental period (30.5–31.5 days) of *C. chinensis* were recorded on four moderately resistant mungbean accessions, LM131, V1123, LM 371 and STY 2633 (Duraimurugan et al. 2014). In another study, the accessions KM-12-5 and P-S-16 were also reported as relatively resistant against *C. analis* (Soumia et al. 2017). Presently, a few mungbean accessions, viz. TC 1966, ACC41,

V2709, V2802, V1128, V2817, are the only known sources of bruchid resistance in mungbean (Sarkar et al. 2011; War et al. 2017). Resistance to bruchid has also been reported in wild black gram, *V. nepalensis*, and ricebean, *V. umbellata*, genotypes (Tomooka et al. 2000). After extensive research in breeding for insect resistance in mungbean, World Vegetable Center has been successful in developing improved mungbean lines with high levels of resistance to *C. maculatus* and *C. chinensis* (Nair et al. 2015).

5.6 Genetic Basis of Resistance

Besides studying the inheritance of resistance to various biotic and abiotic stresses, investigations have been carried out to understand the genetics of quantitative and qualitative traits in mungbean (Table 5.5). Kumar et al. (2006) thoroughly discussed the inheritance pattern of various economically important traits in this crop. The first report on genetic studies in mungbean was made by Bose (1932), who reported that the colour of

Table 5.5 Inheritance pattern, resistance genes/loci and associated markers of major mungbean diseases and storage pest

Disease	Resistant accessions	Resistant genes/alleles/locus	Linkage group	Associated marker	Inheritance pattern
Yellow mosaic disease	NM-6-68-2 KMG189, BM6, TM-99-37, BARI Mung-6, NM-12-1, VC6372 (45-8-1)	qYMIV1 and qYMIV2 qYMIV3, qYMIV4 and qYMIV5 qYMIV7	LG2, LG3 LG4, LG7 LG9A	CM9, CM815, MYMVR-683 (SCAR), CEDG180 cp02662, DMB-SSR158	Independent major recessive genes with additive effects (Akbar et al. 2018; Aski et al. 2014; Chen et al. 2015; Kitsanachandee et al. 2013; Alam et al. 2014). Monogenic inheritance of single recessive genes (Khattak et al. 2000; Sai et al. 2017)
Powdery mildew	V4718, RUM5, VC1210A VC6468-11-1A, V4758, V4785, VC3980A	Qpmr-1 Qpmr-2 Pm1, Pm2	LG2, LG4, LG9	CEDG282, CEDG191, CEDG166, MB-SSR238 (SSR)	Additive and dominant gene action of >2 major and minor genes (Kasettranan et al. 2010; Chankaew et al. 2013; Humphry et al. 2003)
Cercospora leaf spot	VC6372 (45-8-1), v4781, HUM-1, ML-1194, ML-1229, ML 820, EC-27087, EC26271-3, BARI Mung-2	qCLS	LG3	Between CEDG117 and VR393	Monogenic inheritance of single dominant gene (Singh et al. 2017; Chauhan and Gupta 2004; Thakur et al. 1977) Quantitative inheritance of multiple genes (Chankew et al. 2013)
Bruchid	V2709, V2802 TC1966, ACC41, VC1973A, Jangan Mung	VrPG1P1, VrPG1P2, Br1	LG5, LG9	MB87, SSR017 SSR037, OPC-06, DMB-SSR-158 (SSR), STSbr1/SMJ44, STSbr2/SMJ64 (STS)	Monogenic inheritance of few dominant major genes with some modifiers (Bhatacharya 2014; Kaewwongwal et al. 2017; Schafleitner et al. 2016) Single dominant gene (Hong et al. 2015; Mahato et al. 2015; Sun et al. 2008)

unripe pod is due to the same gene responsible for flower colour. Later, numerous studies were conducted on the inheritance pattern of morphological traits, viz. plant type, plant colour, leaf type, flower colour, inflorescence type, pod pubescence, shape and colour, shattering habit, seed coat colour and surface, hard-seededness, resistance/tolerance to biotic and abiotic stresses (Singh 1982).

There are many reports on the inheritance of resistance to MYMD in mungbean; however, most of these reports are contradictory. The discordance in the nature of inheritance of MYMD could be ascribed to species non-specificity as most of these are silent on the exact species of MYMD-causing virus. In most of the reports, the allelic relationships have been studied in MYMD, which suggest that the resistance is controlled by a single recessive gene (Malik et al. 1986; Saleem et al. 1998; Reddy and Singh 1995; Sudha et al. 2013), dominant gene (Sandhu et al. 1985), two recessive genes (Pal et al. 1991; Ammavasai et al. 2004) and complementary recessive genes (Shukla and Pandya 1985). Thus, a more extensive study is needed to finalize the mode of inheritance of the resistance of MYMD in mungbean. The studies on the genetic basis of resistance to MYMV in F_1, F_2 and F_3 progenies indicated that a single recessive gene is responsible for its resistance, and the expression of the major gene responsible for MYMD resistance/susceptibility is affected by modifying genes (Khattak et al. 2000). These modifying genes caused variation in the degree of MYMD resistance/susceptibility in the progenies derived from a single cross. It has been further reported that the inheritance of MYMV resistance occurs through a major recessive gene without any maternal effect (Khan et al. 2007). Though MYMD resistance has been suggested to be monogenic (Pal et al. 1991; Gupta et al. 2013), some reports consider this resistance as digenic (Singh 1980; Verma and Brar 1996; Shukla and Pandya 1985; Ammavasai et al. 2004). Its control has also been reported due to digenic inhibitory gene interaction (Verma and Singh 1986; Solanki et al. 1982). In addition to complex inheritance mechanisms of this disease, non-uniform and fluctuating distributions of whitefly populations in the field always reduce the accuracy in evaluating the resistance and lead to errors in the selection of resistant genotypes. Developing MYMD-resistant varieties through conventional approaches remains difficult due to the explosion of new isolates and complex mechanisms of MYMD resistance (Selvi et al. 2006). In this situation, molecular marker technology can increase the efficiency of breeding through marker-assisted selection (MAS), in which phenotypic selection is carried out using DNA markers associated with the trait of interest. The marker–trait association and gene tagging have shown that single dominant gene is involved in governing MYMIV in black gram (Gupta et al. 2013) and soya bean, whereas five QTLs were identified till date in mungbean for MYMIV (Kitsanachandee et al. 2013). Among other diseases, powdery mildew was reported to be quantitatively inherited with high heritability and predominantly additive gene action (Kasettranan et al. 2010).

Resistance to bruchids has been reported to be controlled by a single gene (Kitamura et al. 1988; Young et al. 1992; Srinives 1996; Miyagi et al. 2004; Lawn and Rebetzke 2006). Sun et al. (2008) reported that the bruchid resistance of resistant mungbean cultivar V2709 was controlled by a single dominant locus named *Br2*. Sarkar et al. (2011) reported that bruchid resistance in Indian *V. sublobata* accession is controlled by a major dominant gene but might have varying degrees of expressivity. Some reports have suggested that resistance to *C. chinensis* in mungbean is dominant and governed by a few major genes (probably two) with some modifiers (Sarkar and Bhattacharyya 2015). The F_1 and F_2 seeds in mungbean showed that the resistance to *C. chinensis* and *Riptortus clavatus* Thunberg is controlled by a single dominant gene (Hong et al. 2015). However, the segregation pattern of reciprocal reaction to each insect in F_2 seeds showed that seeds were susceptible to both the insects.

5.7 Breeding Methods and Strategies

The last 3 decades have seen improvement in mungbean breeding with the focus on the development of short duration, widely adaptable, synchronous and photo- and thermo-period-tolerant varieties. Of late, the major emphasis is on development of mungbean varieties having resistance to multiple diseases as well as insect pests. Germplasm resources and improved mungbean lines are being deployed to develop more stable and resilient varieties. To develop high yielding and biotic stress-resistant cultivars in mungbean, the common methods that have been deployed are the introduction, selection, hybridization and mutation breeding (Pratap et al. 2012).

5.7.1 Selection

Selection from indigenous and exotic germplasm as well as landraces has always been an important tool in the development of superior cultivars. For a successful hybridization programme, selection of diverse parents with useful traits is the prime requisite as the breeding materials developed from genetically divergent parents are likely to produce more heterotic effects and also lead to the recovery of better segregants in the subsequent segregating generations. Selection from indigenous and exotic germplasm as well as landraces has played an important role in the development of superior cultivars of pulse crops. Before 1950, virtually all the mungbean varieties were developed by a selection of superior genotypes from the collected samples of local cultivars. Some of the varieties were also developed from the exotic materials. The desirable plants were selected, and the superior pure lines were established after their progeny testing (Pratap et al. 2012, 2013; Nair et al. 2013). The pure lines were evaluated for yield, yield traits and reaction to diseases, and the best pure line was released for cultivation (Pratap et al. 2012; Nair et al. 2013).

The earliest efforts to collect landraces were made from all over India and Burma as early as 1925 (Bose 1932). These collections were utilized to isolate pure lines from the stocks, and selections were made on the basis of colour of the stem, flowers, ripe pods, seed colour and texture and other morphological features. As a result, a few varieties were developed. However, most of the early selections such as Jalgaon Local, China Mung 1/49, Kopargaon, Krishna II, Gwalior 3, Khachrod 5, Bhilsa Green 16, BR 5, BR 6, BR 7 were susceptible to MYMD and powdery mildew although better in a few other traits such as uniformity, bold seeds and longer pods. Later, a few selections such as KM 1, Co2, R 288-8, T 150, Utkal 2, selections 196, 697, 855, 932, 946, T 1630 and T 2105 which were either tolerant or moderately resistant to MYMD were carried out. Most of these selections were taken before the 1960s, and a few of them such as T 44, Sona mung, Kopargaon, Co 1 and Co 2 are still cultivated in small pockets in interiors of India. Large-seeded varieties of mungbean, Pusa Vishal, Pant moong 5 and SML 668 were developed from the selection of AVRDC material, and these became highly popular among the farmers.

5.7.2 Hybridization

Most of the biotic stress-resistant varieties of mungbean, especially in the last 4 decades, have been developed through hybridization. Resistance sources have been utilized to combine good agronomic base and disease/pest resistance to develop such varieties.

5.7.3 Intraspecific Hybridization

Studies have been carried out on the development of short duration, photo- and thermo-period-insensitive varieties of mungbean coupled with resistance to major biotic stresses, viz. MYMD and powdery mildew (Pratap et al.

2014a, 2015; Singh et al. 2017). In most of the varieties developed through hybridization, the pedigree method of breeding has been followed. Until now, >100 varieties have been developed in India through intraspecific hybridization.

With the development of varieties such as HUM 6, HUM 12, Meha (IPM 99-125), HUM 16 and MH 2-15, the mungbean production and productivity witnessed a big boost in India. ML 1265, a whitefly-resistant variety, was released as a commercial cultivar in India (Cheema et al. 2017). The variety PKV AKM 4 developed from a cross between BM4 X PS 16 was recommended for two zones, viz. central zone and south zone of India. In more recent times, IPM 02-3, currently the most popular variety of mungbean in India, was developed using IPM 99-125 and Pusa Bold 2 and recommended for both *spring* and *kharif* seasons. This variety is highly resistant to MYMD as well as to other major diseases of mungbean and has high yield potential, medium–large, shining and green seed and wider adaptability. However, this variety recorded high (82.52%) pre-harvest sprouting value (Lamichhane et al. 2017) making it prone to pre-harvest sprouting during the rainy season. Another highly popular variety, MH 421, developed from the cross Muskan X BDYR 2 is also highly resistant to MYMD and has a share of about 15–18% in breeder seed indent. Recently, the ICAR-Indian Institute of Pulses Research, Kanpur, released IPM 410-3 (Shikha) variety for entire northern, western and central India and covers the majority of the mungbean area in the country. This variety is also highly resistant to MYMD and powdery mildew, and moderately resistant to CLS. IPM 205-7 is an early duration mungbean variety that matures in <55 days and is suitable for summer cultivation (Pratap et al. 2013). Developed from the cross IPM 02-1 X EC 398889, this variety is most suitable to be grown as a catch crop after the harvest of *rabi* crops and before the onset of monsoon and best utilizes the short-season window of 60–70 days available during the summer season. This variety is also highly resistant to MYMD and powdery mildew and moderately resistant to CLS. IPM 2-14 was released for spring cultivation in south zone of the country and is gaining popularity. The varieties DGGV-2 developed from the cross between China mung x TM-98-50 and Pusa 0672 developed from the cross between 11/395 x ML 267 were released for south zone of India. The varieties such as KM 2241, HUM 16, MH 2-15 and TMB 37 were developed through intraspecific hybridization and became very popular among the farmers in short time. Table 5.6 illustrates the popular mungbean varieties developed in India in the last 10 years. Mungbean genotypes/improved lines developed in countries other than India are shown in Table 5.7.

5.7.4 Interspecific Hybridization

Planned utilization of exotic and wild genetic resources of mungbean can result in yield improvement, plant type and several other characters, such as resistance to biotic and abiotic stresses (Pratap et al. 2015). Wild relatives of cultivated *Vigna* species offer new sources of variability for a number of traits, viz. resistance to biotic stresses such as powdery mildew (Tomooka et al. 2006), MYMV (Pandiyan et al. 2008), bruchids (Tomooka et al. 1992; Somta et al. 2006) (Table 5.8), abiotic stresses such as photo- and thermo-insensitivity (Pratap et al. 2014a; Basu et al. 2019) and agronomic traits (Tomooka et al. 2001), which are hitherto not found in the cultivated species and therefore provide additional avenues of selection for agronomic traits (Pratap et al. 2014a, b). While mungbean has erect growth habit, a large number of seeds/pod, early maturity and desired quality traits, to further improve its branching, synchronous maturity, non-shattering pods and durable resistance to CLS, urdbean can be utilized as a donor (Singh 1990). Likewise, traits such as the number of clusters/plants, longer pods with a large number of seeds, durable resistance to MYMV, CLS, powdery mildew and bruchids may be transferred from ricebean.

Crossability barriers create complications for making successful inter-species gene transfer in mungbean (Pratap et al. 2018). These barriers

Table 5.6 List of mungbean varieties identified/released by all India Coordinated Research Project on MULLaRP (Project Coordinators Report 2018)

Name of variety	Pedigree	Year of release	Average yield (q/ha)	Reaction to major disease
KM 2241	Samrat X PDM 54	2008	10–11	Resistant to MYMD
IPM 02-3	IPM99-125 X Pusa Bold 2	2009	11.0	Resistant to MYMD
PKV AKM 4	BM4 X PS16	2009	10.0	Resistant to MYMD
Pusa 0672	11/395 X ML 267	2009	10.0	Resistant to MYMD
IPM 02-14	IPM99-125 X Pusa Bold 2	2010	11.0	Resistant to MYMD
MGG 347	K-851 X PDM-54	2009	13–15	Tolerant to thrips, stem fly, MYMD, CLS
VBN (Gg)3	K 1 X Vellore Local	2009	9.75	Moderately resistance to MYMD
Basanti	Asha X PDM 90-1	2010	12–15	Resistance to MYMD
Pairymung	TARM 1 X J 781	2010	12	Tolerant to MYMD and resistance to PM
TM-2000-2	JL-781 X TARM-2	2010	10.9	Resistant to PM
SML 832	SML 302 X Pusa Bold 1	2010	11.6	Tolerant to MYMD and thrips
DGGV-2	China Mung X TM-98-50.	2012	11–14	Moderately resistant to PM, tolerant to apion beetle
Shalimar moong-2	PS-7 X Larkipora Local	2013	10.0	Resistant to CLS, moderately resistant to aphid
CO (Gg) 8	COGG923 X VC 6040	2013	–	Resistance to MYMD
MH 421	Muskan X BDYR 2	2014	10–12	Resistant to MYMD
IPM 410-3 (Shikha)	IPM 03-1 X NM 1	2016	11–12	Resistant to MYMD
IPM 205-7 (Virat)	IPM 02-1 X EC 398889	2016	10–11	Resistant to MYMD
SML 1115	SML 134 X SML 715	2016	11–12	Moderately resistant to MYMD
MH 318	CCS HAU, Hisar	2016	12–14	Resistance to MYMD
Pant Mung 8 (PM 09-6)	PM 3 X NDM 99-3	2016	10–11	Resistant to MYMD, CLS and PM
MSJ 118	Mutant of K 851	2016	7–8	Moderately resistant to MYMD
RMG 975	ML 613 X ML 1189	2016	8–9	Moderately resistant to MYMD and tolerant to root knot nematode
KM 2328	KM 2241 X HUM 16	2018	10–12	Resistant to MYMD, CLS, WB, MB and anthracnose

(continued)

Table 5.6 (continued)

Name of variety	Pedigree	Year of release	Average yield (q/ha)	Reaction to major disease
Pusa 1431	Pusa 9531 X IPM 02-19	2018	12–14	Resistant to MYMD, CLS, anthracnose, web blight and ULCV
Kanika	Pant Mung 4 X EC398897	2018	12–13	Highly resistant to MYMD and CLS and resistant to leaf crinkle and leaf curl
Varsha	EC398885 X PDM 139	2018	12–13	Highly resistant to MYMD and PM and moderately resistant to CLS

Table 5.7 List of mungbean varieties/advanced breeding lines developed in different countries

Genotype/improved lines	Pedigree	Specific feature	Country	References
Improved BC_3F_3 lines	CN60 x TC1966	Resistant against bruchid damage (*Callosobruchus chinensis, C. maculatus*)	Thailand	Tomooka et al. (1992)
Improved F_{12} RILS	TC1966 X NM92	Tolerant to bruchid damage	Taiwan	Schafleitner et al. (2016)
Improved F_7 RILS and subsequent advanced generation	V2802 X NM94	Resistant to bruchid damage	Taiwan	Schafleitner et al. (2016)

Table 5.8 Successful transfer of resistance to biotic stresses through distant hybridization

Useful traits	Interspecific crosses	References
MYMD	*Vigna radiata* × *Vigna mungo*	Lekhi et al. (2018)
	Vigna radiata ×13 wild *Vigna* species	Pandiyan et al. (2010)
	V. radiata × *V. sublobata* and *V. mungo*	Pal et al. (2000)
	V. radiata × *V. umbellata*	Bhanu et al. (2018), Bharathi et al. (2006)
MYMD and CLS	VC1482E × NM 20-21 10-43 (NM89) and 10-12 (NM88) (Pakistan)	Ali et al. (1997)
Pest resistance	*V. glabrescens* × *V. radiata*	Chen et al. (1989)
Bruchid beetles	Chainat 60' ('CN60'), *Vigna radiata* × *V. radiata* var. *sublobata* (Thailand)	Tomooka et al. (2006)

may express through reduced fertilization, reduction in the number of hybrid seeds or development of abnormal—shrivelled, small or non-viable seeds, retarded development of hybrid endosperm leading to embryo death or hybrid sterility (Pratap et al. 2015). Several measures have been suggested such as the deployment of embryo rescue, hormonal manipulations and use of mentor pollen, for increasing the success of interspecific crosses. By using sequential embryo rescue, the reciprocal hybrids between *V. mungo* and *V. radiata* are successfully obtained (Gosal and Bajaj 1983; Verma and Singh 1986). *V. mungo* has also been reported to cross successfully with *V. glabrescens* (Dana 1968; Krishnan and De 1968), *V. trilobata* (Dana 1966), and *V. dalzelliana* (Chavan et al. 1966). Similarly, *V. radiata* × *V. umbellata* crosses were generated to transfer resistance to MYMV and other desirable traits into mungbean (Verma and Brar 1996). Derivatives from mungbean × urdbean crosses have been reported to exhibit a higher level of MYMD resistance caused by MYMV (Gill et al. 1983; Lekhi 2017). Useful disease-resistant genes were also identified from amphidiploids of mungbean x ricebean crosses (Dar et al. 1991). Similarly, progenies from mungbean x ricebean and mungbean × *V. radiata* var. *sublobata* crosses were also recovered which exhibited a high degree of resistance to MYMV (Verma and Brar 1996). Singh et al. (2003) produced successful hybrids between *V. radiata* and *V. umbellata*, and the hybrids possessed intermediate morphology with MYMV resistance. One accession of wild mungbean (*Vigna radiata* var. *sublobata*) exhibited complete resistance to adzuki bean weevils and cowpea weevils (Fujii et al. 1989), which was successfully used in a breeding programme (Tomooka et al. 1992).

Despite numerous attempts of hybridization between cultivated mungbean genotypes and wild genetic resources, the actual release of new cultivars from distant crosses has remained limited. Three mungbean cultivars, viz. HUM 1, Pant Moong 4 and IPM 99-125, have been developed from mungbean x urdbean crosses in India. All these have been highly popular among the farmers and possess improved plant types in addition to high levels of MYMD resistance. There are a few reports of the development of advanced breeding lines and genetic resources utilizing the wild genetic resources indirectly. For example, using IPM 99-125 as one of the parents, genotypes IPM 02-1 and IPM 03-1 were developed at ICAR-IIPR, Kanpur, which were further used in the development of two extra early mungbean genotypes, IPM 205-7 and IPM 409-4 that mature in 50–55 days (Pratap et al. 2013). Currently, a few advanced lines derived from interspecific crosses are under multilocation evaluation in AICRP for their possible release as a cultivar.

5.7.5 Mutation Breeding

Induced mutation using physical and chemical mutagens is one of the many ways to develop new cultivars with improved traits and better characteristics. While most of the mutants usually have one or a few traits improved, such characters may be incorporated in other cultivated varieties through backcross breeding, besides releasing the developed material directly as a variety. Mutation breeding has been used successfully to develop improved cultivars in mungbean possessing resistance to a few biotic stresses (Table 5.9). Mutations were induced in two mungbean varieties, K-851 and PS-16, using EMS and gamma rays. Selection studies were conducted to improve the yield and to generate genetic variability in different quantitative traits, viz. fertile branches per plant, pods per plant and seed yield per plant (Khan and Goyal 2009). Other varieties developed through mutation breeding include Pant Mung 2, Co 4, TMB 37, Dhauli, BM4 and MUM 2. In Pakistan also, the popular varieties NM 51 and NM 54 were developed which were large-seeded varieties resistant to MYMD. These varieties were developed by hybridization and irradiation of F_1 seeds. Bean fly, *O. phaseoli*, is a key pest of mungbean in Thailand (Srinives 1991). To control bean fly, apart from insecticide spraying, the induced mutation was used to improve the resistance of mungbean variety Khampang Saen 2 (KPS2) (Ngampongsai et al. 2009) and an insect tolerant selection Chai Nat 72 (CN 72) was isolated.

5.8 Impact of Resistance Breeding

Besides the development of more than 100 improved varieties in mungbean, remarkable progress was also made in collection, evaluation, characterization and documentation of germplasm resources. There has also been a notable success in transferring disease and insect resistance alleles from wild *Vigna* relatives to

Table 5.9 Varieties/advanced breeding materials developed through mutation breeding

Variety name	Resistance to disease	Radiation used	Area/season of cultivation	References
TARM-2	PM	X-rays, gamma rays, ethyl methyl sulphonate	Southern and central zone, summer season, rice fallow	D'souza et al. (2009)
TARM-18	PM			
TM-96-2	PM			
TMB-37	PM, YMV			
TJM-3	PM, YMV, Rhizoctonia root-rot disease			
NIAB Mung 2006	CLS, MYMV	Induced mutation and hybridization	Pakistan	Haq (2009)
M4-2	CLS (moderately resistant), PM (moderately resistant), bean fly (tolerance)	500 Gy (gamma rays and treated with 1% EMS	Thailand	Ngampongsai et al. (2004)
M5-1				
M5-5				
CN 36				
KPS 32				
Mutant SML-668	YMV	600 Gy gamma rays (M1 generation) 500 Gy gamma rays (M3 generation)	Summer season, India	Reddy (2009)

PM=Powdery mildew, YMV: Yellow Mosaic Virus, CLS: Cercospora leaf spot

cultivated mungbean backgrounds. The impact of biotic stress-resistant varieties has been realized well in production as well as in productivity of mungbean, which showed continuous increase despite fluctuations in its area. The area, production and productivity of mungbean in India have seen a consistent upward trend since the 1960s, and the production increased from 0.60 million tonnes in 1964–65 to about 2.17 million tonnes in 2016–17 (PC Report, 2018, AICRP on MULLaRP). During the corresponding period, productivity also increased from about 280 to >500 kg/ha. While increased irrigation facilities, better inputs and crop management had a role in increasing productivity, deploying biotic stress-resistant cultivars for cultivation had a definite role to play in enhancing productivity. Significant growth in mungbean area and production was witnessed in non-traditional niches, especially in summer, spring and rice fallow cultivation during the last decade (Gupta and Pratap 2016; Singh et al. 2017). It is noteworthy that about 80% of the mungbean breeder seed indent is shared by top ten varieties in India including IPM 02-3, MH 421, GM 4, HUM 16, SML 668, IPM 2-14, Samrat, Pant Moong 5 and Meha (Singh et al. 2017). Among these, IPM 02-3 alone contributes to about 25% of breeder seed indent. Most of these varieties are highly resistant to major diseases and insect pests of mungbean.

5.9 Future Outlook

Mungbean has a distinct advantage of being a short duration and widely adaptable crop that can fit well in several cropping systems. The reduced maturity duration and synchronous maturity in new cultivars have made it an attractive option as a catch crop in short-season windows between two crops and also as a non-competing intercrop in cash crops like sugar cane. Therefore, it has a tremendous scope of vertical as well as horizontal expansion in all major mungbean-growing ecologies of the world. Mungbean plays an important role in food and nutritional security of several countries including India, Pakistan, Bangladesh, Myanmar and several African countries. In India, it has been projected as one of the major crops for vertical and horizontal expansion to achieve self-sufficiency in pulses in policy documents (Vision 2030, ICAR-IIPR). Therefore, there is a need of a major research boost to this crop to make it a key component of pulse revolution. Biotic stresses are the major constraints in realizing the actual yield potential of a cultivar and ultimately affect the productivity and quality of mungbean to a great extent. Therefore, a major thrust is required on combining pre-harvest sprouting and bruchid resistance and pyramiding genes for resistance to major insect pests (thrips, Jassids and pod borer) and diseases (MYMD, powdery mildew and CLS) utilizing resistance sources in cultivated and wild germplasm.

While several improved cultivars have been developed with enhanced resistances to yellow mosaic, powdery mildew, CLS and a few more diseases, only those problems for which resistance sources are known have been addressed till date. Stem fly and bruchids are serious pests worldwide, and the resistance sources are either limited or genes difficult to utilize for breeding resistance to these pests. Such traits remain untouched and need major attention of breeders. Marker-assisted breeding has been successfully deployed in other pulses such as chickpea (Varshney et al. 2014; Pratap et al. 2017; Mannur et al. 2019), and this technology needs to be put to use for breeding for complex traits in mungbean as well. Molecular markers are now available for powdery mildew and CLS, which require to be utilized in breeding programmes. Root rot and anthracnose are other important diseases, and more attention is required towards the development of molecular markers for these stresses. There is a strong need for generating additional genomic resources to fully utilize the potential of marker technology. One such mission has been recently launched by the Department of Biotechnology (DBT), Government of India, in minor legumes including mungbean urdbean, moth bean, cowpea and horse gram, where numerous genomic and genetic resources will be developed for various biotic stresses. A fine map on the distribution of MYMD-causing viruses will be developed covering all mungbean-growing ecologies of India besides preparing a differential set of mungbean genotypes to identify the prevalence of MYMD-causing species of viruses.

Germplasm has played an important role in the development of many cultivars in mungbean and collection, evaluation and characterization of trait-specific germplasm need a systematic investment of time and money so that potential germplasm can be deployed to best use in filling the gaps related to traits of interest. Mungbean minicore collection (Schafleitner et al. 2015) has been made available to partner countries of the Australian Centre for International Agricultural Research (ACIAR)-funded International Mungbean Improvement Network (IMIN). This germplasm needs to be thoroughly screened for a host of biotic stresses and deployed to introgression breeding for developing biotic stress-resistant cultivars. Breeding materials have already been developed at ICAR-Indian Institute of Pulses Research (IIPR), Kanpur; Department of Agricultural Research (DAR), Myanmar; and Bangladesh Agricultural Research Institute (BARI), Bangladesh, besides World Vegetable Center, Hyderabad, utilizing promising mungbean minicore accessions. While the development of biotic stress-resistant cultivars is important, capacity building of mungbean farmers is also equally significant. Cultivating only resistant cultivars, adopting good practices of crop management and raising a clean crop need to be taught to the poor and marginal farmers as to make mungbean a mainstream pulse crop towards providing a vegetarian solution to global protein and calorie malnutrition.

References

Abate T (1990) Studies on genetic, cultural and insecticidal controls against bean fly, *Ophiomyia phaseoli* (Tryon) (Diptera: Agromyzidae) in Ethiopia. Dissertation, Simon Fraser University, pp 220

Abate T, Gebremedhin T, Ali K (1982) Arthropod pests in grain legumes in Ethiopia—their importance and distribution. Institute of Agriculture Research, Addis Ababa, Ethiopia, p 63

Afzal M, Ahmad T, Bashir M (2002) Relative toxicity of different insecticides against whitefly, *Bemisia tabaci* (Genn.) and black thrips, *Caliothrips indicus* on nm-92 mung bean, *Vigna radiata* (L.). Pak J Agri Sci 39(3):224–225

Ahmad M, Harwood RE (1973) Studies on whitefly transmitted yellow mosaic disease of cowpea (*Vigna unguiculata*). Plant Dis Rep 62:224–226

Akbar W, Aslam M, Maqbool MA, Ali M (2018) Inheritance pattern of mungbean yellow mosaic disease resistance and gene action for different traits in mungbean (*Vigna radiata* (L.) Wilczek) under protected and unprotected field conditions. Plant Breed. https://doi.org/10.1111/pbr.12626

Alam AKMM, Somta P, Srinives P (2014) Identification and confirmation of quantitative trait loci controlling resistance to mungbean yellow mosaic disease in mungbean [*Vigna radiata* (L.) Wilczek]. Mol Breed 34:1497–1506

Ali M, Malik IA, Sabir HM, Ahmad B (1997) The mungbean green revolution in Pakistan. Asian Vegetable Research and Development Center, Shanhua, Taiwan. AVRDC Technical Bulletin No. 24. Publication No. 97–459

Ammavasai S, Phogat DS, Solanki IS (2004) Inheritance of resistance to *mungbean yellow mosaic virus* (MYMV) in green gram (*Vigna radiata* L. Wilczek). Indian J Genet 64:145–146

Asawalam EF, Constance EE (2018) Control of field insect pests of mung bean (*Vigna radiata* L. Wilczek) using some plant extracts in Umudike, Nigeria. J Med Plants Econ Develop 2(1):a27

Aski M, Dikshit HK, Singh D, Singh A, Prapti P (2014) Inheritance of resistance to Mungbean Yellow Mosaic India Virus (MYMIV) in mungbean (*Vigna radiata*). Indian J Agric Sci 85:1144–1147

AVRDC (1998) Diseases and insect pests of mungbean and blackgram: a bibliography, Shanhua, Taiwan, Asian Vegetable Research and Development Centre, 1999. VI. P. 254

AVRDC (2012) Asian Vegetable Research and Development Center, Mungbean, AVRDC Progress Report 2012 (Shanhua)

Bashir M, Alam SS, Qureshi SH, Malik BA (1985) Control of mungbean anthracnose by foliar fungicides. Pakistan J Agric Res 6:173–175

Basu PS, Pratap A, Gupta S, Sharma K, Tomar R, Singh NP (2019) Physiological traits for shortening crop duration and improving productivity of greengram (*V. radiata* (L.) Wilczek) under high temperature. Front. Plant Sci 10:1508. https://doi.org/10.3389/fpls.2019.01508

Bharathi A, Selvaraj KSV, Veerabadhiran P, Lakshm BS (2006) Crossability barriers in mungbean (*Vigna radiata* L. Wilczek): with its wild relatives. Indian J Crop Sci 1(1–2):120–124

Bhattacharya S (2014) Inheritance of bruchid resistance and morphological traits in greengram. Ind J Genet Plant Breed 74(1):98

Bose RD (1932) Studies in Indian pulses mung or greengram (*Phaseolus radiatus Linn.*). Indian J Agric Sci 2:607–624

Chankaew S, Somta P, Isemura T, Tomooka N, Kaga A, Vaughan DA et al (2013) Quantitative trait locus mapping reveals conservation of major and minor loci for powdery mildew resistance in four sources of resistance in mungbean [*Vigna radiata* (L.) Wilczek]. Mol Breed 32:121–130

Chauhan MP, Gupta RP (2004) Genetics of Cercospora leaf spot disease resistance in mungbean [*Vigna radiata* (L.) Wilcek]. Legume Res 27:155–156

Chavan VM, Patil GD, Bhapkar DG (1966) Improvement of cultivated *Phaseolus* species-need for interspecific hybridization. Indian J Genet Plant Breed 26:152–154

Cheema HK, Singh RS, Taggar GK (2015) Screening of some advanced mungbean genotypes against sucking insect-pests. In: Abstracts of the brain storming meeting on promotion pulses in Indo-Gangetic plains of India. Punjab Agricultural University, Ludhiana, pp 63–64

Cheema HK, Pratap A, Sujayanand GK, Arora R, Sandhu S (2017) Breeding for insect resistance in mungbean and urdbean breeding insect resistant crops for sustainable agriculture. In: Arora R. Sandhu S (eds) Breeding insect resistant crops for sustainable agriculture. Springer Singapore, Singapore, pp 353–385

Chen H, Wang L, Wang S, Liu C, Blair MW, Cheng Z (2015) Transcriptome sequencing of mung bean (*Vigna radiata* L.) genes and the identification of EST-SSR markers. PLoS ONE 10(4):e0120273

Chen HK, Muk MC, Shenmugasundaram S, Muk DWS (1989) Interspecific hybridization between *Vigna radiata* L. Wilczek. and *Vigna glabrescens*. Theor Appl Genet 78:641–647

Chhabra KS (2001) Advances in pest management in grain legume crops—*Vigna* spp. Indian J Entomol 47 (1):103–105

Chhabra KS, Kooner BS (1985a) Losses in summer mungbean due to insect pest in Punjab. Indian J Entomol 47:103–105

Chhabra KS, Kooner BS (1985b) Problem of flower shedding caused by thrips, Megalurothrips distalis (Karny) on summer mungbean, *Vigna radiata* (L.) Wilczek and its control. Trop Pest Manag 31:186–188

Chhabra KS, Kooner BS (1998) Insect pest management in mungbean and blackgram—status and strategies. In: Upadhyay R, Mukerji KG, Rajak RL (eds) Pulses, vol 4. IPM system in agriculture, Aditya Books Publishing Pvt. Ltd, New Delhi, pp 233–310

Chhabra KS, Kooner BS, Brar JS (1980) Resistance behaviour of mungbean, *Vigna radiata* (L.) Wilczek cultivars against insect-pest complex and yellow mosaic virus. Indian J Ecol 7:276–280

Chhabra KS, Kooner BS, Sharma AK, Saxena AK (1988) Sources of resistance in mungbean (*Vigna radiata*) to insect pests and mungbean yellow mosaic virus. In: Shanmugasundaram S, McLean BT (eds) Proceedings of the second international mungbean symposium. The World Vegetable Center (AVRDC), Shanhua, pp 308–314

Chiang HS, Talekar NS (1980) Identification of sources of resistance to bean fly and two other agromyzid flies in soybean and mungbean. J Eco Entomol 73:197–199

Choragudi SR, Eswari KB, Sudarshanam A (2012) Reaction of greengram (*Vigna radiata* L.) OVT entries against major insect pests in *rabi* season. Andhra Agric J 59(1):87–92

Clarry S (2016) The rise and rise of mungbeans. In Ground cover supplement issue 125 November–December. Australia: Grains Research and Development Corporation, 15. Accessed 22 Apr 2019

D'Souza SF, Reddy KS, Badigannavar AM, Manjaya JG, Jambhulkar SJ (2009) Mutation breeding in oilseeds and grain legumes in India: accomplishments and socio-economic impact. In: Shu QY (ed) Induced plant mutations in the genomic era food and agricultural organization of the United Nations, p 55

DAFF (2010) Mungbean-production guide. Department of Forestry and Fisheries, Republic of South Africa, p 24

Dana S (1966) Cross between *Phaseolus aureus* and *P. mungo*. Genetica 37:259–274

Dana S (1968) Hybrid between *Phaseolus mungo* and tetraploid *Phaseolus species*. Japan J Genet 43:153–155

Dar GM, Verma MM, Gosal SS and Brar JS (1991) Characterization of some interspecific hybrids and amphidiploids in Vigna. In: Sharma B (ed) Proceedings Golden Jubilee symposium on grain legumes. Indian Soc Genet Plant Breed, pp 73–88

Das SB (1999) Outbreak of til hawk moth *Acherontia styx* Westwood on green gram. Insect Environ 5(2):71

Diatloff A, Imrie BC (2000) Inoculation techniques for evaluating resistance to *Curtobacterium flaccumfaciens* pv. *flaccumfaciens* in mungbean cultivars. Austr Plant Pathol 29:24–28

Duraimurugan P, Mishra A, Pratap A, Singh SK (2014) Toxicity of spinosad to the pulse beetle, *Callosobruchus chinensis* (Coleoptera: Bruchidae) and its parasitoid, *Dinarmus basalis* (Hymenoptera: Pteromalidae). The Ecoscan 8(1–2):17–21

Duraimurugan P, Tyagi K (2014) Pest spectra, succession and its yield losses in mungbean and urdbean under changing climatic scenario. Legumes Res 37(2):212–222

Ebert AW, Hidayat IM, de los Santos EB (2013) Cultivar trials of indigenous vegetables in Indonesia and community-based seed conservation and multiplication in the Philippines. In: Massawe, F, Mayes, S, Alderson, P (eds) Proceedings of the 2nd international symposium on underutilized plant species: crops for the future—beyond food security, vol 2. International Society for Horticultural Sciences (ISHS): Korbeek-Lo, Belgium, pp 341–348

Fernandez GCJ, Talekar NS (1990) Genetics and breeding for bruchid resistance in Asiatic *Vigna* species. In: Fujii K, Gatehouse AMR, Johnson CD, Mitchell R, Yoshida T (eds) Bruchids and legumes: economics, ecology and coevolution. Kluwer Academic Pub, Dordrecht, Netherlands, pp 209–217

Fujii K, Ishimoto M, Kitamura K (1989) Patterns of resistance to bean weevils (Bruchidae) in *Vigna radiata-mungo-sublobata* complex inform the breeding of new resistant variety. Appl Ent Zool 24:126–132

Fujii K, Miyazaki S (1987) Infestation resistance of wild legumes (*Vigna sublobata*) to azuki bean weevil, *Callosobruchus chinensis* (L.) (Coleoptera: Bruchidae) and its relationship with cytogenetic classification. Appl Ent Zool 22:229–230

Ganapathy N, Durairaj C (1995) Pest status of pulses in Pudukkottai district, Tamil Nadu. Madras Agri J 82 (4):322

Gentry J (2010) Mungbean management guide, 2nd edn. Department of Employment, Economic Development and Innovation, Queensland Government, p 24

Gill AS, Verma MM, Dhaliwal HS, Sandhu TS (1983) Interspecific transfer of resistance to mungbean yellow mosaic virus from *Vigna mungo* to *Vigna radiata*. Curr Sci 52:31–33

Gosal SS, Bajaj YPS (1983) Interspecific hybridization between *Vigna mungo* and *Vigna radiata* through embryo culture. Euphytica 32:129–137

Gupta S, Gupta DS, Anjum TK, Pratap A, Kumar J (2013) Inheritance and molecular tagging of MYMIV resistance gene in blackgram (*Vigna mungo* L. Hepper). Euphytica 193:27–37

Gupta S, Pratap A (2016) Mungbean: summer cultivation in India (Pocket Guide), AICRP on MULLaRP, ICAR-Indian Institute of Pulses Research, Kanpur-208024. Extension Bulletin, p 42

Haq MA (2009) Development of mutant varieties of crop plants at NIAB and the impact on agricultural production in Pakistan. In: Shu QY (ed) Induced plant mutations in the genomic era. Food and Agricultural Organization of the United Nations, p 61

Haque MF, Mukherjee AK, Mahto RN, Jha DK, Chakraborty M, Srivastava G, Pand Prasad D (1997) Birsa Urid-1—a new variety for Chotanagpur region of Bihar. J Res 9:177–178

Hartman GL, Wang TC, Kim D (1993) Field evaluation of mungbean for resistance to Cercospora leaf spot and powdery mildew. Int J Pest Manag 39:418–421

Hegde VS (1999) Evaluation of mungbean genotypes for resistance to powdery mildew. Indian J Plant Gen Res 12(3):408–409

Hong MG, Kim KH, Ku JH, Jeong JK, Seo MJ, Park CH, Kim YH, Kim HS, Kim YK, Baek SH, Kim DY, Park SK, Kim SL, Moon JK (2015) Inheritance and quantitative trait loci analysis of resistance gene to bruchid and bean bung in mungbean (*Vigna radiata* L. Wilczek). Plant Breed Biotechnol 3:39–46

Hossain MA, Ferdous J, Sarkar MA, Rahman MA (2004) Insecticidal management of thrips and pod borer in mungbean. Bangladesh J Agri Res 29(3):347–356

Humphry SME, Magner T, McIntyre CL, Aitken EA, Liu CL (2003) Identification of major locus conferring resistance to powdery mildew (*Erysiphe polygoni* D. C.) in mungbean (*Vigna radiata* L. Wiczek*)* by QTL analysis. Genome 46:738–744

Indiati SW (2015) Thrips pest management through integrated pest management approach. J Agric Res 34(2):51–60

Indiati SW, Bejo Rahayu M (2017) Diversity of mung bean insect pests and their natural enemies in farmers' fields in East Java, Indonesia. Biodiversitas 18 (4):1300–1307

Iqbal SM, Zubair M, Haqqani AM (2004) Resistant in mungbean to cercospora leaf spot disease. Int J Agric Biol 6:792–793

Iqbal U, Iqbal MS, Afzal R, Jamal A, Farooq MA, Zahid A (2011) Screening of mungbean germplasm against mungbean yellow mosaic virus (MYMV) under field conditions. Pak J Phytopathol 23(1):48–51

Islam W, Ahmed KN, Joarder 0I (1984) Timing and extent of damage caused by insect pests of green gram, *Vigna radiata* (L.) Wilczek in Bangladesh. Crop Protec 3(3):343–348

Jhamaria SL, Sharma OP (2002) Management of web blight of mungbean through chemicals and plant product. Indian Phytopath 55:526

Kaewwongwal A, Chen J, Somta P, Kongjaimun A, Yimram T, Chen X, Srinives P (2017) Novel alleles of two tightly linked genes encoding polygalacturonase-inhibiting proteins (VrPGIP1 and VrPGIP2) associated with the *Br* locus that confer bruchid (*Callosobruchus* spp.) resistance to mungbean (*Vigna radiata*) Accession V2709. Front Plant Sci 8:1692

Kasettranan W, Somta P, Srinives P (2010) Mapping of quantitative trait loci controlling powdery mildew resistance in mungbean (*Vigna radiata* (L.) Wilczek). J Crop Sci Biotechnol 13:155–161

Kashiwaba K, Tomooka N, Kaga A, Han OK, Vaughan DA (2003) Characterization of resistance to three bruchid species (*Callosobruchus* spp. Coleoptera, Bruchidae) in cultivated rice bean (*Vigna umbellata*). J Econ Entomol 6:207–213

Kaur L, Chadha ML, Kuo G, Gowda CLL (2007) Multiple disease resistant sources of mungbean. Acta Horti 752:423–426

Khan KSH, Shuaib M (2007) Identification of sources of resistant in mungbean (*Vigna radiata* L.) against charcoal rot *Macrophomina phaseolina* (Tassi) Goid. Afr Crop Sci Conf Proc 8:2101–2102

Khan MG, Ahmad W, Khattak GSS, Siraj-ud-Din Ahmad H (2007) Mode of inheritance of resistance to mungbean yellow mosaic virus (MYMV) in mungbean (*Vigna radiata* (L.) Wilczek). Sarhad J Agric 23:1071–1074

Khan S, Goyal S (2009) Improvement of mungbean varieties through induced mutations. African J Plant Sci 3(8):174–180

Khattak GSS, Haq MA, Ashraf M, Elahi T (2000) Genetics of mungbean yellow mosaic virus in mungbean (*Vigna radiata* (L.) Wilczek.). J Genet Breed 54:237–243

Khattak MK, Shafqat A, Chisti JI (2004) Varietal resistance of mungbean (*Vigna radiata* L.) against whitefly (*Bemisia tabaci* Genn.), jassid (*Amrasca devastans* Dist.), and thrips (*Thrips tabaci* Lind.). Pak Entomol 26:9–12

Kitamura K, Ishimoto M, Sawa M (1988) Inheritance of resistance to infestation with azuki bean weevil in *Vigna sublobata* and successful incorporation to *V. radiata*. Jap J Breed 38:459–464

Kitsanachandee R, Somta P, Chatchawankanphanich O, Akhtar KP, Shah TM, Nair RM, Bains TS, Sirari A, Kaur L, Srinives P (2013) Detection of quantitative trait loci for mungbean yellow mosaic India virus (MYMIV) resistance in mungbean (*Vigna radiata* (L.) Wilczek) in India and Pakistan. Breed Sci 63:367–373

Kooner BS (1998) Identification of sources of resistance to whitefly, jassids and MYMV in kharif mungbean. In: Proceedings of national symposium on management of biotic and abiotic stresses in pulse crops, vol 40. Indian Society of Pulses Research and Development, Kanpur

Kooner BS, Cheema HK (2007) Screening of mungbean germplasm against whitefly, *Bemisia tabaci* and MYMV. Acta Hort 752:307–310

Kooner BS, Cheema HK, Kaur R (2006) Insect pests and their management. In: Ali M, Kumar S (eds) Advances in mungbean and urdbean. Indian Institute of Pulses Research, Kanpur, pp 335–401

Kooner BS, Chhabra KS (1980) Pest of pulse crop and their control. In: Gill KS (ed) Breeding methods for the improvement of pulse crops. Ludhiana, India: Punjab Agricultural, pp 132–141

Kooner BS, Chhabra KS, Arora BS (1997) Resistant sources in mungbean to manage whitefly, jassids and yellow mosaic virus. In: Proceedings of third agricultural science congress, March 12–15, vol 2. PAU Ludhiana, India

Kooner BS, Malhi BS, Cheema HK (2005) Insect pest management of mungbean. In: Shanmugasundaram S (ed) Proceedings of the final workshop and planning meeting DFID-mungbean project (2002–2004), improving income and nutrition by incorporating mungbean in cereal fallows in the Indo-gangetic plains of South Asia, pp 214–235

Kooner BS, Singh K, Singh H, Singh KB (1977) Field screening of mungbean germplasm against whitefly

(*Bemisia tabaci* Genn.) and yellow mosaic virus. J Res Punjab Agric Univ 14:75–76

Krishnan R, De DN (1968) Cytological studies in *Phaseolus*, I. Autotetraploid *Phaseolus aureus* × a tetraploid species of *Phaseolus* and the backcrosses. Indian J Genet Plant Breed 28:12–22

Kumar R, Shamshad A, Rizvi SMA (2006) Screening of mungbean genotypes for resistance against white fly, *Bemisia tabaci* and mungbean yellow mosaic virus. Ind J Pulses Res 19:135–136

Lal SS (1987) Insect pests of mungbean, urd, cowpea, and pea and their management. In: Rao VM, Sithanantham S (eds) Plant protection in field crops. Plant Protection Association of India, Hyderabad, pp 185–202

Lambrides CJ, Imrie BC (2000) Susceptibility of mungbean varieties to the bruchid species *Callosobruchus maculatus* (F.), *C. phaseoli* (Gyll.), *C. chinensis* (L.), and *Acanthoscelide sobtectus* (Say.) (Coleoptera: Chrysomelidae). Austr J Agric Res 51:85–90

Lamichhane JR, Dürr C, Schwanck AA et al (2017) Integrated management of damping-off diseases. Agron Sustain Dev 37:10

Lawn RJ, Rebetzke GJ (2006) Variation among Australian accessions of the wild mungbean (*Vigna radiata* ssp. *sublobata*) for traits of agronomic, adaptive, or taxonomic interest. Aus J Agric Res 57:119–132

Lekhi P, Gill RK, Kaur S, Bains TS (2018) Generation of interspecific hybrids for introgression of mungbean yellow mosaic virus resistance in *Vigna radiata* (L.) Wilczek. Legume Res Int J, Article ID: LR-3808

Lekhi P (2017) Identification of molecular markers linked to MYMV resistance in mungbean (*Vigna radiata* (L.) Wilczek). Masters thesis, Punjab Agriculture University, Ludhiana, p 69

Mahato S, Jena M, Mohanty P, Behera C, Senapati N, Dash GB, Nayak PK, Lenka D and Tripathy SK (2015) Bruchid resistance in mungbean: an overview. e-planet 13(2):10–15

Maheshwari SK, Krishna H (2013) Field efficacy of fungicides and bio-agents against *Alternaria* leaf spot of mungbean. Ann Plant Prot Sci 21:364–367

Malik LA, Sarwar G, Ali Y (1986) Genetic studies in mungbean (*Vigna radiata* (L.) Wilczek). Inheritance of tolerance to mungbean yellow mosaic virus and some morphological characters. Pak J Bot 18:189–198

Malik SPS (1990) Comparative resistance of summer mungbean genotype to the thrips, *Megalurothrips distalis* (Karny.). M.Sc. thesis, Punjab Agricultural University, Ludhiana

Mandhare VK, Suryawanshi AV (2008) Dual resistance against powdery mildew and yellow mosaic virus in green gram. Agri Sci Digest 281:39–41

Manivannan N, Sethuraman K, Natarajan S (2001) Screening of greengram (*Vigna radiata* (L.) Wilczek) germplasm for yellow mosaic resistance. Legume Res 24:268–271

Mannur DM, Babbar, A, Thudi, M, Sabbavarapu MM, Roorkiwal M, Yeri S B, Bansal VP, Jayalakshmi S K, Yadav S S, Rathore A, Chamarthi S K, Mallikarjuna B P, Gaur P M, Varshney R K (2019) Super Annigeri 1 and improved JG 74: two Fusarium wilt-resistant introgression lines developed using marker-assisted backcrossing approach in chickpea (*Cicer arietinum* L.). Mol Breed 39:2

Mansoor-Ul-Hassan AR, Akbar R, Latif A (1998) Varietal response of mung and mash beans to insect attack. Pak J Entomol 20:43–46

Marwoto S (2008) Strategy and control of armyworm (*Spodoptera litura* Fabricius) on soybean. J Agric Res 27(4):131–136

Mbeyagala KE, Amayo R, Obuo J P, Pandey AK, War AR, Nair RM (2017) A manual for mungbean (greengram) production in Uganda. National Agricultural Research Organization (NARO), 32 pages

Mclean DM (1941) Studies on mosaic disease of cowpea. Phytopathol 31:420–430

Miyagi M, Humphry M, Ma ZY, Lambrides CJ, Bateson M, Liu CJ (2004) Construction of bacterial artificial chromosome libraries and their application in developing PCR-based markers closely linked to a major locus conditioning bruchid resistance in mungbean (*Vigna radiata* L. Wilczek). Theor Appl Genet 110:15–56

Mohan S, Sheeba A, Murugan E, Ibrahim SM (2014) Screening of mungbean germplasm for resistance to mungbean yellow mosaic virus under natural condition. Indian J Sci Tech 7(7):891–896

Nadeem S, Hamed M, Asghar MJ, Abbas G, Saeed NA (2014) Screening of mungbean (*Vigna radiata* (L.) Wilczek) genotypes against sucking insect pests under natural field conditions. Pak J Zool 46(3):863–866

Nair RM, Schalfleitner R, Kenyon L, Srinivasan R, Easdown W, Ebert A, Hanson P (2013) Genetic improvement of mungbean. SABRAO J Breed Gen 44:177–190

Nair RM, Yan MR, Srinivasan R, Schafleitner R (2015) Developing bruchid resistant mungbean varieties. In: Paper presented at the SABRAO 13th congress and international conference IPB, 14–16 Sept 2015. International Convention Center, Bogor, Indonesia

Nair RM, Götz M, Winter S, Giri RR, Boddepalli VN, Sirari A, Bains TS, Taggar GK, Dikshit HK, Aski M, Boopathi M (2017) Identification of mungbean lines with tolerance or resistance to yellow mosaic in fields in India where different begomovirus species and different *Bemisia tabaci* cryptic species predominate. Eur J Plant Path 149(2):349–365

Naraini TK (1960) Yellow mosaic of mung (*Phaseolus aureus* L.). Indian Phytopath 13:24–29

Ngampongsai S, Chotechuen S, Wongpiyasatid A (2004) Induced mutations in mungbean breeding programme. In: Proceedings of the ninth national mungbean research conference, 5–6 Mar 2004. Lampangwiengthong Hotel Lampang, pp 111–118 (in Thai)

Ngampongsai S, Watanasit A, Srisombun S, Srinives P, Masari A (2009) In: Induced plant mutations in the genomics era. Food and Agriculture Organization of the United Nations, Rome, pp 355–357

Noble TJ, Young A, Douglas CA, Williams B, Mundree S (2019) Diagnosis and management of halo blight in

Australian mungbeans: a review. Crop Pasture Sci (in press)
Pal SS, Dhaliwal HS, Bains SS (1991) Inheritance of resistance to yellow mosaic virus in some *Vigna* spp. Plant Breed 106:168–171
Pal SS, Singh JJ, Singh I (2000) Transfer of YMV resistance in cultivar SML32 of *Vigna radiata* from other related *Vigna* species. Plant Disease Res 15: 67–69
Pandey AK, Burlakoti RR, Kenyon L, Nair RM (2018) Perspectives and challenges for sustainable management of fungal diseases of mungbean [*Vigna radiata* (L.) R. Wilczek var. *radiata*]: a review. Front Environ Sci 6:53. https://doi.org/10.3389/fenvs.2018.00053
Pandiyan M, Senthil N, Ramamoorthi N, Muthiah AR, Tomooka N, Duncan V, Jayaraj T (2010) Interspecific hybridization of *Vigna radiata* × 13 wild *Vigna* species for developing MYMV donor. Elect J Plant Breed 1(4):600–610
Pathak AK, Jhamaria SL (2004) Evaluation of mungbean (*Vigna radiata L.)* varieties to yellow mosaic disease (YMD). J Mycol Plant Pathol 34:64–65
Patel PN, Jindal JK (1972) Bacterial leaf spot and halo blight disease of mungbean and other legumes in India. Indian Phytopathol 25:526–529
Pratap A, Basu PS, Gupta S, Malviya N, Rajan N, Tomar R, Latha M, Nadarajan N, Singh NP (2014) Identification and characterization of sources for photo- and thermo-insensitivity in *Vigna* species. Plant Breed 133:756–764
Pratap A, Chaturvedi SK, Tomar R, Rajan N, Malviya N, Thudi M, Saabale PR, Prajapati U, Varshney RK, Singh NP (2017) Marker-assisted introgression of resistance to fusarium wilt race 2 in Pusa 256, an elite cultivar of *desi* chickpea. Molec Genet Genom 292:137–1245
Pratap A, Gupta DS, Rajan N (2012) Mungbean. In: Bharadwaj D (ed) Breeding Indian field crops. Agrobios Publishers, New Delhi, pp 208–227
Pratap A, Gupta DS, Singh BB, Kumar S (2013) Development of super early genotypes in greengram (*Vigna radiata* L. Wilczek). Legume Res 36:105–110
Pratap A, Gupta S, Basu S, Tomar R, Dubey S, Rathore M, Prajapati U, Singh P, Kumari G (2019) Towards development of climate-smart mungbean: challenges and opportunities In: Kole C (ed) Genomic designing of climate smart pulse crops. Springer Nature, New York, pp 235-264
Pratap A, Kumar J (2014b) Alien gene transfer in crop plants: an introduction. In: Pratap A, Kumar (eds) Alien gene transfer in crop plants. Innovations, methods and risk assessment, vol 1. Springer, New York, NY, USA, pp 1–23
Pratap A, Gupta S, Malviya N, Tomar R, Maurya R, Joseph JK, Singh NP (2015) Genome scanning of Indian *Vigna* species through microsatellite variation for genetic diversity and population structure analysis. Mol Breed 35:1–3
Pratap A, Prajapati U, Singh CM, Gupta S, Rathore M, Malviya N, Tomar R, Gupta AK, Tripathi S, Singh NP (2018) Potential, constraints and applications of in vitro methods in improving grain legumes. Plant Breed 137:235–249
Project Coordinators Report (2018) All India Coordinated Research Project on MULLaRP (Mungbean, Urdbean, Lentil, Lathyrus, Rajmash, Fieldpea). ICAR-Indian Institute of Pulses Research, Kalyanpur, Kanpur, p 46
Rahman MM, Mannan MA, Islam MA (1981) Pest survey of major summer and winter pulses in Bangladesh. In: Kaul AK (ed) The proceedings of the national workshop on pulses, pp 265–273
Rajarathinam S, Natarajan E, Thyagarajan K, Arjunan G, Sundera RPV, Rathinasamy R (1990) Vamban 1-A, yellow mosaic resistant black gram variety. Madras Agric Jour 77:73–76
Rana DS, Anchal Dass, Rajanna GA, Ramanjit Kaur (2016) Biotic and abiotic stress management in pulses. Indian J Agron 61(4th IAC Special issue):S238–S248
Reddy KR, Singh DP (1995) Inheritance of resistance to mungbean yellow mosaic virus. Madras Agric J 88:199–201
Reddy KS (2009) A new mutant for yellow mosaic virus resistance in mungbean (*Vigna radiata* (L.) Wilczek) variety SML-668 by recurrent gamma-ray irradiation In: Shu QY (ed) Induced plant mutations in the genomic era Food and Agricultural Organization of the United Nations, p 361
Reddy KS, Kao KC, Reddy MS (1992) Evaluation of some new fungicides against *Rhizoctonia solani* Kühn, the incident of damping-off in mungbean. Indian J Plant Protect 20:37–42
Ryley MJ, Tatnell JR (2011) Management of the major foliar diseases of mungbeans and peanuts in Australia. In: 4th Asian conference on plant pathology and the 18th biennial Australasian plant pathology society conference (ACPP/APPS 2011): new frontiers in plant pathology for Asia and Oceania, 26–29 Apr 2011, Darwin, Australia
Sahoo BK, Hota AK (1991) Field screening of greengram germplasm against insect pest and disease complex. Madras Agric J 78(1–4):84–86
Sahoo BK, Patnaik NC (1994) Insect pests in green gram and blackgram in the South Costal region of Orissa with notes on their seasonal activity. Orissa J Agri Res 7(Supp 1):74–76
Saleem M, Haris WAA, Malik IA (1998) Inheritance to yellow mosaic virus in mungbean (*Vigna radiata* L. Wilczek). Pak J Photopath 19:30–32
Sandhu TS (1978) Breeding for yellow mosaic virus resistance in mungbean. Punjab Agricultural Univ., Ludhiana (India). Dept. of Plant Breeding
Sandhu TS, Brar JS, Sandhu SS, Verma MM (1985) Inheritance of resistance to mungbean yellow mosaic virus in greengram. J Res Punjab Agric Univ 22:607–611

Sarkar S, Bhattacharyya S (2015) Screening of greengram genotypes for Bruchid (*Callosobruchus chinensis* L.) Resistance and selection of parental lines for hybridization programme. Legume Res 38(5):704–706

Sarkar S, Ghosh S, Chatterjee M, Kalita PD, Lahari T, Maji A, Mondal N, Pradhan KK, Bhattacharyya S (2011) Molecular markers linked with bruchid resistance in *Vigna radiata* var. *sublobata* and their validation. J Plant Biochem Biotechnol 20:155–160

Satyagopal K, Sushil SN, Jeyakumar P, Shankar G, Sharma OP et al (2014) AESA based IPM package for blackgram and greengram. National Institute of Plant Health Management, Rajendranagar, Hyderabad, p 43

Schafleitner R, Huang SM, Chu SH Yen JY, Lin CY, Yan MR, Krishnan B, Liu MS, Lo HF, Chen CY, Long-fang OC (2016) Identification of single nucleotide polymorphism markers associated with resistance to bruchids (*Callosobruchus* spp.) in wild mungbean (*Vigna radiata* var. *sublobata*) and cultivated V. radiata through genotyping by sequencing and quantitative trait locus analysis. BMC Plant Biol 16:159

Selvi R, Muthiah AR, Manivannan N, Raveendran TS, Manickam A, Samiyappan R (2006) Tagging of RAPD marker for MYMV resistance in mungbean (*Vigna radiata* (L.) Wilczek). Asian J Plant Sci 5:277–280

Sharma I, Dubey GS (1984) Control of urdbean leaf crinkle virus through heat treatment, chemotherapy and resistance. Indian Phytopathol 37:26–30

Shukla GP, Pandya BP (1985) Resistance to yellow mosaic in greengram. SABRAOJ 17:165–171

Singh A, Awadhiya G K, Gupta B (2002) Effect of varieties and sowing dates on powdery mildew severity and yield of field pea (*Pisum sativum* L.). J Mycol Plant Pathol 32:286

Singh BB, Solanki RK, Chaubey BK, Verma P (2011) Breeding for improvement of warm season food legumes. In: Pratap A, Kumar J (eds) Biology and breeding of food legumes. CAB International, Oxfordshire, pp 63–80

Singh DP (1980) Inheritance of resistance to yellow mosaic virus in blackgram (*Vigna mungo* (L.) Hepper). Theor Appl Genet 57:233–235

Singh DP (1982) Genetics and breeding of blackgram and greengram. G.B. Pant University of Agriculture and Technology, Pantnagar, India. Res Bull 109:68

Singh DP (1990) Distant hybridization in genus Vigna: a review. Ind J Genet 50(3):268–276

Singh DP, Singh BB, Pratap A (2017) Genetic improvement of mungbean and urdbean and their role in enhancing pulse production in India. Indian J Genet 76:550–567

Singh RA, De RK (2006) Major diseases and their management. In: Ali M, Kumar S (eds) Advances in mungbean and urdbean. IIPR, Kanpur, pp 283–334

Singh RA, De RK, Gurha SN, Gosh A (2003) Yellow mosaic virus of mungbean and urdbean. In: Upadhyay RK, Arora DK, Dubey OP (eds) IPM system in agriculture, key pathogens and diseases, vol 8. New Delhi: Aditya Books Pvt. Ltd., pp 395–408

Singh RA, Gurha SN (2005) Search for sources of resistance to powdery mildew in mungbean and urdbean. In: Proceedings of the second global conference on plant health-global wealth, MPUAT, Udaipur, India, 25–29 Nov 2005, vol 256. ISMPP

Singh SK, Singh PS (2014) Screening of mungbean (*Vigna radiata*) genotypes against major insects. Curr Adv Agric Sci 6(1):85–87

Singh VP, Chhabra A, Kharb RPS (1988) Production and utilization of mungbean in India. In: Proceedings of the second international symposium 16–20 Nov 1987. Bangkok, Thailand, AVRDC, Shanhua, Tainan, p 588

Solanki IS, Dahiya BS, Waldia RS (1982) Resistance to mungbean yellow mosaic virus in blackgram. Indian J Genet 43:240–242

Somta P, Ammaranan C, Ooi PAC, Srinives P (2007) Inheritance of seed resistance to bruchids in cultivated mungbean (*Vigna radiata* L. Wilczek). Euphytica 155:47–55

Somta P, Kaga A, Tomooka N, Kashiwaba K, Isemura T, Chaitieng B, Srinives P, Vaughan DA (2006) Development of an interspecific *Vigna* linkage map between *Vigna umbellata* (Thunb.) Ohwi & Ohashi and *V. nakashimae* (*Ohwi*) Ohwi & Ohashi and its use in analysis of bruchid resistance and comparative genomics. Plant Breed 125:77–84

Soumia PS, Chitra S, Dikshit HK, Guru Pirasanna PG (2017) Screening for resistance against pulse beetle, *Callosobruchus analis* (F.) in greengram (*Vigna radiata* (L.) Wilczek) accessions. Proc Natl Acad Sci India Sect B Biol Sci 87(2):551–558

Srinives P (1991) Breeding for resistance to diseases and pests of mungbean in Thailand. In: Green SK, Kim D (eds) Proceedings international workshop on mungbean yellow mosaic virus, 2–3 July 1991. Bangkok, Thailand, pp 59–67

Srinives P (1996) Mungbean breeding: past, present and future. In: Srinives P et al (eds) Mungbean germplasm: collection, evaluation and utilization for breeding program. Proceedings of workshop on mung bean germplasm, 17 Aug 1995. Thailand, pp 73–83

Sudha M, Karthikeyan A, Anusuya P, Ganesh NM, Pandiyan M, Senthil N, Raveendran M, Nagarajan P, Angappan K (2013) Inheritance of resistance to mungbean yellow mosaic virus (MYMV) in inter and intraspecific crosses of mungbean (*Vigna radiata*). Am J Plant Sci 4:1924–1927

Suman S (2015) Genetic and molecular analysis with respect to yellow vein mosaic virus resistance in mungbean (*Vigna radiata*). PhD thesis, Rajendra Agricultural University, Pusa (Samastipur), p 164

Sun K, Cheng XZ, Wang SH, Wang LX, Liu CU, Li MEI, Ning XU (2008) Heredity analysis and gene mapping of bruchid resistance of a mungbean cultivar V2709. Agric Sci China 7(6):672–677

Sun S, Zhi Y, Zhu Z, Jin J, Duan C, Wu X, Xiaoming W (2017) An emerging disease caused by *pseudomonas syringae* pv. *phaseolicola* threatens mungbean production in China. Plant Dis 101:95–102

Swaminathan R, Singh K, Nepalia V (2012) Insect pests of greeng gram *Vigna radiata* (L.) Wilczek and their management. In: Aflakpui G (ed) Agricultural science, pp 197–222

Talekar NS (1990) Agromyzid flies of food legumes in the tropics. Wiley Eastern Limited, New Delhi, p 299

Thakur PR, Patel PN, Verma JP (1977) Genetic relationship between reactions to bacterial leaf spot, yellow mosaic and Cercospora leaf spot diseases of mungbean (*Vigna radiata* L.). Euphytica 26:765–774

Tomooka N, Kaga A, Vaughan DA (2006) The Asian *Vigna* (*Vigna* Subgenus *Ceratotropis*) biodiversity and evolution. In: Sharma AK, Sharma A (eds) Plant genome: biodiversity and evolution. CRC Press, pp 87–126

Tomooka N, Kashiwaba K, Vaughan D, Ishimoto M, Egawa Y (2000) The effectiveness of evaluating wild species, searching for sources of resistance to bruchid beetle in the genus *Vigna* subspecies ceratotropis. Euphytica 115:27–41

Tomooka N, Vaughan DA, Xu RQ, Kashiwaba K, Kaga A (2001) Japanese native *Vigna* genetic resources. Jap J Agri Res Q 35:1–9

Tomooka NC, Lairungreang R, Nakeeraks P, Egawa Y (1992) Development of bruchid resistant mungbean line using wild mungbean germplasm in Thailand. Plant Breed 109:60–66

Varshney RK, Mohan SM, Gaur PM, Chamarthi SK, Singh VK, Srinivasan S, Swapna N, Sharma M, Pande S, Singh S, Kaur L (2014) Marker-assisted backcrossing to introgress resistance to fusarium wilt race 1 and ascochyta blight in C 214, an elite cultivar of chickpea. Plant Genome 7(1):1–11

Verma MM, Brar JS (1996) Breeding approaches for increasing yield potential of mungbean. In: Asthana AN, Kim DH (eds) Recent advances in mungbean research. Indian Society of Pulses Research and Development, Kanpur, India, pp 102–123

Verma RPS, Singh DP (1986) The allelic relationship of genes giving resistance to mungbean yellow mosaic virus in blackgram. Theor Appl Genet 72:737–738

War AR, Murugesan S, Boddepalli VN, Srinivasan R, Nair MR (2017) Mechanism of resistance in mungbean [*Vigna radiata* (L.) R. Wilczek var. *radiata*] to bruchids, *Callosobruchus* spp. (Coleoptera: Bruchidae). Front Plant Sci 20

Wongpiyasatid A, Chotechuen S, Hormchan P, Srihuttagum M (1999) Evaluation of yield and resistance to powdery mildew, cercospora leaf spot and cowpea weevil in mungbean mutant lines. Kasetsart J 33:204–215

Yadav DK, Singh SK (2006) Forecast model of major insect pests of mung bean. Annals Plant Protec Sci 14 (2):323–328

Yadav DL, Pandey RN, Jaisani P, Gohel NM (2014) Sources of resistance in mungbean genotypes to cercospora leaf spot disease and its management. Afr J Agric Res 9(41):3111–3114

Yadav GS, Dahiya B (2000) Screening of some mungbean genotypes against major insect-pests and yellow mosaic virus. Ann Agric Bio Res 5:71–73

Yadav MS, Brar KS (2010) Assessment of yield losses due to mungbean yellow mosaic India virus and evaluation of mungbean genotypes for resistance in South-West Punjab. Ind Phytopathol 63:318–320

Young ND, Kumar L, Menancio-Hautea D, Danesh D, Talekar NS, Shanmugasundaraum S, Kim DH (1992) RFPL mapping of a major bruchid resistance gene in mungbean (*Vigna radiata* L. Wilczek). Theor Appl Genet 84:839–844

Zhimo VY, Panja BN, Saha J, Nath R (2013) Evaluation of mungbean genot ypes for resistance against cercospora leaf spot and yellow mosaic diseases under field condition. J Mycopathol Res 51:273–278

Breeding Progress and Future Challenges: Abiotic Stresses

6

Col Douglas, Aditya Pratap, Bindumadhava Hanumantha Rao, B. Manu, Sonali Dubey, Parikshit Singh and Rakhi Tomar

Abstract

Mungbean is a short-season tropical grain legume grown on some six million hectares each year. Though predominantly a crop of smallholder farmers and subsistence agriculture mungbean is increasingly seen as a high value crop for international markets with broad acre production under modern farming systems established in Australia, South America, West Asia and Africa. Key benefits of mungbean are its nutritional and monetary value. It provides a short duration, flexible disease break when fit into intensive wheat, rice and summer cereal rotations and its self-sufficiency for nitrogen. The short growing season of 55–100 days places a ceiling on productivity which is further impacted by the traditional low-input farming systems where mungbean is most frequently produced; global yield averages are 0.5 tonnes per hectare though 3 tonnes per hectare is considered achievable under favourable conditions. Increased reliability of mungbean in subsistence systems has been achieved by developing shorter duration, more determinate ideotypes and by the manipulation of sowing time. The strategy of reducing exposure to risk was very successful in transforming mungbean rather than identifying and breeding inherent resilience. The major abiotic stresses of mungbean presented here are drought, heat, waterlogging, low temperatures and salinity. Sources of tolerance identified for all of these stresses have been identified in the germplasm collections of cultivated mungbean as well as wild relatives. Future research efforts must combine known sources of genetic variation with the investigation into the biochemical and physiological processes in order to understand and breed for tolerance to abiotic stress in mungbean.

C. Douglas (✉)
Department of Agriculture and Fisheries, Hermitage Research Facility, 604 Yangan Road, Warwick Qld 4370, Australia
e-mail: col.douglas@daf.qld.gov.au

A. Pratap · B. Manu · S. Dubey · P. Singh · R. Tomar
ICAR-Indian Institute of Pulses Research, Kalyanpur, Kanpur 208024, Uttar Pradesh, India

B. H. Rao
World Vegetable Center, South Asia, ICRISAT Campus, Patancheru 502324, Greater Hyderabad, Telangana, India

6.1 Introduction

Mungbean or green gram [*Vigna radiata* (L.) Wilczek var. *radiata*] is a short duration tropical legume that was domesticated in India before spreading to Southeast Asia and China. Today the vast majority of production is still in Asia where mungbean is grown by smallholder farmers for a ready source of income and as

R. M. Nair et al. (eds.), *The Mungbean Genome*, Compendium of Plant Genomes,
https://doi.org/10.1007/978-3-030-20008-4_6

a nutritious staple food. Mungbean has been grown under mechanized agriculture in Australia since the late 1960s and commercial production has more recently spread to West Asia, Africa and South America thanks to growing international demand. As a short-season legume mungbean aids soil fertility and is a significant contributor to the flexibility and overall sustainability and productivity of many tropical and sub-tropical farming systems.

Globally, mungbean is grown on around 6 million hectares annually yet average production of only 3 million tonnes reflects that its status as an orphan crop left behind by the green revolution. Mungbean is predominantly grown in traditional, low-input farming systems and subject to a myriad of biotic and abiotic stresses. Increased reliability of mungbean in wheat–rice cropping systems has been achieved by selecting for short duration types with a more determinate growth habit (Hanumantharao et al. 2016).

There is a considerable gap between anecdotal reports of farmer yields and crop simulation models (Chauhan and Rachaputi 2014) with the productivity varying from 3 t/ha for favourable rainfed crop to an average yield of 1 t/ha in Australia and 0.5 t/ha in Asia. In India, its production has increased to almost 3 million tonnes from 1.17 million tonnes in 2007–08 (PC Report 2018). In Australia, a fledgling industry supported by public sector research and development has seen production triple to an average volume of 100,000 tonnes per year. Empirical breeding has delivered yield gains of 20–30% and stepwise improvements in disease resistance have been made from germplasm introductions and the first locally bred varieties. The current crop research landscape is characterized by intense competition for resources; investors and donors looking for maximum impact in productivity, food security and nutrition. Researchers must be more connected and work more collaboratively. As a relatively new crop, there are encouraging and easy-within-reach productivity gains for mungbean. There are established and successful research models from the major crops. Increased power and affordability of genomic and high throughput phenotyping has made these technologies more attainable to crops such as mungbean. The mungbean genome was published in 2014 (Kang et al. 2014) and two complementary diversity panels have subsequently been assembled and genotyped (Schafleitner et al. 2015) laying the foundation for researchers and breeders to understand abiotic stresses in mungbean and develop more productive, more resilient varieties.

6.2 Major Abiotic Stresses

Abiotic stresses are those environmental factors that impact on a plant's biochemical and physiological processes to constrain productivity. Either individually or more often in combination, abiotic stresses prevent crops from reaching their genetic yield potential. Of the major abiotic stresses that plants are subject to our focus in this chapter has been on temperature (high and low) and water availability (drought and waterlogging). Other abiotic factors causing stress to mungbean crops include salinity, mineral toxicity, UV-radiation, greenhouse gases triggering climate change.

Mungbean has nine growth stages; the progression and the phenological development of the plant are driven by temperature (Carberry 2007; Chauhan et al. 2010). Key temperatures for mungbean are 7.5 °C (baseline), 30 °C (optimum) and 40 °C (maximum). Kaur et al. (2015) described optimum temperatures for mungbean as 28–30 °C but clearly, mungbean is subjected to the conditions that are far in excess of this and faces still greater heat stress conditions from the context of climate change.

Patriyawaty et al. (2018) described reduced root weight and pollen viability as key contributors impacting reproductive function among commercial varieties and breeding lines of mungbean under heat stress. On the physiological level differences in SPAD value, stomatal conductance, stomatal density and leaf temperature were proposed as quantifiable traits that may useful to researchers and breeders to discern between susceptible and tolerant genotypes with respect to heat stress. While evaluating the heat stress response of two extra-short duration genotypes at >40/25 °C, Kaur et al. (2015) found reduced biomass, pod set, seed number and seed yield per plant but that there was no effect on phenology. Underlying reproductive functions were pollen viability, germination and pollen tube growth as well as pollen load and receptivity of the stigma. Heat stress was also linked to impacts on stomatal conductance, photosynthetic efficiency and a decrease in sucrose concentrations and its synthesis. Using the late sowing approach, Sharma et al. (2016) were able to accelerate phenology through elevated temperatures and documented a reduction in leaf area, biomass, number of flowers and number of pods. Reduced reproductive function manifested as flower and pod abortion. Fives lines were identified with purported heat stress tolerance for donors in the breeding program and to elucidate the biochemical and physiological processes involved in heat stress mode of action. Bindumadhava et al. (2015) reported that heat stress research on mungbean has been sparse and suggested that progress could be made by examining the functional viability of pollen cells produced under heat stress.

6.2.1 Temperature Stress (High and Low)

High-temperature stress has been reported to negatively affect reproductive development in mungbean (Tzudir et al. 2014; HanumanthaRao et al. 2016). It affects all reproductive traits viz., flower initiation, pollen viability, stigma receptivity, ovule size and viability, fertilization, pod set, grain filling as well as seed quality (Wassman et al. 2009; Barnabas et al. 2008). Flower shedding is very common in mungbean crop and has been reported up to 79% (Kumari and Varma 1983). Heat stress severely affects flower bud initiation and this sensitivity may prevail for 10–15 days (Bita and Gerats 2013). While terminal high-temperature stress is a serious problem in spring and summer grown mungbean early-stage heat stress is observed to occur in *kharif* (rainy season) crop (Pratap et al. 2019). Temperatures exceeding 42 °C during summer cause hardening of mungbean seeds due to incomplete sink development. Terminal heat stress is a severe problem in India (Singh and Singh 2011) especially during spring and summer seasons. Even during early kharif season, temperatures may remain as high as 38 °C and may cause drastic reduction in seed yield due to pollen inviability, incomplete fertilization and significant flower drop. Abscission of reproductive organs has been reported as one of the primary symptoms of reduced yield under heat stress in annual grain legumes (Rainey and Griffiths 2005).

Screening and selection of mungbean genotypes which can retain maximum number of flowers and produce productive pods during high temperature (>40 °C) are essential to increase productivity but only limited basic information is available about mungbean flower shedding and no work has been carried out for breeding of mungbean for maximum flower retention under high temperature (>40 °C).

Increased flower shedding under high temperature, precipitation and desiccating winds during the flowering period in legumes has been reported by different researchers (Sinha 1977; Tickoo et al. 1996; Rainey and Griffiths 2005). Significant flower shedding above 40 °C in mungbean was reported by Tickoo et al. (1996). Khattak et al. (1999) reported absence of resistance to flowers shedding under high temperature in mungbean. In this crop flowers are borne in clusters of 10–20 in axillary or terminal racemes and come in different flushes. Generally a higher daily mean temperature hastens flowering, while a lower mean temperature is expected to delay it at all photoperiods. However, this relationship may not hold true for all strains of mungbean as observed by Aggarwal and Poehlman (1977). Rawson and Craven (1979) reported temperature-flowering interactions in particular groups of strains with high mean temperatures (24–28 °C) and long photoperiods (15–16 h).

In field experiments at Hisar the mungbean cultivar 'K 851' was sown in the summer season of 1989 (Singh et al. 1991). The leaf, flower and pod growth showed a linear response to cumulative heat units and were initiated at 140, 630 and 725 °C, respectively. A strong linear correlation was also observed between cumulative heat units and dry weight, although dry matter gain was slow until 300 °C of cumulative heat units. Perennial accessions of *V. radiata* var. *sublobata* have been found to be tolerant to low temperatures (Lawn et al. 1988).

Low or cooler temperatures are one of the important environmental conditions that hamper growth and yield of many cool-season legumes. Though mungbean is a warm-season legume, lower temperatures (9–11 °C) that prevail during *Rabi* growing season (January–February months in the rice fallow) do impede growth rates as well as the productivity. A field-level low-temperature phenotyping assay was carried out by the World Vegetable Center research team in 30 mungbean lines in five districts of Odisha (eastern state of India), during Rabi season of 2017 (WorldVeg Report 2017) to select low-temperature tolerant lines. The cold tolerance scoring index (CTSI) an empirical index (derived based on leaf appearance, plant morphology and architecture characters), was applied (1—Highly tolerant, 2—Tolerant, 3—Moderately tolerant/sensitive, 4—Sensitive, 5—Highly sensitive) for a large set of mungbean lines. Further, the biometric parameters such as plant height, days to 50% flowering and pod and seed yield were recorded for mungbean in all these five districts and mapped with CTSI to shortlist promising low-temperature tolerant lines. Among the 30 mungbean lines phenotyped, a few lines (e.g., PUSA 9704, VI001400AG and KPS-1) showed quite a similar cold tolerance response across the districts.

6.2.2 Water Stress (Deficit and Excess)

Mungbean requires a fair moisture regime in soil during its entire growing period, while excess moisture or waterlogging even for a short period of time, is detrimental. Drought is one of the major stresses of mungbean, affecting growth and development by reducing vegetative growth, flower initiation and pod set (Morton et al. 1982). As mungbean is mostly grown under rainfed situations it is reported to be more susceptible to water deficits as compared to many other food legumes (Pandey et al. 1984). Spring and summer grown crops are more prone to water stress as rainfall is unpredictable and farmers mostly prefer to grow this crop on residual moisture. Therefore cultivating short-duration cultivars may help in escaping terminal moisture stress (Pratap et al. 2013a).

Among the abiotic stresses, waterlogging also stands prominent in water-dense areas which receive heavy downpours within a short time span (where soil drainage is poor coupled with maximum water holding). Mungbean cultivation may incur severe yield loss where the annual rainfall is >1000 mm (Fernandez and Shanmugasundaram 1988). Though the impacts of waterlogging in mungbean are well recognised information on the physiological responses of mungbean is scarce.

Mungbean is highly sensitive to waterlogging and cannot withstand even a few hours of excess soil moisture, in particular during the early stages of growth (Tickoo et al. 2006). Flooding around the roots of submerged plants leads to restricted aeration thereby reducing nodule activity and nitrogen fixation (Singh and Singh 2011). In later growth stages heavy rains coupled with strong winds may lead to plant damage and consequently heavy losses. Even if the plants survive they may further be severely affected by fungal diseases and also insect–pests (Tickoo et al. 2006). Haqqani and Pandey (1994) suggested that decrease in leaf area index and increase in specific leaf weight, leaf water potential and root density are some of the drought avoidance mechanisms during the reproductive phase of the plants when the moisture is generally depleted in the soil.

Duong et al. (1988) reported that 48 h of waterlogging reduced plant height from 36 to 76%, leaf area from 12 to 46% and dry matter production from 25 to 57%. The cultivar differences to waterlogging were also observed. Musgrave and Vanhoy (1989) conducted a growth analysis of waterlogging damage and illustrated the interrelationships between root and shoot carbon budgets during the response to the stress of waterlogging. Pre-harvest sprouting has been reported as a serious problem in tropical regions (Fernandez and Shanmugasundaram 1988). Varma and Rao (1975) reported that excess soil moisture (75 and 100% of soil dry weight) was more detrimental to seed yield and nodulation than the limited soil moisture (25% of soil dry weight) in mungbean. Despite being protected inside the pod mungbean seeds are susceptible to pre-harvest sprouting following rainfall due to lack of fresh seed dormancy, this deteriorates the quality of the seed and grain produced. Therefore the development of mungbean cultivars with a short (10–15 days) period of fresh seed dormancy has become important to curtail losses incurred by pre-harvest sprouting (Lamichaney et al. 2017). Long duration cultivars with more reproductive flushes would give more stable yields because flowering would be continued over the longer period (Chowdhury and Haque 1977). However, such cultivars would require additional pickings and would be prone to lodging, shattering and sprouting.

To select suitable mungbean lines that withstand excess water conditions a field screening study was conducted with 40 mungbean lines jointly by World Vegetable Center and Department of Crop Physiology, at the University of Agricultural Science, GKVK Campus, Bengaluru, India. Specially designed 'Field Root Structures' waterlogging were used to screen mungbean lines for waterlogging tolerance (Rameshreddy et al. 2019; WorldVeg Report 2017). This study offered growth and yield responses of these mungbean lines for waterlogging stress, helps in exploring genetic variability within a species and to identify efficient waterlogging tolerant genotypes. The study also determined the genotypic response to waterlogging stress at various soil depths.

The first treatment of waterlogging was applied at 30 days after sowing DAS for five days and the number of plants in each row was recorded before and after the treatment. The second flooding treatment was applied at flowering (43 DAS) and the third treatment applied at 60 DAS.

After each treatment, the biometric parameters such as plant height, number of branches, number of leaves, fruiting points, number of pods, pod and seed yield per plant along with harvest index, were measured through destructive plant sampling method. Based on the observations from seed yield and total biomass (TDM) at different depths at various growth stages, a standardized Z-distribution was plotted. The following lines that showed higher seed yield as well as high TDM were classified as tolerant (AVMU 1001, AVMU 1201, VO6381A-G, KPS-1, ML 1628, PDM 139, IPM 02-14). Similarly, lines having low TDM and seed yield were classified as susceptible (EC 693362, VC 3960-88, VC 6153 B20 and EC 693370) (WorldVeg Report 2017).

6.2.3 Salinity

Adverse effects of salinity on plant growth may be due to ion cytotoxicity (mainly due to Na^+, Cl^- and SO_4^-) and osmotic stress (Zhu 2002). Most mungbean cultivars tolerated salt to an extent of 9–18 m mhos/cm and are tolerant to salinity at germination stage. Paliwal and Maliwal (1980) reported that mungbean seeds could tolerate 6 m mhos/cm salinity with a progressive reduction in germination and seedling development, nodulation and nitrogen fixation was observed with an increase in salinity. In general legumes are highly salt-sensitive crops (Hanumantharao et al. 2016). Salinity affects crop growth and yield in three ways: (a) osmotic stress, (b) ion toxicity, and (c) reduced nodulation and therefore reduced nitrogen-fixing ability (Pratap et al. 2019). Salinity causes a significant reduction in yield (Abd-Alla et al. 1998; Saha et al. 2010) as it has several pronounced symptoms in mungbean viz., enhanced chlorosis, necrosis and decreased the content of chlorophyll and carotenoids (Gulati and Jaiwal 1993; Wahid et al. 2004). Several symptoms viz., decrease in total leaf area and stomatal opening (Chakrabarti and Mukherji 2002) and reduction in total chlorophyll content, sugar, starch and peroxidise enzyme activity in roots and shoots (Arulbalachandran et al. 2009) have been reported due to salinity stress in *Vigna* species.

Friedman et al. (2006) observed that NaCl stress has a more deleterious effect on roots than shoots as there is a reduction in root growth associated traits. In seedlings high salt concentration was reported to cause increased H_2O_2 content in both roots and leaves (Saha et al. 2010). Mudgal et al. (2010) reported that salt stress significantly affected initiation, weight and nitrogen-fixing ability of the root nodules and also leads to inhibition of root colonization by Rhizobium. Delayed and reduced flowering and yield of crop plants as a result of salinity stress were reported by Maas (1986). Excessive salt was reported to lead injury to leaves and subsequently to reduced photosynthesis (Hossain and Fujita 2010).

6.2.4 Other Abiotic Stresses

Mungbean may also suffer from prolonged rainy season resulting in excess soil and atmospheric moisture leading to pre-harvest sprouting and also increased the incidence of insect pests and diseases. Pod shattering is another constraint which affects mungbean production. While most of the modern cultivars with synchronous maturity are shattering resistant, those cultivars with prolonged flowering and pod set are still prone to shattering. Among other stresses, mungbean plants exposed to UV-B radiation showed a significant reduction in plant height, leaf area total biomass, photosynthetic rate, chlorophyll, soluble proteins and carbohydrate (Pal et al. 1999). However, UV-B radiation increased the content of flavanoids, anthocyanins, total free amino acids and PAL activity. Among the *Vigna* species maximum reduction for net photosynthesis was observed with increased UV-B radiation, whereas the greater reduction of nitrogenase activity was observed in mungbean (Singh 1997). Tolerance to aluminium toxicity as measured by root length, plant height, and plant dry matter varied with cultivars tested (Duong et al. 1988).

6.3 Sources of Resistance

Mungbean genotypes including germplasm lines, breeding materials and commercial cultivars have been screened for resistance to abiotic stresses and sources having resistance to heat, drought, salinity, flooding, etc. have been reported (Table 6.1). At the World Vegetable Centre (AVRDC), twenty genotypes were subjected to 75–100 centibars of soil moisture tension for eight days from flowering (AVRDC 1979). Consequently, the genotypes 'V 1281', 'V 2013' and 'V 3372' were identified as drought tolerant. Simultaneously, the accessions 'V 1968, 'V 2984 m', 'V 3092' and 'V 3372' were reported to have varying levels of tolerance to flooding (AVRDC 1979). Fernandez and Shanmugasundaram (1988) identified the breeding lines 'VC 1163 D', 'VC 2570A', 'VC 2754 A' and 'VC 2768 A' as drought tolerant based on minimum reduction in yield, total dry matter and plant height in stressed environment as compared to the non-stressed environment. He et al. (1988) reported the high yielding lines 'V 1381' and 'VC 2778' as tolerant to drought and flood besides being resistant to lodging. Efforts have also been undertaken to identify tolerant cultivars and the selections such as 'V 1968', 'V 2984', 'V 3092' and 'V 3372' were made (Tickoo et al. 2006).

High-temperature stress directly affects flower retention and pod formation in mungbean. Khattak et al. (2006) evaluated 14 commercial varieties and 24 advanced genotypes of mungbean developed through hybridization for maximum flower retention capability under high temperature (above 40 °C). They also evaluated 77 mutants derived from 'NM 92' and 51 recombinants selected from three crosses viz., 'VC1560D' × 'NM92', 'VC1482C' × 'NM92' and 'NM98' × 'VC3902A'. Under high temperature none of the genotypes showed absolute tolerance to flower shedding. In another study Khattak et al. (1999) evaluated 242 recombinants/

Table 6.1 Sources of resistance for abiotic stresses in mungbean

Source of resistance/tolerance	Abiotic stress	Reference
K-851	Drought tolerance	Dutta and Bera (2008) and Dutta et al. (2016)
EC693357, EC693358, EC693369, Harsha and ML1299	Heat tolerance	Sharma et al. (2016)
TCR 20	Drought tolerance	Tripathy et al. (2016)
SML-1411, SML-1136	Drought tolerance	Kaur et al. (2017)
ML 267	Drought tolerance	Swathi et al. (2017)
VC2917 (seedling stage) and Zhangjiakouyinggelv (adult stage)	Drought tolerance	Wang et al. (2014, 2015a, b)
'V 1281', 'V 2013' and 'V 3372'	Drought tolerance	AVRDC (1979)
V 1968', 'V 2984', 'V 3092' and 'V 3372	Flooding	AVRDC (1979)
AVMU 1001, AVMU 1201, VO6381A-G, KPS-1, ML 1628, PDM 139, IPM 02-14	Waterlogging	Rameshreddy et al. (2019)
VC 1163 D', 'VC 2570A', 'VC 2754 A' and 'VC 2768 A	Drought tolerance	Fernandez and Shanmugasundaram (1988)
V 1381' and 'VC 2778 A'	Drought and Flood	He et al. (1988)
PUSA 9074, VI001400AG, KPS-1	Low temperature	WorldVeg Report (2017)
Perennial accessions of *V. radiata* var. *sublobata*	Low temperature	Lawn et al. (1988)
AVMU 1001, AVMU 1202, VC6512-6A, TV 0193A-G, ML 1299	Salinity	Manasa et al. (2017)
S 72','H 45', 'No. 525', 'Madira' and 'RS-4	Salt tolerance	Maliwal and Paliwal (1982)
Accessions of *V. radiata* var. *sublobata*	Alkaline and calcareous soil	Lawn et al. (1988)
V. radiata var. *sublobata*	Hard seededness (Pre-harvest sprouting)	Singh et al. (1983)
Pant Moong-1	Non-shattering	Singh and Sharma (1984)
Chamu 4	Pre-harvest sprouting tolerance	Lamichaney et al. (2017)

mutants. Among them 163 were moderately tolerant (10–20% flower drop) and 79 were susceptible with more than 40% flowers shedding from terminal raceme on the main stem. Based on multilocation evaluation at Kanpur, Vamban (Tamilnadu) and Durgapura (Rajasthan) and further reconfirmation based upon sucrose synthase activity and protein profiling as biochemical markers the candidate genotypes were validated by repeated field trial across diverse agro-climatic zones prone to be affected by recurrent high-temperature stress. Twelve promising mungbean genotypes (IPM 02-16, IPM 9901-10, IPM 409-4, IPM 02-3, PDM 139, IPM 02-1, IPM 2-14, IPM 9-43-K, PDM 288, EC 470096, IPM 2K14-9, IPM 2K14-5) were identified.

Maliwal and Paliwal (1982) studied 42 cultivars of mungbean and black gram for tolerance to salinity at five levels (3–18 m mhos/cm) in 1/5 Hoagland nutrient solution and observed that germination of all the cultivars was delayed and decreased with an increase in salinity level. Seedling height also decreased significantly with salinity. Some varieties of mungbean viz., 'S 72', 'H 45', 'No. 525', 'Madira' and 'RS-4' were reported to be more salt-tolerant (Maliwal and Paliwal 1982). Some of the accessions of *V. radiata* var. *sublobata* showed no symptoms of

Table 6.2 Potential sources of alien variation in *Vigna* spp

Character	Species	References
Low trypsin inhibitor activity	*V. tenuicaulis*	Konarev et al. (2002)
Chymotrypsin absent	*V. grandiflora*	Konarev et al. (2002)
High methionine content	*V. radiata* var. *sublobata*	AVRDC (1979), Babu et al. (1988)
High photosynthetic efficiency and drought tolerance	*V. radiata* var. *sublobata*	Ignacimuthu and Babu (1987)
Drought tolerance	*V. aconitifolia*	Jain and Mehra (1980)
Heat tolerance	*V. aconitifolia*	Tomooka et al. (2001)
	V. riukinensis	Egawa et al. (1999)
High tolerance to saline and alkaline soils	*V.radiata* var. *sublobata*	Lawn et al. (1988)
Photo–thermo insensitivity	*V. umbellata, V. glabrescens*	Pratap et al. (2014)

chlorosis when grown on extremely alkaline (pH > 8.5) calcareous soils (Lawn et al. 1988).

Pod shattering is another important abiotic stress in mungbean. This is mainly due to the indeterminate flowering habit where flowering and pod maturity occur on the same plant over the entire reproductive phase. As a result the earliest maturing pods may shatter while later developing pods may still be green. Most modern-day commercial mungbean cultivars are resistant to shattering. Pant Moong-1 has been reported to be tolerant of shattering and its harvesting can be delayed by 7–10 days so as to allow the maturity of pods from the second flush of flowers (Singh and Sharma 1984). Excessive moisture especially during seedling stage and at the time of maturity also poses threats to mungbean production. In a comparison of two mungbean cultivars during a short period of overwatering Bagga et al. (1984) reported that the cultivar Pusa Baisakhi utilized the additional water more efficiently than cultivar PS 16. Islam (1994) also observed that leaf area, CGR, NAR, crop growth rate, net assimilation rate, leaf canopy structure, dry matter production and grain yield in mungbean were all more susceptible to waterlogging than to drought. Lamichaney et al. (2017) evaluated 163 mungbean genotypes for tolerance to pre-harvest sprouting and found 14 genotypes tolerant to this trait; Chamu 4 was reported as highly tolerant.

Wild species that are easily crossable with cultivated species are important genetic resources for the improvement of cultivated legumes (Hamdi and Erskine 1996; Chen et al. 2006, Hillocks et al. 2006; Pratap et al. 2014). A number of wild *Vigna* accessions have been identified as possessing abiotic stress resistance traits such as high photosynthetic activity and drought tolerance in *V. radiata* var. *sublobata* (Ignacimuthu and Babu 1987) (Table 6.2), drought tolerance in *V. aconitifolia* (Jain and Mehra 1980) and heat tolerance in *V. aconitifolia* and *V. riukinensis* (Egawa et al. 1999). Pratap et al. (2014) reported two wild accessions, *V. umbellata* (IC 251442) and *V. glabrescens* (IC 251472) possessing photo–thermo period insensitivity.

6.4 Genetic Studies

Attempts have been made to study the inheritance of several morphological traits including plant type, leaf characters, flower colour, pod pubescence, pod colour, shattering habit, seed coat colour, etc. in mungbean (Singh et al. 1982, 2017; Table 6.3). Most of the earlier cultivars had a twining habit which was considered as a survival trait. For this trait, Pathak and Singh (1963) reported a single recessive gene to be responsible while a single dominant gene (*T*) was reported by Khattak et al. (1999). Erect plant

Table 6.3 Genetic studies to study inheritance of major morphological traits related to abiotic stress resistance

Trait	Importance	Inheritance	Reference
Hypocotyl pigmentation	Genetic purity	Single dominant/recessive gene, anthocyanin in hypocotyl governed by two supplementary genes	Pathak and Singh (1963), Mukherjee and Pradhan (2002)
Plant type and growth habit	Synchronous maturity, shattering resistance	Single dominant/recessive gene, semi-spreading is dominant over erect habit	Sen and Ghosh (1959), Pathak and Singh (1963), Khattak et al. (1999)
Pubescence	Insect resistance	Single dominant gene	Murty and Patel (1973), Sen and Ghosh (1959)
Pod shattering	Seed yield	Single dominant gene	Verma and Krishi (1969)
Seed coat colour	Market preference	One or few genes; mottling governed by single gene	Khattak et al. (1999), Chen and Liu (2001), Lambrides et al. (2004)
Seed coat surface	Market preference	Two complementary genes	Bose (1939), Sen and Ghosh (1959), Murty and Patel (1973)
Hard seededness	Grain quality, market preference, cooking quality	One or few dominant genes involved	Lambrides (1996), Singh et al. (1983), Humphry et al. (2005)
Pre-harvest sprouting	Grain quality, seed viability	Additive and non-additive gene action; high G x E interaction	Durga and Kumar (1997)

types are more suited to mechanical harvesting and more number of erect plants can be accommodated in a unit area as compared to spreading or semi-erect plants. Semi-spreading habit was observed dominant over erect habit and it was reported to be governed by a single dominant gene (Pathak and Singh 1963). Mukherjee and Pradhan (2002) indicated that anthocyanin pigmentation in the hypocotyl is controlled by two supplementary genes ('*Sh*' and '*Ph*') with recessive epistatic interaction. Dwivedi and Singh (1985) reported that purple pigmentation on stem, petiole and veins of the leaves was conditioned by a single dominant gene '*Ppp1*' with pleiotropic effect. Singh and Singh (2011) reported that indeterminacy is responsible for pod shattering as pod development is spread over a longer period of time in the plant and the pods which ripe earlier are prone to shattering by the time the late-developing pods mature, consequently leading to significant yield loss. A single dominant gene inherited independently from leaf shape was reported to be responsible for indeterminate growth habit (Talukdar and Talukdar 2003). Among the leaf traits, the trifoliate leaf was dominant over the entire leaf and this trait was reported to be governed by a single dominant gene (Chhabra 1990; Talukdar and Talukdar 2003). Sareen (1985) reported two dominant genes, '*Tlb1*' and '*Tlb2*', with duplicate gene action for trilobed leaves. Dwivedi and Singh (1985) reported narrow lanceolate leaf to be controlled by two recessive genes, '*nl1*' and '*nl2*'. Pod pubescence was reported to be dominant over non-pubescence and governed by independent duplicate genes (Khadilkar 1963). For seed coat colour, Khattak et al. (1999) reported monogenic inheritance while for hard seededness Humphry et al. (2005) reported four loci to be responsible through QTL analysis among which two QTLs for hard seededness were reported to be co-localized with the loci conditioning seed weight. Chen and Liu (2001) suggested that the inheritance of black and green seed colours was controlled by a single gene (*B*), black being dominant over green. Verma and Krishi (1969) showed that shattering is completely dominant to non-shattering.

6.5 Breeding

The initial phase of breeding in mungbean saw selections from locally adapted germplasm, mainly for biotic stresses resistance and high yield. While selecting for abiotic stress resistance was not practiced directly selection for yield, plant type and adaptation-related traits indirectly lead to selection for abiotic stress resistance as well. Selection has been a useful strategy to identify superior cultivars with significant drought tolerance. Before breeding for drought tolerance per se, the first step is to determine the type of drought. Warm-season food legumes generally encounter two types of drought stresses: (i) terminal drought which is more prominent in summer/spring crops, usually coinciding with late reproductive stage and increasing towards generative stage, and (ii) intermittent drought, which may occur anytime during vegetative growth and results due to a break in rainfall or insufficient rains at the vegetative stage. The ranking of warm-season food legumes in increasing order of drought resistance was soybean, followed by blackgram, mungbean, groundnut, Bambara nut, lablab bean and cowpea (Singh et al. 1999). Fernandez and Kuo (1993) used stress tolerance index to select genotypes with high yield and tolerance to temperature and water stresses in mungbean. Singh (1997) described the plant type of mungbean suitable for kharif (rainy) as well as dry (spring/summer) seasons. He emphasized the need of developing mungbean genotypes with semi-determinate, medium statured plants (70–75 cm) having large inflorescence, more bunches/plants, 3–4 branches, an average pod length with 8–10 seeds and average seed size of 30–40 g/1000 seeds, tolerance to shattering and moderate seed dormancy. Pratap et al. (2013b) also suggested development of short-duration cultivars for spring/summer cultivation so that these escape terminal heat and drought stress. Cultivars with 60–65 days crop duration, determinate growth habit, high harvest index, reduced photoperiod sensitivity, fast initial growth, longer pods with more than 10 seeds/pod and large seeds are more suitable to summer season. Keeping this in mind a number of early maturing mungbean lines have been selected and released as commercial cultivars. Samrat and SML 668 are two such selections which remained highly popular varieties for the spring/summer seasons in northern part of India.

While most of the varieties released before 1990 were developed through selection, hybridization has been most often used to develop varieties in the last three decades. More than 100 varieties have been developed in India through hybridization, among which about 20 varieties are suited to more than one season and agro-climatic zone, indicating their relatively higher level of stability, photo–thermo period insensitivity and tolerance to abiotic stresses (Table 6.4). Among these, IPM 02-3, IPM 410-3, PKV AKM 4, MH 421 and MH 3-18 are highly popular among farmers. The variety PKV AKM 4 developed from a cross between BM4 X PS 16 was recommended for two zones, viz., Central Zone and South Zone of India. In more recent times, IPM 02-3 was developed using IPM 99-125 and Pusa Bold 2 and recommended for both spring and *kharif* seasons. IPM 410-3 (Shikha) has been released for entire northern, western and central India covering majority of the mungbean area in the country. Sehrawat et al. (2014) attempted hybridization between previously selected salt-resistant genotypes (EC 528960 (wild) and JP31300 (cultivated) of mungbean and highly salt-sensitive cultivar, i.e. IC10492. The inter- and intra-specific hybrids of two different types of crosses were assessed for morphological characterization and hybrid purity using azukibean specific SSR markers. The SSR (CEDG149) produced reproducible band of 188 bp in male parent JP31300 (salt-tolerant); however, it was absent in female parent IC10492 (salt susceptible). In this study, F_1 hybrids of mungbean were developed successfully for salt tolerance.

Development of early genotypes is of utmost importance in mungbean improvement because such genotypes fit well in different crop rotations and multiple cropping systems (Ali and Kumar 2006). While comparatively longer duration genotypes (>75 days) are more suited to rainy season, short-duration cultivars (<60 days) are preferred in spring/summer season as well as specific niches (Pratap et al. 2013b). Cultivation

Table 6.4 List of release varieties of mungbean in India

S. no.	Name of variety	Pedigree	Year of release	Area of adaptation	Season
1.	Narendra Mung 1	G 65 X UPM 79-3-4	1992	UP	*Kharif* and *Spring*
2.	Warangal 2 (WGG2)	W 75-70 X Pusa 101	1995	AP	*Kharif*, *Rabi* and *Summer*
3.	PDM 84-178	–	1996	AP	*Summer* and early *Kharif*
4.	SML 134	V 2164 X ML 258	1996	Punjab	*Spring*/Summer season
5.	HUM 1 (Malviya Jyoti)	BHUM 1 X Pant U 30	1999	CZ and SZ	*Kharif* season
6.	RMG 268	R288-8 X J 781	1997	Rajasthan	*Kharif* and *Summer* season
7.	CO 6	WGG 37 X CO5	1999	TN	Suitable for all seasons
8.	HUM 2 (Malviya Jagrati)	Sel. from local germplasm TVCM-3	2000	UP and Uttarakhand	Spring/Summer season
9.	PDM 139	ML 20-20/19 X ML 5	2001	UP and plains of Uttrakhand	Spring/Summer
10.	Pant Mung 5	Selection from VC 6368	2002	UP and plains of Uttrakhand	–
11.	IPM 02-3	IPM 99-125 X Pusa bold2	2009	NWPZ	*Kharif* and *Spring* season
12.	PKV AKM 4	BM4 X PS16	2009	CZ and SZ	*Kharif* season
13.	MH 421	Muskan X BDYR 2	2014	NWPZ	Summer/*Spring*
14.	IPM 410-3 (Shikha)	IPM 03-1 X NM 1	2016	NWPZ/CZ	Summer/*Spring*
15.	IPM 205-7 (Virat)	–	2016	Entire India	Summer
16.	Yadadri (WGG 42)	–	2016	Telangana	*Kharif*/*Rabi* and summer cultivation
17.	Sri Rama (MGG 351)	–	2016	Telangana	*Rabi*/summer and rice fallow cultivation
18.	MSJ 118 (Keshvanand mung 2)	Mutant of K 851	2016	Rajasthan	*Kharif*/*Spring* cultivation
19.	SGC 16 (Rupohi)	PDM 91-243 X WGG 62	2018	Assam	*Summer* and *kharif* season
20.	GAM 5	Sel. From local germplasm VM 6	2018	Gujarat	Summer and *kharif* season

NWPZ North-West Plain Zone of India, *CZ* Central Zone, *SZ* South Zone, *UP* Uttar Pradesh

of shorter duration genotypes not only helps by escaping from drought and heat but also saves at least one irrigation and a spray of insecticides, leading to considerable cost savings to farmers. Such genotypes escape terminal moisture stress and therefore are less affected by heat and drought. Samrat is a short-duration cultivar which matures in 55–60 days and has been one of the most preferred mungbean varieties in India for cultivation during spring and summer seasons. More recently Pratap et al. (2013a, b) developed extra early mungbean cultivar IPM 205-7 which matures in 52–55 days. It has been released for cultivation in Punjab, Haryana, Uttar Pradesh, Bihar, Jharkhand, Odisha, Tamil Nadu and Madhya Pradesh and fits well in the short-season window of 60–65 days after the harvest of wheat and before the sowing of rice during summer season. Another genotype, IPM 409-4, is also a short-duration genotype, recently identified for cultivation in Uttar Pradesh during the spring/summer season. Both these genotypes will help in horizontal expansion of mungbean cultivation in India, especially during the spring/summer season in northern India as well as in rice fallows in peninsular India. Recognizing their potential IPM 205-7 was also registered as INGR 11043 and IPM 409-4 as INGR 11044 by the National Bureau of Plant Genetic Resources (ICAR), New Delhi, both for extra-early maturity (Pratap et al. 2013b).

Naidu et al. (1996) reported several factors including small pod beak and angle, thick pod wall, less moisture absorption by pod wall, hard seediness and high cuticular wax content on pod wall that determine resistance of mungbean cultivars to pre-harvest sprouting.

6.6 Future Prospects

Mungbean is a short-duration crop that can be grown across seasons, environments and soil types and therefore faces a host of biotic and abiotic stresses. Nevertheless, owing to its multifarious uses in human nutrition and its capacity to fit well in a number of cropping systems it has a tremendous scope for expansion in area as well as production volume. To make it more versatile and climate resilient new cultivars need to be photo- and thermo-period insensitive, widely adaptable and resistant/tolerant to biotic and abiotic stresses. While a large number of improved cultivars have been bred to date most breeding efforts have remained focussed on development of disease-resistant and high yielding cultivars.

It is now time to prioritize the importance of abiotic stress tolerance in short duration crops like mungbean and initiate trait-specific breeding. Increasing osmotic stress tolerance, introducing photo- and thermo-period insensitive genes, manipulating maturity duration and dissecting physiological and biochemical pathways will pave the way to development of abiotic stress-tolerant cultivars. There is a need to explore modification of biochemical and physiological processes which may lead to development of heat tolerance through acclimation. Work must be initiated by developing root systems that help plants to withstand moisture deficits by drawing water from deeper in the soil. Screening for various abiotic stresses must be precise and stringent to identify robust donors for these traits. The identified donors need to be further put to use by the plant breeders at a faster pace. Plant types having a deep root system, maturity period of 55–60 days for the spring/summer season and 70–75 days for rainy season, erect plants with sympodial pod bearing, multiple pods per cluster and longer pods, and 45–60 cm plant height with many node and shorter internodes will help in withstanding heat and drought-related stresses. Selection of the plants having thick leaves that allow the penetration of light to the lower part of the plant will help better photosynthesis. Utilization of germplasm resources will be a key to selecting and introgressing such traits. For this, the collections such as minicore, exotic germplasm, landraces and wild genetic resources must be pooled and screened thoroughly. There is a strong need for systematic investments towards utilization of germplasm and biotechnological approaches to harness desirable genes from genetic resources. Molecular approaches revealing tolerance mechanisms will help in modifying mungbean plants to suit changing climates and also prevailing stresses. There must be systematic efforts towards exploring physiological and biochemical regulations of abiotic stresses and studying whole profile of genes, proteins and metabolites imparting stress tolerance in resistant genotypes so that the same can be employed to develop improved cultivars.

References

Abd-Alla MH, Vuong TD, Harper JE (1998) Genotypic differences in nitrogen fixation response to NaCl stress in intact and grafted soybean. Crop Sci 38:72

Aggarwal VD, Poehlman JM (1977) Effects of photoperiod and temperature on flowering in mungbean (Vigna radiata (L.) Wilczek). Euphytica 26:207–219

Ali M, Kumar S (2006) Mungbean and urdbean: retrospect and prospects

Arulbalachandran D, Sankar GK, Subramani A (2009) Changes in metabolites and antioxidant enzyme activity of three Vigna species induced by NaCl stress. Amer Eur J Agron 2:109–116

AVRDC (1979) AVRDC Progress Report for (1978). Asian Vegetable Research and Development Centre, Shanhua, Tainan, p 173

Babu CR, Sharma SK, Chaterjee SR, Abrol YP (1988) Seed protein and amino acid compositions of wild *Vigna radiata var. sublobata* (Fabaceae) & two cultigens *V. mungo and V. radiata* Econ Bot 42:54–61

Bagga AK, Bela M, Tomar OPS (1984) Effect of short duration of waterlogging on water use efficiency of two mungbean (*Vigna radiata* (L.) Wilczek) varieties. Indian J Physiol 27:159–165

Barnabas B, Jäger K, Fehér A (2008) The effect of drought and heat stress on reproductive processes in cereals. Plant Cell Environ 31:11–38

Bindumadhava H, Nair RM, Easdown W (2015) Physiology of mungbean accessions grown under saline and high temperature conditions. AVRDC-Annual Report, No. 74199. Weihai Shanhua Carpet Group Co Ltd, Tainan, pp 1–46

Bita CE, Gerats T (2013) Plant tolerance to high temperature in a changing environment: scientific fundamentals and production of heat stress tolerant crops. Front Plant Sci 4:273

Bose RD (1939) Studies in Indian pulses. IX. Contributions to the genetics of mung (*Phaseolus radiatus* Linn. Syn. Ph. aureus Roxb.) Indian J Agri Sci 9:575–594

Carberry (2007) Crop development models. In: Encyclopedia of water science, 2nd edn. Encyclopedia of science. CRC Press, Boca Raton, FL, pp 122–124

Chakrabarti N, Mukherji S (2002) Growth regulator mediated changes in leaf area and metabolic activity in mungbean under salt stress condition. Indian J Plant Physiol 7:256–263

Chauhan YS, Rachaputi RCN (2014) Defining agro-ecological regions for field crops in variable target production environments. A case study on mungbean in the northern grains region of Australia. Agric For Meteorol 194:207–217

Chauhan YS, Douglas CA, Rachaputi RCN, Agius P, Martin WD, King K, Skerman A (2010) Physiology of mungbean and development of the mungbean crop model. In: Proceedings of the 1st Australian summer grains conference 2010

Chen H, Liu X (2001) Inheritance of seed color and lustre in mungbean (Vigna radiata). Agric Sci Technol 2:8–12

Chen W-M, James EK, Coenye T, Chou J-H, Barrios E, de Faria SM, Elliott GN, Sheu SY, Sprent JI, Vandamme P (2006) Burkholderia *mimosarum* sp. nov., isolated from root nodules of *Mimosa* spp. from Taiwan and South America. Int J Syst Evol Microbiol 56:1847–1851

Chhabra AK (1990) Inheritance of lobed and pentafoliate leaf in mungbean. Indian J Pulses Res 3(1):69–72

Chowdhury SK, Haque MN (1977) Stability performance of some green gram varieties. Indian J Agric Sci 47:303–304

Duong M, Nauyen PB, Vo MO, Nguyen HT, Duong NV, Nguyen CH (1988) Mungbean varietal development for problem soils in the Mekong Delta of Vietnam. TVIS Newslett 3:4–9

Durga KK, Kumar SS (1997) Screening for pre-harvest sprouting in pulses. Legume Res 20:193–197

Dutta P, Bera AK (2008) Screening of mungbean genotypes for drought tolerance. Legume Res 31:145–148

Dutta P, Bandopadhyay P, Bera AK (2016) Identification of leaf based physiological markers for drought susceptibility during early seedling development of mungbean. Am J Plant Sci 7:1921–1936

Dwivedi S, Singh DP (1985) Inheritance of resistance to yellow mosaic virus in a wide cross of black gram (*Vigna mungo* (L.) Hepper). Z Pflanzenzuchtg 95:281–284

Egawa Y, Takeda H, Suzuki K (1999) Research plan on crop heat tolerance at the crop introduction and cultivation laboratory. Jpn Int Res Center Agric Sci Working Rep 14:103–107

Fernandez GCJ, Kuo CG (1993) Effective selection criteria for assessing plant stress tolerance. In: Kuo CG (ed) Adaptation of food crops to temperature and water stress. Proceedings of the international symposium, Tainan, pp 257–270. 13–18 Aug 1992

Fernandez GCJ, Shanmugasundaram S (1988) The AVRDC mungbean improvement program. The past, present and future. In: Fernandez J, Shanmugsundaram S (eds) Mungbean. Asian Vegetable Research and Development Centre, Shanhua, Tainan, pp 58–70

Friedman R, Altman A, Nitsa L (2006) The effect of salt stress on polyamine biosynthesis and content in mung bean plants and in halophytes. Physiol Plant 76:295–302

Gulati A, Jaiwal PK (1993) In vitro selection of salt resistant Vigna radiata (L.) Wilczek plants by adventitious shoot formation from cultured cotyledon explants. J Plant Physiol 142:99–102

Hamdi A, Erskine W (1996) Reaction of wild species of the genus Lens to drought. Euphytica 91:173–179

Hanumantharao B, Nair RM, Nayyar H (2016) Salinity and high temperature tolerance in mungbean (Vigna radiata (L.) Wilczek) from a physiological perspective. Front Plant Sci 7:957

Haqqani AM, Pandey RK (1994) Response of Mungbean to aterstress and irrigation at various growth stages and plant densities: II. Yield and yield components. Tropical Agric 71(4):289–294

He X, He T, Xiong Y, Jiao C (1988) Research and use of mungbean germplasm resources in Hubei, China. In: Fernandez J, Shanmugsundaram (eds) Mungbean. Asian Vegetable Research and Development Centre, Shanhua, Tainan, pp 35–41

Hillocks RJ, Catherine SM, Rowland C, Eliainery MM, Msolla S (2006) Phaseolus bean improvement in Tanzania, 1959–2005. Euphytica 150:215–231

Hossain MA, Fujita M (2010) Evidence for a role of exogenous glycine betaine and proline in antioxidant defense and methyl glyoxal detoxification systems in mungbean seedlings under salt stress. Physiol Mol Biol Plants 16:19–29

Humphry ME, Lambrides CJ, Chapman SC, Aitken EAB, Imrie BC, Lawn RJ, McIntyre CL Liu CJ (2005) Relationships between hard-seededness and seed weight in mungbean (Vigna radiata) assessed by QTL analysis. Plant Breed 124:292–298

Ignacimuthu S, Babu CR (1987) Vigna radiata var. sublobata (Fabaceae): Economically useful wild relative of urd and mung beans. Econ Bot 41:418–422

Islam MT (1994) Eco-physiological studies on photosynthesis and dry matter production in mungbean (*Vigna radiata* (L.) Wilczek). Ph D Thesis. Kyushu University, Fukuoka, Japan

Jain HK, Mehra KL (1980) Evaluation, adaptation, relationship and cases of the species of Vigna cultivation in Asia. In: Summerfield RJ, Butnting AH (eds) Advances in legume science. Royal Botanical Garden, Kew, London, vol 273, pp 459–468

Kang YJ, Kim SK, Kim MY, Lestari P, Kim KH, Ha B, Jun TH, Hwang WJ, Lee T, Lee J, Shim S, Yoon MY, Jang YE, Han KS, Gwag JG, Moon JK, Lee YH, Park BS, Bombarely A, Doyle JJ, Jackson SA, Schafleitner R, Srinives P, Varshney RK, Lee SH (2014) Genome sequence of mungbean and insights into evolution within Vigna species. Nat commun 5:5443

Kaur R, Bains TS, Bindumadhava H, Harsh N (2015) Responses of mungbean genotypes to heat stress. Effects on reproductive biology, leaf function and yield traits. Sci Hortic 197:527–541

Kaur R, Kaur J, Bains TS (2017) Screening of mungbean genotypes for drought tolerance using different water potential levels. J Adv Agric Technol 4:2

Khadilkar BT (1963) Inheritance of two characters in green gram (Phaseolus aureus Roxb. Piper). Akola Agr Coll Mag 3:29–32

Khattak GSS, Haq MA, Rana SA, Srinives P, Ashraf M (1999) Inheritance of resistance to mungbean yellow mosaic virus (MYMV) in mungbean (Vigna radiata (L.) Wilczek). Thai J Agri Sci 32:49–54

Khattak GSS, Saeed I, Muhammad T (2006) Breeding for heat tolerance in mungbean (*Vigna radiata* (L.) Wilczek). Pak J Bot 38(5):1539–1550

Khattak GSS, Saeed I, Muhammad T (2009) Flowers shedding under high temperature in mungbean (*Vigna radiata* (L.) Wilczek). Pak J Bot 41(1):35–39

Konarev AV, Tomooka N, Vaughan DA (2002) Proteinase inhibitor polymorphism in the genus Vigna Savi subgenus Ceratotropis and its biosystematic implications. Euphytica 123:165–177

Kumari P, Varma SK (1983) Genotypic differences in flower production/shedding and yield in mungbean (Vignaradiata). Indian J Plant Physiol 26:402–405

Lambrides CJ (1996) Breeding for improved seed quality traits in mungbean (*Vigna radiata* L. Wilczek) using DNA markers. PhD Dissertation, University of Queensland, Brisbane, Australia

Lambrides CJ, Godwin ID, Lawn RJ, Imrie BC (2004) Segregation distortion for seed testa color in mungbean (*Vigna radiata* L. Wilczek). J Hered 95:532–535

Lamichaney A, Katiyar P, Laxmi V, Pratap A (2017) Variation in pre-harvest sprouting tolerance and fresh seed germination in mungbean (*Vigna radiata* L.) genotypes. Plant Genet Resour Charact Util 16 (5):437–445

Lawn RJ, Williams RW, Imrie BC (1988) Potential of wild germplasm as a source of tolerance to environmental stresses in mungbean. In: Fernandez J, Shanmugsundaram S (eds) Mungbean. Asian Vegetable Research and Development Centre, Shanhua, Tainan, pp 136–145

Maas EV (1986) Salt tolerance of plants. Appl Agric Res 1:12–26

Maliwal GL, Paliwal KV (1982) Salt tolerance of some mungbean (*Vigna radiata*), urdbean (Vigna mungo) and guar (*Cyamopsis tetragonoloba*) varieties at germination and early stages. Legume Res 5:23–30

Manasa R, Rameshreddy, Bindumadhava H, Nair RM, Prasad TG, Shankar AG (2017) Screening mungbean (*Vigna radiata* L.) lines for salinity tolerance using salinity induction response technique at seedling and physiological growth assay at whole plant level. Int J Plant Anim Enviro Sci 7(4):1–12

Morton F, Smith RE, Poehlman JM (1982) The mungbean. University of Puerto Rico, Mayaguez, p 136

Mudgal V, Madaan N, Mudgal A (2010) Biochemical mechanisms of salt tolerance in plants: a review. Intl J Bot 6:136–143

Mukherjee A, Pradhan K (2002) Genetics of lobation in trifoliate leaf of mungbean. Abstracted Perspect Cytol Genet 11:88

Murty BK, Patel GJ (1973) Inheritance of some morphological characters in mungbean. Bansilal Amrital Coll Agri Mag 25:1–9

Musgrave ME, Vanhoy MA (1989) A growth analysis of waterlogging damage in mungbean (*Phaseolus aureus*). Canadian J Bot 67:2391–2395

Naidu NV, Seenaiah P, Satyanarayana A (1996) Mechanism of resistance to pre-harvest sprouting in mungbean (*Vigna radiata* (L.) Wilczek). Ann Agric Res 17:167–170

Pal M, Sengupta UK, Srivastava AC, Jain V, Meena RC (1999) Changes in growth and photosynthesis of mungbean induced by UV-Bv radiation. Indian J Plant Physiol 4(2):79–84

Paliwal KV, Maliwal GL (1980) Growth and nutrient uptake relationship of some crops in saline substrate. Annals of the Arid Zone 19:251–253

Pandey RK, Herrera WAT, Villegas AW, Penletion JW (1984) Drought response of grain legumes under irrigation gradient. III. Plant Growth Agron J 76:557–560

Pathak GN, Singh B (1963) Inheritance studies in green gram. Indian J Genet 23:215–218

Patriyawaty NR, Rachaputi RCN, George D, Douglas CA (2018) Genotypic variability for tolerance to high temperature stress at reproductive phase in Mungbean [*Vigna radiata* (L.) Wilczek]. Sci Hortic 227:132–141

Pratap A, Gupta DS, Singh BB, Kumar S (2013a) Development of super early genotypes in greengram (*Vigna radiata* L. Wilczek). Legume Res 36:105–110

Pratap A, Gupta DS, Singh BB, Kumar S (2013b) IPM 205-7 (IC0589309-IC0589310; INGR11043-INGR11044), a mungbean (*Vigna radiata* (L.) Wilczek) germplasm with super early maturity. Indian J Plant Genet Resour 26:89–90

Pratap A, Basu PS, Gupta S, Malviya N, Rajan N, Tomar R, Latha M, Nadarajan N, Singh NP (2014) Identification and characterization of sources for photo- and thermo-insensitivity in Vigna species. Plant Breed 133:756–764

Pratap A, Gupta S, Basu PS, Tomar R, Dubey S, Rathore M, Prajapati U, Singh P Kumari G (2019) Towards development of climate-smart mungbean: challenges and opportunities. In: Kole C (ed) Genomic designing of climate smart pulse crops. Springer Nature, New York (In press)

Project Coordinators Report (2018) All India coordinated research project on MULLaRP (Mungbean, Urdbean, Lentil, Lathyrus, Rajmash, Fieldpea). ICAR-Indian Institute of Pulses Research, Kalyanpur, Kanpur, Kalyanpur, Kanpur, p 46

Rainey KM, Griffiths PD (2005) Differential response of common bean genotypes to high temperatures. J Am Soc Horti Sci 130:18–23

Rameshreddy BH, Shankar AG, Boddepalli N, Nayak R, Tripathy N, Nair RM (2019) Assessing waterlogging tolerance in mungbean (*Vigna radiata*) and Urdbean (*Vigna mungo*) under managed field conditions (unpublished data, under preparation)

Rawson HM, Craven CL (1979) Variation between short duration mungbean cultivars *(Vigna radiata (L.)* Wilczek) in response to temperature and photoperiod. Indian J Plant Physiol 22:127–136

Saha P, Chatterjee P, Biswas AK (2010) NaCl pretreatment alleviates salt stress by enhancement of antioxidant defense and osmolyte accumulation in mungbean (*Vigna radiata* L. Wilczek). Indian J Exp Biol 48:593–600

Sareen PK (1985) Further genetic analysis of the trilobate leaf mutants in mungbean (*Vigna radiata* var. *aureus* (L.) Wilczek). Curr Sci 54:930–931

Schafleitner R, Nair RM, Rathore A, Wang Y-W, Lin C-Y, Chu S-h, Lin P-Y, Chang J-C, Ebert AW (2015) The AVRDC—The World Vegetable Center mungbean (*Vigna radiata*) core and mini core collections. BMC Genom 16:344

Sehrawat N, Bhat KV, Kaga A, Tomooka N, Yadav M and Jaiwal PK (2014) Development of new gene-specific markers associated with salt tolerance for mungbean (*Vigna radiata* L.Wilczek). Spanish J Agric Res 12(3):732–741

Sen NK, Ghosh AK (1959) Genetic studies in green gram. Indian J Genet 19:210–227

Sharma L, Priya M, Bindumadhava H, Nair RM, Nayyar H (2016) Influence of high temperature stress on growth, phenology and yield performance of mungbean (*Vigna radiata* (L.) Wilczek) under managed growth conditions. Sci Hortic 213:379–391

Singh DP (1997) Tailoring the plant type in pulse crops. Plant Breed 67:1213–1220

Singh DP, Sharma BL (1984) New varieties of mungbean. Indian Farm 33(4):31

Singh DP, Singh BB (2011) Breeding for tolerance to abiotic stresses in mungbean. J Food Legumes 24 (2):83–90

Singh BR, Sing M, Yadav MD, Dingar SM (1982) Yield loss in mungbean due to Cercospora leaf spot yellow mosaic. Sci Cult 48(12):435–436

Singh DP, Sharma BL, Dwivedi S (1983) Inheritance of hard seeds in interspecific cross of mungbean. Indian J Genet Plant Breed 43:378–379

Singh S, Rao VUM, Singh D, Singh S and Singh D (1991)Thermal requirement of summer moong *(Vigna radiata (L.)* Wilczek) Crop J Res 21:295–297

Singh BB, Kondomi YM, Terao T (1999) Relative drought tolerance of major rainfed crop of the semi-arid tropics. Indian J Genet Plant Breed 59:1–8

Singh DP, Singh BB, Pratap A (2017) Genetic improvement of mungbean and urdbean and their role in enhancing pulse production in India. Indian J Genet 76:550–567

Sinha SK (1977) Food legumes: Distribution adaptability and biology of yield. In: Food and agriculture organization of the United Nations Plant Production and Protection Rome, p 124

Swathi L, Reddy DM, Sudhakar P, Vineela V (2017) Screening of Mungbean (*Vigna radiata* L. Wilczek) Genotypes against Water Stress Mediated through Polyethylene Glycol. Int J Curr Microbiol App Sci 6 (10):2524–2531

Talukdar LT, Talukdar D (2003) Inheritance of growth habit and leaf-shape in mungbean [*Vigna radiata* (L.) Wilczek.]. Indian J Genet Plant Breed 63(2)

Tickoo JL, Grajraj R, Matho M, Manji C (1996) Plant type in mungbean. In: Asthana N, Kum H (eds) Proceedings of recent advances in mungbean. Indian Society of Pulses Research, Kanpur, India, pp 197–213

Tickoo SK, deParalta-Venturia MN, Harik LR, Worcester HD, Salama ME, Young AN, Moch H, Amin MB (2006) Am J Surgical Pathol 30:141–153

Tomooka N, Vaughan DA, Xu RQ, Kashiwaba K, Kaga A (2001) Japanese native *Vigna* genetic resources. Jpn Agri Res Q 35:1–9

Tripathy S, Mohanty P, Jena M, Dash S, Lenka D, Mishra D, Nayak P, Swain D, Ranjan R, Pradhan K, Senapati N, Mohapatra P, Dash G (2016) Identification of seed storage protein markers for drought tolerance in mungbean. Res Biotechnol 7

Tzudir L, Bera PS, Chakraborty PK (2014) Impact of temperature on the reproductive development in mungbean (Vigna radiata) varieties under different dates of sowing. Intl J Bioresour Stress Manage 5:194–199

Varma AK, Rao NSS (1975) Effect of different levels of soil moisture on growth, yield and some physiological aspects of nodulation in greengram. Indian J Agric Sci 45:11–16

Verma SNP and Krishi JN (1969) Inheritance of some qualitative characters in greengram (*Phaseolus aureus* Roxb.). Indian J Hered 1:105–106

Wahid A, Hameed M, Ejaz R (2004) Salt-induced injury symptom, changes in nutrient and pigment composition and yield characteristics of mungbean. Int J Agric Biol 6:1143–1152

Wang LF, Jing WU, Jing RL, Cheng XZ, Wang SM (2014) Drought resistance identification of mungbean germplasm resources at bud stage. J Plant Genet Resour 15:498–503

Wang LF, Jing WU, Jing RL, Cheng XZ, Wang SM (2015a) Drought resistance identification of mungbean germplasm resources at seedlings stage. Acta Agron Sin 41:145–153

Wang LF, Jing WU, Jing RL, Cheng XZ, Wang SM (2015b) Identification of mungbean germplasm resources resistant to drought at adult stage. Acta Agron Sin 41:1287–1294

Wassman R, Jagadish S, Sumfleth K, Pathak H, Howell G, Ismail A (2009) Regional vulnerability of climate change impacts on Asian rice production and scope for adaptation. Adv Agron 102:91–133

WorldVeg Report (2017) Improve mungbean and urdbean productivity in Odisha State (RKVY project funded by Govt of India), pp 1–24. Jan 2016–May 2017

Zhu JK (2002) Salt and drought stress signal transduction in plants. Annu Rev Plant Biol 53(1):247–273

7 Breeding Progress and Future Challenges—Nutritional Quality

Ramakrishnan Nair

Abstract

Mungbean is a good source of protein, carbohydrates and important minerals like iron and zinc. However, there is a good scope for improving the nutritional quality of currently available mungbean varieties. Variation for protein, carbohydrates, iron, zinc and in anti-nutritional factors like phytic acid have been reported. Even though some efforts have been made to improve the protein quality, further concerted efforts required to developed stable varieties. Screening of the recently developed mungbean mini-core collection would help in unravelling the extent of variation for the important quality traits and for their potential use in varietal improvement of commercial varieties. The sprouting industry has special demands, and special varieties have to be developed to cater to this premium market segment. The growing need for plant protein foods also provides an opportunity to further investigate on the functional food properties of mungbean protein and starch.

7.1 Introduction

Mungbean grains and sprouts produced from currently available varieties provide significant amounts of protein (240 g/kg) and carbohydrate (630 g/kg) and a range of micronutrients in diets. However, there is lot of scope for improving the nutritional quality of mungbean, as very little work has been done in this area. The need to expand mungbean breeding research to nutritional and food processing properties of mungbean was pointed out in an exhaustive review done by Dahiya et al. (2013). The nutrient content depends on several factors including the variety used, the location where the crop is grown, agronomic practices adopted and the storage conditions. In addition, postharvest processes such as sprouting, dehulling, soaking, boiling, autoclaving and microwave cooking can affect the composition of nutritional and anti-nutritional factors of mungbean (see review by Nair et al. 2013). In this chapter, a review on the progress made in improving mungbean for nutrients such as protein, carbohydrates, lipids, vitamins, minerals like iron and zinc, and other factors, including anti-nutrients like phytic acid and their implications have been discussed.

R. Nair (✉)
World Vegetable Center, South Asia, ICRISAT Campus, Patancheru, Hyderabad, Telangana, India
e-mail: ramakrishnan.nair@worldveg.org

R. M. Nair et al. (eds.), *The Mungbean Genome*, Compendium of Plant Genomes,
https://doi.org/10.1007/978-3-030-20008-4_7

7.2 Protein

Mungbean seeds contain 20.97–31.32% protein content (Itoh et al. 2006). There is high variation of crude protein content in mungbean, which has been attributed to the varietal differences (Yohe and Poehlman 1972; Thakare et al. 1988; Das et al. 2015) and different analytical methods employed (Dahiya et al. 2013). Polymorphism in protein profiles was reported by Naik and Kole (2001) in their study with improved mungbean varieties and local land races from the state of Odisha, India. Kudre et al. (2013) reported that the total protein content in mungbean protein isolates (MBPI) was 87.8%, with a total amino acid content of 800.2 mg/g. The average amino acid composition of mungbean is presented in Table 7.1. The constitution of the essential amino acids was 43.5%, while the sulphur-containing amino acids (methionine and cysteine) constituted about 1.6% of the total MBPI. A negative correlation between protein content and methionine content in mungbeans has been reported (Yi-Shen et al. 2018). Interspecific hybridisation has been attempted to transfer high methionine content from black gram, also known as urd bean, *Vigna mungo* L. into mungbeans (Nair et al. 2013). In mature mungbean seeds, the major storage protein is 8S globulin. Methionine and cysteine residues have been introduced into the 8S globulin through protein engineering technology, which resulted in improving the amino acid score from 41 to 145% (Toria et al. 2011). The same authors introduced free sulfhydryl groups and disulphide bonds to generate cysteine-modified mungbean 8Sα globulin to improve protein quality. Variation for trypsin inhibitor among mungbean varieties has been reported by Chattopadhyay et al. (2009); 1324.26 TIU/g to 1502 TIU/g by Das et al. (2015). Bioactivities for proteins and hydrolysed peptides, including angiotensin-converting enzyme inhibitory activity, anti-fungal activity and trypsin inhibitory activity, have been reviewed by Yi-Shen et al. (2018), all of them with potential commercial applications.

7.3 Carbohydrates

Carbohydrates in mungbean include starch components—available and resistant starch; and fibres—lignin, cellulose; monosaccharides—maltose, glucose, xylose; oligosaccharides—raffinose, stachyose, verbascose. Among carbohydrates in mungbean, starch is the major constituent and utilised by the food industry and for noodle preparation. The mungbean starch granule can be oval, round or bean-shaped and 7–26 μm in diameter (Hoover et al. 1997) and is characterised by its high-cross linkage properties (Lii et al. 1988). Starch separated from ten mungbean varieties widely cultivated in China possesses different physicochemical characteristics and diverse processing properties (Li et al. 2011). It has been reported that the starch granule size (5–40 μm), total starch content (55–58%), amylose content (40–42%), solubility (14–18%),

Table 7.1 Amino acid (g/16 g of nitrogen) composition of mungbean

Amino acid	Alanine	Arginine	Aspartic acid	Cysteic acid	Glutamic acid	Glycine	Histidine	Isoleucine	Leucine
Mean	4.1	5.8	13.0	13.5	18.3	3.6	3.2	4.3	7.6
Amino acid	Lysine	Methionine	Phenylalanine	Proline	Serine	Threonine	Tryptophan	Tyrosine	Valine
Mean	6.5	1.2	5.4	4.5	4.9	3.2	1.2	2.7	5.1

Source Adapted from Dahiya et al. (2013)

Table 7.2 Digestibilities of starch and starch fractions in different legume species

Legume	Resistant starch (%)	Readily digestible starch (%)	Slowly digestible starch (%)	Hydrolysis index	Glycemic index
Black gram	60.9 ± 1.1^{d}	9.5 ± 0.6^{d}	29.6 ± 0.8^{b}	16.5 ± 0.8^{c}	48.7 ± 0.3^{c}
Chickpea	54.3 ± 0.9^{b}	10.9 ± 0.4^{e}	34.8 ± 0.6^{cd}	18.4 ± 0.4^{d}	49.8 ± 0.2^{d}
Field pea	58.0 ± 1.4^{c}	8.1 ± 0.5^{c}	33.9 ± 0.4^{c}	16.8 ± 0.5^{c}	48.9 ± 0.4^{c}
Lentil	65.2 ± 0.9^{e}	5.2 ± 0.6^{b}	29.7 ± 0.5^{b}	13.0 ± 0.7^{b}	46.8 ± 0.5^{b}
Mungbean	50.3 ± 1.3^{a}	9.7 ± 0.6^{d}	40.0 ± 0.8^{d}	20.0 ± 0.7^{e}	50.7 ± 0.3^{d}
Pigeon pea	78.9 ± 0.8^{f}	4.2 ± 0.3^{a}	16.9 ± 0.7^{a}	8.2 ± 0.3^{a}	44.2 ± 0.6^{a}

Means followed by the same superscript within a column do not differ significantly ($p < 0.05$). Data (±standard deviation) are the mean of triplicate analyses (reproduced from Sandhu and Lim 2008)

swelling power (17–21%), hydration coefficient (52–84%), degree of gelatinisation (63–95%) and hardness (26–112 N). In a study of 20 mungbean varieties released in China, the total starch content ranged from 40.6 to 48.9% of seed (Shi et al. 2016). The above authors also found that the resistant starch accounted for 16.1–22.3% of total carbohydrates. The comparison of the starch and starch fractions between mungbean and other legumes is presented in Table 7.2. Resistant starch has attracted interest for its potential to improve gut microbiota composition (Nielsen et al. 2015). Nair et al. (2013) pointed out that exploring more varieties and evaluating the relationship between the structural and functional properties of mungbean starch would help develop new processing applications for the crop. Oligosaccharides, such as raffinose, stachyose and verbascose, are associated with flatulence after consumption of beans. Mungbeans cause less flatulence compared to other legumes (Goel and Verma 1981).

7.4 Lipids

The fatty acids (g/kg of seed) present in mungbean are palmitic (2.8–4), stearic (1.4–1.7), oleic (2.1–2.9), linoleic (3.4–4.6, linolenic (1.9–2.4), arachidic (0.23–0.25) (Anwar et al. 2007). Adsule et al. (1986) and Abdel-Rehman et al. (2007) recorded that linoleic acid was the most predominant one and that lauric acid was the lowest one. Zia-Ul-Haq et al. (2008) recorded that the oil content in mungbean seeds is relatively low (2.1–2.7%). The total tocopherol content of mungbean (12.5 mg/100 g) was reported to be higher than in other legumes like black gram: 6.7 mg/100 g), chickpea (*Cicer arietinum* L.): 11.4 mg/100 g and horse gram (*Macrotyloma uniflorum* (Lam.) Verdc): 7.4 mg/100 g) (Gopala Krishna et al. 1997). The scope for genetic improvement is limited owing to the low oil content (Nair et al. 2013).

7.5 Vitamins

The amount and distribution of carotenoids in 20 mungbean varieties differing in seed size and colour were studied by Harina and Ramirez (1978). They found that these carotenoids were present in the form of β-carotene and xanthophylls. Carotenoid content in cotyledons of mungbean (0.5–0.8 mg/100 mg) differs slightly between green and yellow varieties, whereas it varies significantly in seed coats (0.07–0.44 mg/100 mg) of green and yellow mungbean varieties. Interestingly, the authors found that grain size has no correlation with the carotenoid content in mungbean. Carotenoids including provitamin A, such as α-carotene and β-carotene, are converted to vitamin A in the human body after absorption. According to USDA (2010), mungbean sprouts (per kg, dry weight basis) have higher (100 μg retinol activity equivalent

(RAE) content of Vitamin A than grains (70 μg RAE). Nisha et al. (2005) recorded the riboflavin content of mungbean to be 0.29 mg/100 g. Wide range (0–10 mg/100 g, dry weight basis) in vitamin C content in mungbean was reported by Prabhavat (1990). Like in the case of vitamin A, the vitamin C content in mungbean sprouts (1.38 g/kg, dry weight) is higher than in mungbean grains (0.05 g/kg, dry weight). Mungbean grains have a folate content of 0.0069 g, compared with 0.0064 g for sprouts (per kg, dry weight basis) (USDA 2010). Rychlik et al. (2007) used the stable isotope dilution assays to quantify folates in legumes and found 5-methyltetrahydrofolate as the predominant vitamin in mungbean. They reported a folate content of 0.0028 g/kg, dry mungbean seeds. Folates are vital nutrients for pregnant women.

7.6 Iron

Mungbean is an important source of minerals like iron in countries where the large portions are consumed as part of the meal. The iron content in mungbean could range from 5.9 to 7.6 mg/100 g (Dahiya et al. 2015). Consumption of improved mungbean varieties with higher iron content had a positive effect on the productivity of female anaemic workers in Pakistan (Weinberger 2003) and among children in India (Vijayalakshmi et al. 2001). The iron content in mungbean lines/varieties grown in India ranged from 3.5 to 8.7 mg/100 g (Nair et al. 2015a, b). The authors opined that daily consumption of 100 g of mungbean could potentially provide 46–109% of RDA for males and 19–48% of RDA for females. The soil available iron may be affected by agronomic factors, soil and weather conditions, thereby, affecting the grain iron content (Thavarajah et al. 2009; Nair et al. 2015a, b). Mungbean varieties (CN 9-5 and Harsha) recorded almost double their iron content when grown in soils with high available iron (Nair et al. 2015a, b). A promising QTL (qFe-11-1) for iron was located on LG 11 map at the position of 113.7 cM by mapping in a recombinant inbred line (RIL) population developed from a cross between ML446 (high iron content) and Sattya (low iron content) by Singh (2013).

7.7 Zinc

Variation (2.1–6.2 mg/100 g) in zinc content among mungbean varieties has been reported (Taunk et al. 2011; Nair et al. 2015a, b). Karuppanapandian et al. (2006) studied the diversity of zinc content in local landraces of mungbean from Tamil Nadu, India, using random amplified polymorphic DNA (RAPD) markers. Taunk et al. (2011) also obtained polymorphism for zinc content in mungbean using RAPD markers. Four promising QTLs (qZn-11-4, qZn-11-5 on LG 11 and qZn-4-1, qZn-4-2 on LG 4) at the map distance of 196.2 cM, 296.3 cM, 13.7 cM and 87.9 cM by mapping in a RIL population developed from a cross between ML446 (high zinc content) and Sattya (low zinc content) by Singh (2013).

7.8 Other Minerals

Singh et al. (1968) examined the distribution of the different minerals in mungbean plant and found that calcium is primarily present in the seed coat (30–50%), i.e. (812 mg/100 g dry weight), iron in the embryo (23 mg/100 g dry weight) and seed coat (17 mg/100 g dry weight), and phosphorus in the embryo (756 mg/100 g dry weight) and cotyledons (341 mg/100 g dry weight). The mungbean lines/varieties commonly grown in South Asia were studied for their variability in their mineral content by Nair et al. (2015a, b). The following ranges were recorded for the other minerals: Ca (1190–1580 mg/kg), Mg (970–1700 mg/kg), Zn (21–62 mg/kg), Cu (7.5–11.9 mg/kg), Mn (9.8–19.6 mg/kg), Se (0.21–0.91 mg/kg), K (8670–14,100 mg/kg) and P (2760–5170 mg/kg). The effect of the environment on the concentration of minerals in mungbean lines was observed. In addition, variation in some of the values reported could be due to the method of determination of these minerals; inductively couple plasma-emission spectrometry (ICP–EMS)—Nair et al. (2015a, b); atomic absorption spectrometry—Barakoti and Bains (2007) and EDTA Titration method —Kadwe et al. (1974).

7.9 Phytic Acid

Anti-nutritional components reported in mungbean include phytic acid, tannins, hemagglutinins, polyphenols, trypsin inhibitor and proteinase inhibitor (Dahiya et al. 2013). Phytic acid, also known as phytate or myo-inositol-1,2,3, 4,5,6-hexakisphosphate (InsP6), is the main seed storage molecule for phosphorus and is essential for seed development and germination. In legumes, phytates are present in the protein bodies of the endosperm. Variation in phytic acid content (1.8–5.8 g/kg dry grain) in mungbean was reported by Sompong et al. (2010a) in a study with 250 accessions. High broad-sense heritability (80%) for phytic acid content was recorded, which suggested that breeding for low phytate content was feasible. The authors found that high phytic acid content was controlled by dominant alleles at two independent loci, showing duplicated recessive epistasis. A few mungbean QTLs with low or moderate effect on phytic acid content were identified by Sompong et al. (2010b), However, the authors have cautioned that some of the QTLs for phytic acid content overlapped with QTLs for seed size, flowering and maturity, and hence, validation of these results may be conducted before applying for marker-assisted selection. Low phytic acid content (2.6–3.8 g/kg) in mungbean varieties/lines was reported by Nair et al. (2015a, b) and was lower than those reported in kidney bean (*Phaseolus vulgaris* L.) (11–17 g/kg), chickpea (4.9–6.1 g/kg), broad bean (*Vicia faba* L.) (10.1–13.7 g/kg) and soybean (*Glycine max* (L) Merr. (10–14.7 g/kg) (Khokhar and Pushpanjali Fenwick 1994; Konietzny and Greiner 2003). Variation in phytic acid content could also depend upon the method of analysis, estimation of the myo-inositol hexaphosphate content by anion exchange HPLC separation (Lestienne et al. 2005) or phytic acid extracted using 0.5 M HNO_3 and determined colorimetrically (Grewal and Jood 2006). Low phytic acid is considered beneficial, as phytic acid can reduce the bioavailability of iron, zinc and other mineral micronutrients. Reduction in phytic acid content should not lead to a detrimental effect on seed germination (Bohn et al. 2008). Phytic acid serves as an antioxidant, shows anticarcinogenic/antineoplastic properties, reduces or prevents kidney stone formation and plays important roles in many physiological processes (Konietzny and Greiner 2003).

7.10 Other Compounds

Haemagglutinin activity was reported in mungbeans (Mubarak 2005). These are the sugar-binding proteins that bind with red blood cells and agglutinate them. They bind with specific receptors at epithelial cells of the intestine causing lesions and improper microvillus development leading to abnormal absorption of nutrients. Philip and Prema (1998) studied five mungbean varieties cultivated in India and recorded trypsin inhibitor activity: 56–98 trypsin inhibitor units (TIU) mg/protein and tannin content ranging from 3.1 to 4 g/kg grain. Trypsin inhibitor activity of mungbean is much lower than that of other legumes like soybean, kidney bean and chickpea (Sumanthi and Pattabiraman 1976; Guillamón et al. 2008). In addition, the potential to reduce both trypsin inhibitor activity and tannins in mungbean is strengthened by the significant variation recorded for these in the varieties studied (Philip and Prema 1988).

Saponins occur in a wide variety of food plants: chickpea (56 g/kg, dry weight), soybean (43 g/kg, dry weight) being relatively rich compared to mungbeans with 5.7 g/kg, dry weight (Fenwick and Oakenfull 1983). Consumption of large quantities of food containing saponins could lead to stomach discomfort. However, moderate amounts of saponins can be beneficial lowering plasma cholesterol and thereby reducing the risk of heart disease (Potter et al. 1980).

The antioxidant properties of flavonoids in mungbean were investigated by Cao et al. (2011). They found that vitexin and isovitexin (more than 96%) present in the seed coat can help reduce injury due to heat stress (dehydration) in humans. Kim et al. (2012) found that the total phenolic and flavonoid levels of extracts of mungbean sprouts were higher (0.167–0.192 g ferulic acid equivalent (FAE)/kg dry weight than that of dry seeds (which ranged from 0.098 to 0.101 g FAE/kg). Yao et al. (2008, 2011) have suggested the potential use of mungbean seeds and sprouts extracts for therapeutic use.

7.11 Scope for Genetic Improvement in Nutritional Quality: Opportunities and Challenges

A study conducted by Ebert et al. (2017) by comparing the level of nutrients in landraces and improved varieties of mungbean found that relatively old mungbean accessions were superior in protein, calcium (Ca), iron (Fe), zinc (Zn), carotenoid and vitamin C content compared to improved mungbean lines at the fully mature stage. The above finding is a timely reminder to mungbean breeders not to ignore nutritional quality improvement in their program.

The discussion from the previous sections suggests that genetic improvement in mungbean is feasible for further enhancing protein quality, starch content and quality, content of minerals like iron, zinc and also in reducing the anti-nutritional compounds like phytic acid. Progress particularly with improving iron and zinc in the identification of suitable parents, development of mapping populations and in the identification of QTLs for marker-assisted selection is encouraging. Validation of the markers developed would enable other breeders to employ them in their breeding programs. Indirectly progress can also be made by tackling other traits such as resistance to bruchids (*Callosobruchus* spp.), which causes huge losses during storage and also leads to a reduction in the nutritional quality of the stored grains. Bruchid infestation during storage in a 6-month period could lead to increases of 25% in trypsin inhibitor activity, 16% in saponin level and 46% in phytic acid content (Modgil and Mehta 1994). Bruchid-resistant mungbean varieties developed (Nair et al. 2015a, b) would help to reduce the risk of losses in the nutritional quality of these varieties during storage.

However, in order to fully explore the potential for further improvement, screening of the mungbean mini-core collection developed by World Vegetable Center (Schafleitner et al. 2015) for nutritional quality traits would be a good start. The seed size of the variety selected is an important criterion among farmers: large/medium/small. A non-significant correlation between seed size and the micronutrient content indicated that nutrient composition is apparently not affected by seed size and that there is no danger of losing the nutritive value of the grain by developing small- or large-seeded varieties (Nair et al. 2015a, b). Several authors have reported the advantage of sprouts over grains for nutritive value (Ebert et al. 2017). Hence, any variety with a better nutrient content in the grains would have increased nutritive value when consumed as sprouts. However, varieties preferred by the growing sprout market segment are very selective (Nair et al. 2015a, b) and would need special attention to seed size and colour, colour of the seedlings and shelf life of the sprouts.

It is promising to see more interest in recent years in identifying compounds in mungbean, which could add value to the crop, for example, identification of aroma volatiles and understanding

2-acetyl-1-pyrroline biosynthetic mechanism in aromatic mungbean (Attar et al. 2017). The need to pay attention to the food processing properties of mungbean was pointed out by Dahiya et al. (2013). In addition, it is important to employ the latest methods with good detection limits so that biological variation can be separated from analytical variation.

References

Abdel-Rahman ESA, El-Fishawy FA, El-Geddawy MA, Kurz T, El-Rify MN (2007) The changes in the lipid composition of mung bean seeds as affected by processing methods. Int J Food Eng 3:1–10

Adsule RN, Kadam SS, Salunkhe DK, Adsule RN, Kadam SS, Salunkhe DK (1986) Chemistry and technology of green gram (*Vigna radiata* [L.] Wilczek). Crit Rev Food Sci Nutr 25:73–105

Attar U, Hinge V, Zanan R, Adhav R, Nadaf A (2017) Identification of aroma volatiles and understanding 2-acetyl-1-pyrroline biosynthetic mechanism in aromatic mung bean (*Vigna radiata* (L.) Wilczek). Physiol Mol Biol Plants 23(2):443–451. https://doi.org/10.1007/s12298-017-0414-2

Barakoti L, Bains K (2007) Effect of household processing on the in-vitro bioavailability of iron in mungbean (*Vigna radiata*). Food Nut Bull 28(1):18–22

Bohn L, Meyer AS, Rasmussen SK (2008) Phytate: impact on environment and human nutrition. A challenge for molecular breeding. J Zhejiang Univ Sci B 9:165–191

Cao D, Li H, Yi J, Zhang J, Che H, Cao J (2011) Antioxidant properties of the mungbean flavonoids on alleviating heat stress. PloS One. https://dx.plos.org/10.1371/journal.pone.0021071

Chattopadhya K, Sarkar HK, Bhattacharyya S (2009) Morphological and nutritional diversity of some mungbean genotypes from Indian subcontinent. Indian J Plant Genet Resour 22(2):155–159

Chen M-X, Zheng S-X, Yang Y-N, Xu C, Liu J-S, Yang W-D et al (2014) Strong seed-specific protein expression from the *Vigna radiata* storage protein 8SG alpha promoter in transgenic *Arabidopsis* seeds. J Biotechnol 174:49–56

Dahiya PK, Linnemann AR, Van Boekel MAJS, Khetarpaul N, Grewal RB, Nout MJR (2015) Mung bean: technological and nutritional potential. Food Sci Nutr 55(5):670–688. https://doi.org/10.1080/10408398.2012.671202

Das S, Dewanjee A, Maji, Sarkar KK (2015) Breeding mungbean [*Vigna radiata* (L.) Wilczek] for development of high yielding protein rich lines with low trypsin inhibitor. Leg Res 38(4):447–451

Ebert AW, Chang CH, Yan MR, Yang RY (2017) Nutritional composition of mungbean and soybean sprouts compared to their adult growth stage. Food Chem 237:15–22. https://doi.org/10.1016/2017.05.073

Fenwick DE, Oakenfull DG (1983) Saponin content of food plants and some prepared foods. J Sci Fd Agr 34:186–191

Goel R, Verma J (1981) Removal of flatulence factor of some pulses by microbial fermentation. Indian J Nutr Diet 18(6):215–217

Gopala Krishna AG, Prabhakar JV, Aitzetmuller K (1997) Tocopherol and fatty acid composition of some Indian pulses. J Am Oil Chem Soc 74:1603–1606

Grewal A, Jood S (2006) Effect of processing treatments on nutritional and antinutritional contents on green gram. J Food Biochem 30:535–546

Guillamón E, Pedrosa MM, Burbano C, Cuadrado C, Sánchez MdC, Muzquiz M (2008) The trypsin inhibitors present in seed of different grain legume species and cultivar. Food Chem 107:68–74

Harina TH, Ramirez DA (1978) The amount and distribution of carotenoids in the mung bean seed (*Vigna radiata*). Philippine J Crop Sci 3:65–70

Hoover R, Li YX, Hynes G, Senanayake N (1997) Physicochemical characterization of mungbean starch. Food Hydrocolloids 11:401–408

Itoh T, Garcia RN, Adachi M, Maruyama Y, Tecson-Mendoza EM, Mikami B et al (2006) Structure of 8Sα globulin, the major seed storage protein of mung bean. Acta Crystallogr D Biol Crystallogr 62(7):824–832

Kadwe RS, Thakare KK, Badhe NN (1974) A note on the protein content and mineral composition of twenty-five varieties of pulses. Indian J Nutr Diet 11:83–85

Karuppanapandian T, Karuppudurai T, Sinha PB, Haniya AK, Manoharan K (2006) Genetic diversity in green gram [*Vigna radiata* (L.)] landraces analyzed by using random amplified polymorphic DNA (RAPD). Afr J Biotechnol 5:1214–1219

Khokhar S, Pushpanjali Fenwick GR (1994) Phytate content of Indian foods and intakes by vegetarian Indians of Hisar Region, Haryana State. J Agric Food Chem 42:2440–2444

Kim DK, Jeong SC, Gorinstein S, Chon SU (2012) Total polyphenols, antioxidant and antiproliferative activities of different extracts in mungbean seeds and sprouts. Plant Food Hum Nutr 67:71–75

Konietzny U, Greiner R (2003) Phytic acid/nutritional impact. In: Caballero B, Trugo L, Finglas P (eds) Encyclopedia of food science and nutrition. Elsevier, London, pp 4555–4563

Kudre TG, Benjakul S, Kishimura H (2013) Comparative study on chemical compositions and properties of protein isolates from mung bean, black bean and bambara groundnut. J Sci Food Agr 93(10):2429–2436. https://doi.org/10.1002/jsfa.6052

Lestienne I, Icard-Verniere C, Mouquet C, Picq C, Treche S (2005) Effects of soaking whole cereal and legume seeds on iron, zinc, and phytate contents. Food Chem 89:421–425

Lii CY, Chu YL, Chang YH (1988) Isolation and characterization of mungbean starch. In: Shanmugasundaram S (ed) Proceedings of 42 second international mungbean symposium. AVRDC, Shanhua 43, pp 528–535

Li W, Shu C, Zhang P, Shen Q (2011) Properties of starch separated from ten mungbean varieties and seeds processing characteristics. Food Bioprocess Technol 4:814–821

Modgil R, Mehta U (1994) Effects of different levels of *Callosobruchus chinensis* L. infestation on proximate principles, true protein, methionine and uric acid contents of greengram and redgram. J Food Sci Technol 31:135–139

Mubarak AE (2005) Nutritional composition and antinutritional factors of mung bean seeds (*Phaseolus aureus*) as affected by some home traditional processes. Food Chem 89(4):489–495

Naik BS, Kole C (2001) Electrophoretic studies on seed protein profile of mungbean. Leg Res 24(3):169–173

Nair RM, Yang RY, Easdown WJ, Thavarajah D, Thavarajah P, Hughes Jd'A, Keatinge, JDH (2013) Biofortification of mungbean (*Vigna radiata*) as a whole food to enhance human health. J Sci Food Agric 93:1805–1813

Nair RM, Yan MR, Srinivasan R, Schafleitner R (2015a) Developing bruchid resistant mungbean varieties. Paper presented at the SABRAO 13th Congress and International conference IPB, International Convention Center, Bogor, Indonesia, 14–16 Sept 2015

Nair RM, Thavarajah D, Thavarajah P, Giri RR, Ledesma D, Yang RY, Hanson P, Easdown W, Hughes Jd'A, Keatinge JDH (2015b) Mineral and phenolic concentrations of mungbean [*Vigna radiata* (L.) R. Wilczek var. *radiata*] grown in semi-arid tropical. Ind J Food Compos Anal 39:23–32

Nisha P, Singhal RS, Pandit AB (2005) A study on degradation kinetics of riboflavin in green gram whole (*Vigna radiata* L.). Food Chem 89:577–582

Philip J, Prema L (1998) Variability in the antinutritional constituents in green gram, *Vigna radiata*. Plant Food Hum Nutr 53:99–102

Potter JD, Illman RJ, Calvert GD, Oakenfull DG, Topping DL (1980) Soya saponins, plasma lipids, lipoproteins and fecal bile acids: a double blind cross-over study. Nutr Rep Int 22:521–528

Prabhavat S (1990) Mungbean utilization in Thailand. In: Proceedings of the second international symposium, Taipei, Taiwan, pp 9–15

Rychlik M, Englert K, Kapfer S, Kirchhoff E (2007) Folate contents of legumes determined by optimized enzyme treatment and stable isotope dilution assays. J Food Compos Anal 20:411–419

Sandhu KS, Lim S-T (2008) Digestibility of legume starches as influenced by their physical and structural properties. Carbohydr Polym 71:245–252

Schafleitner R, Nair RM, Rathore A (2015) The AVRDC—The World Vegetable Center mungbean (*Vigna radiata*) core and mini core collections. BMC Genom 16(1):344. https://doi.org/10.1186/s12864-015-1556-7

Singh V (2013) Genotypic response and QTL identification for micronutrient (iron and zinc) contents in mungbean [*Vigna radiata* (L.) Wilczek]. Department of Genetics and Plant Breeding, College of Agriculture, CCS Haryana Agricultural University, Hisar-125 004, p 96

Sompong U, Kaewprasit C, Nakasathien S, Srinives P (2010a) Inheritance of seed phytate in mungbean (*Vigna radiata* (L.) Wilczek). Euphytica 171:389–396

Sompong U, Somta P, Raboy V, Srinives P (2010b) Mapping of quantitative trait loci for phytic acid and phosphorus contents in seed and seedling of mungbean (*Vigna radiata* (L.) Wilczek). Breed Sci 62:87–92. https://doi.org/10.1270/jsbbs.62.87

Sumanthi S, Pattabiraman TN (1976) Natural plant enzyme inhibitors of seeds. Indian J Biochem Biophys 13:52–56

Taunk J, Yadav NR, Yadav RC, Kumar R (2011) Molecular markers for assessment of genetic diversity for zinc content among green gram [*Vigna radiata* (L.) Wilczek] genotypes. In: Thezincrops 2011. Available via http://www.zinccrops2011.org/presentations/2011_zinccrops2011

Thakare RG, Gadgil JD, Mitra R (1988) Origin and evolution of seed protein genes in Vigna mungo and *Vigna radiata*. In: Proceedings of the 2nd international symposium on mungbean, Asian Vegetable Research and Development Centre, Taipei, Taiwan, 1–6 Aug 1988

Thavarajah P, Thavarajah D, Sarkar Vandenberg A (2009) Lentils (Lens culinaris Medik Subspecies culinaris): a whole food for increased iron and zinc intake. J Agri Food Chem 57:5413–5419

United States Department of Agriculture (USDA) (2010) National Nutrient Database. http://www.nal.usda.gov/fnic/foodcomp/search/. Accessed 10 Jan 2012

Vijayalakshmi P, Amirthaveni M, Devadas RP (2001) Possibilities of increasing bioavailability of iron from mungbean and study on the effects of its supplementation on children and women. Project Report, Avinashilingam Institute for Home Science and Higher Education for Women, Coimbatore, India

Weinberger K (2003) Impact analysis on mungbean research in south and south-east Asia. AVRDC Processing No. 99.9117.5, Shanhua, Taiwan

Yao Y, Chen F, Wang M, Wang J, Ren G (2008) Antidiabetic activity of mungbean extracts in diabetic KK-Ay mice. J Agric Food Chem 56:8869–8873. https://doi.org/10.1021/jf8009238

Yao Y, Cheng X, Ren G (2011) Contents of D-chiro-inositol, vitexin, and isovitexin in various varieties of mungbean and its products. Agric Sci China 10:1710–1715. https://doi.org/10.1016/s1671-2927-11-60169-7

Yohe JM, Poehlman JM (1972) Genetic variability in the mung bean, *Vigna radiata* (L.) Wilczek. Crop Sci 12:461–464

Yi-Shen Zhu, Shuai Sun, FitzGerald Richard (2018) Mung bean proteins and peptides: nutritional, functional and bioactive properties. Food Nutr Res 62:1290. https://doi.org/10.29219/fnr.v62.1290

Zia-Ul-Haq M, Ahmad MandIqbal S (2008) Characteristics of oil from seeds of 4 mungbean [*Vigna radiata* (L.) Wilczek] cultivars grown in Pakistan. J Am Oil Chem Soc 85:851–856

Molecular Marker Resources and Their Application

8

Roland Schafleitner

Abstract

Molecular markers are DNA fragments or sequence tags that are associated with a certain location of the genome of an organism. Markers have been used in mungbean to analyze the genetic diversity among germplasm accessions and cultivars and to map important traits, including resistance to pests and diseases. Early studies were performed with isoenzyme and RAPD markers. Microsatellite markers derived from various *Vigna* species were efficiently used for generating genetic maps and map traits such as bruchid resistance in segregating populations. The advent of the mungbean whole genome sequence has strongly improved the access to molecular markers for this crop. Large numbers of single-nucleotide markers have been produced by genotyping by sequencing and whole genome re-sequencing, and the generated information has been used to assess the diversity and population structure of mungbean collections and for mapping traits in segregating populations and germplasm panels. Markers for agro-morphological traits as well as for disease and pest resistance are available for marker-assisted selection in mungbean breeding programs.

R. Schafleitner (✉)
World Vegetable Center, Shanhua, Tainan, Taiwan
e-mail: roland.schafleitner@worldveg.org

8.1 Introduction—The Most Important Molecular Marker Types for Plant Science

Molecular markers can be considered to be landmarks in the genome. A molecular marker is either a DNA fragment or a DNA sequence associated with a specific genomic location. In the past, also protein markers have been used. For being useful, a molecular marker needs to show differences among the genotypes under investigation, by either sequence or fragment length. Subsequently, the most important marker types used in plant sciences are briefly described.

Restriction fragment length polymorphism (RFLP) markers were invented by AJ Jeffries in the late 1970s (Zagorski 2006) and were first applied in human genetics. In the late 1980, this technique was introduced to study genome architecture in major crops such as wheat (Sharp et al. 1989), tomato and potato (Bonierbale et al. 1988). RFLP marker detection involves digestion of genomic DNA with restriction enzymes, labeling of specific DNA fragments—usually with the radioactive isotope ^{32}P—and then using the fragments one by one as a probe in Southern blot analyses (Williams 1989). RFLP markers are usually designed to detect both alleles in a heterozygous sample. These markers were a breakthrough for genetic fingerprinting, but the technique is time consuming, needs large amounts of high-quality DNA, and involves handling of a radioactive substance, making the analysis laborious and expensive. Therefore, this

R. M. Nair et al. (eds.), *The Mungbean Genome*, Compendium of Plant Genomes,
https://doi.org/10.1007/978-3-030-20008-4_8

technology is not used anymore, but many published RFLP markers referring to important pest and disease resistances are still relevant to research and breeding.

The advent of polymerase chain reaction (PCR) opened the path for a number of different marker types. Random amplified polymorphic DNA (RAPD) markers rapidly have became popular, as they are easy and cheap to use on virtually any organism. RAPD applies combinations of short random primers in PCRs to amplify different genomic regions. The obtained DNA fragments are resolved according to their size by agarose or polyacrylamide electrophoresis. The banding patterns can vary among genotypes, resulting in potentially polymorphic markers. The drawbacks of RAPD markers are that they require relatively high-quality DNA, are dominant, and have low reproducibility. Consequently, they are difficult to be compared among experiments and among laboratories. In addition, the multiple bands produced in PCR by RAPD primer pairs make identification of alleles difficult. Converting RAPD markers to more robust sequence characterized amplified regions (SCAR) markers enhances reproducibility and allele identification (Bhagyawant 2016). For developing SCAR markers, polymorphic RAPD fragments are cloned and sequenced, and primers are designed to specifically amplify the polymorphic RAPD fragments. Generally, the primer pairs are made to amplify a single RAPD band (Paran and Michelmoore 1993). Polymorphism of SCAR markers is either scored as presence or absence of the amplified band, or as length polymorphisms in the case of co-dominant SCAR markers.

Amplified fragment length polymorphism (AFLP) uses genomic DNA digested with restriction enzymes that are ligated to adapters with known sequence (Vos et al. 1995). Primers complementary to the ligated adapters are used to amplify DNA fragments. Complexity reduction is achieved by adding one or a few specific bases at the 3′ end of the primers to amplify only a subset of the restriction fragments. Presence and absence of specific fragments are scored after separation of the fragments according to their size on a gel or on a fragment analyzer. DNA bands are visualized either through autoradiography or fluorescence methods. AFLP can produce a large number of markers and is at least partly amenable to automatization. Polymorphic bands can be converted to SCAR markers (see above). Today, AFLP markers are not broadly used any more.

Microsatellite markers (SSR marker) contain simple sequence repeats (SSRs) of 1–8 base pairs (Hamada and Kakunaga 1982). SSR motifs are hot spots for mutations, where DNA polymerase adds or eliminates one or more repeat units during DNA replication. SSR motifs can be present in coding and non-coding sequences. They can be amplified by PCR using flanking sequence-specific primers. Size polymorphisms of specific SSR fragments among different genotypes are scored after size fractionation by electrophoresis or on fragment analyzers. SSR markers are abundant in the genome and are generally co-dominant; however, the degree of their polymorphism varies among species and populations.

Single-nucleotide polymorphism (SNP) markers consist of single-nucleotide changes observed by comparing the DNA of different genotypes. They are very abundant in the genome, but their generation generally requires sequence information. As improvements in sequencing technology made sequence information readily available at low cost, SNP markers have become the marker species of choice.

Strictly speaking, markers like AFLP or RFLP that do not require sequence information also score (beside indels and structural variations) SNPs, but only if they are present at restriction enzyme cutting sites. Likewise, the Diversity Array Technology (DArT) also scores SNPs (and indels and structural variants) by testing for the presence or amount of a specific DNA restriction fragments in a representation derived from the total genomic DNA of different individuals or populations (Jaccoud et al. 2001). The DArT technology is fast and cost-effective, but produces dominant markers.

There are various methods to obtain SNP markers. Comparing genome or transcriptome sequences among individuals is a relatively simple method to produce SNP information, but cost-effective genotyping of these SNPs in a large population needs specialized technologies based on fluorescence detection in PCR format (Tapp et al. 2000; Semagn et al. 2014), or arrays (Elbasyoni et al. 2018). Smaller SNP sets can be analyzed in large populations by cleaved amplified polymorphic sequence (CAPS) markers (Thiel et al. 2004), high-resolution melting (Liew et al. 2004), mass spectroscopy (Storm and Darnhofer-Patel 2003), or other methods. Dependent of the number of loci to be genotyped, establishing the SNP resource for a population may be a costly investment. Multiplexing SNP assays is an often practiced way to improve cost efficiency (Fan et al. 2003).

SNP genotyping technologies such as genotyping by sequencing (Elshire et al. 2011) and the single-primer enrichment technology (SPET) became very popular. They are based on sequencing a fraction of the genome. Complexity reduction is achieved either by restriction enzyme digestion or by multiplex PCR amplification of target sequences. Both methods can be scaled to obtain a greater or lesser number of SNPs. DNA barcoding allows for pooling of many samples in one sequencing reaction, which leads to dramatic cost reductions (Elshire et al. 2011). GBS and SPET combine SNP discovery and genotyping in one step, but the reproducibility and precision of these methods are inferior to array-based genotyping (Elbasyoni et al. 2018). Both GBS and SPET are patented technologies (patent numbers WO2013009175A1 and US20130231253A1, respectively).

8.2 Molecular Markers in Plant Breeding

Plant breeding consists of crossing the best parents and subsequent identification and recovery of the progeny that outperforms the parents (Moose and Mumm 2008). Genetic gain is defined by (i) the phenotypic variation present in the breeding population, (ii) the probability that a trait phenotype will be transmitted from parents to offspring (heritability), (iii) the proportion of the population that is selected as parents for the next generation (selection intensity), and (iv) by the time necessary to complete a cycle of selection. All four key factors for genetic gain can be positively impacted by using molecular markers.

One factor affecting genetic gain is the available phenotypic variation in the population. Measuring the phenotypic diversity is affected by the environment. Exotic material may not be adapted to the selection environment and thus not show the real potential for breeding. Phenotypic diversity is positively associated with genetic diversity. Molecular markers are cheap and efficient tools to characterize genetic diversity in populations. They contribute to understand population structure and inform about the presence of heterotic groups in a germplasm set or in breeding populations and facilitate the exploitation of heterosis for producing hybrids and improved populations (Van Inghelandt et al. 2010; Barata and Carena 2006).

Heritability depends on the number of genes affecting a trait, the magnitude of their effects, and the type of gene action associated with the phenotype (Moose and Mumm 2008). Molecular marker technologies facilitate the definition of loci associated with a trait of interest. For traits with low heritability such as yield, molecular markers associated with loci influencing the trait often account for a greater proportion of additive genetic effects than the phenotype alone. Knowledge of the genetic architecture underlying the trait can be exploited to add or eliminate specific alleles that contribute to the breeding value. If linkage drag or epistasis among loci with antagonistic effects on a trait limits the genetic gain, information on loci associated with the traits can be used to break these undesirable allelic relationships (Moose and Mumm 2008).

Selection intensity in conventional plant breeding relies on phenotypic selection. Environmental variability, genotype by environment interaction, and evaluation errors add complexity to phenotypic selection. Multi-environment evaluation improves selection accuracy, but is time consuming and expensive. Some traits require destructive sampling or exposure of the population to diseases and pests, which affects the recovery of the desired genotypes. Pest and disease resistance screening in natural environments is particularly challenging, as it depends on the presence and activity of the pathogen (and its vector) or the pest. Molecular markers can make selection more precise and increase the selection intensity.

Some traits, including those associated with yield or stress resistances appear at late developmental stages and only can be measured on mature plants. Therefore, large testing populations have to be cultivated up to maturity for selection. Molecular markers associated with traits of interest can allow selecting for these traits at early stages, reducing the time and costs required for plant cultivation and testing. Markers linked to traits of interest make the selection environment independent, allow for selection in off season nurseries, and permit accommodating multiple selection rounds in a year, therefore shortening the time required for completing a selection cycle. The advantages of molecular marker-based selection have been realized by plant breeders, and the technology is now applied on a broad range of crops, including legumes (Varshney et al. 2018).

8.3 Molecular Markers of Mungbean—A Brief History

Before molecular markers became available, genetic studies relied on studying morphological traits, such as flower color in pea (Mendel et al. 1993). Morphological traits that are controlled by a single gene can be used as genetic markers, but their number is limited, and without progeny tests, it is impossible to distinguish heterozygous from homozygous individuals. With the advent of isoenzyme markers, first genetic maps were constructed (Mahmoud et al. 1984). Protein markers were developed based on differences in mobility of certain proteins among different accessions. Mobility differences of seed proteins were used for cultivar detection (Mohanty et al. 2011). Pattnaik and Kole (2002) found protein markers that were polymorphic between MYMV-resistant and susceptible genotypes. The phylogenetic relationship between *Vigna* species was also addressed with protein markers (Kole and Panigrahi 2001). The limited number of polymorphic protein markers favored the development of more polymorphic DNA-based markers. About one decade after the first RFLP marker studies on major crop species, the technology was applied also on "orphaned" crops such as mungbean, and in the early 1990s, the first reports using restriction fragment length polymorphic (RFLP) markers for mapping traits in mungbean were published (Fatokun et al. 1992; Young et al. 1992, 1993). A genetic map based on RFLPs comprising 171 loci on 14 linkage groups was produced (Menancio-Hautea et al. 1993). Reports on molecular marker in mungbean became more frequent with the advent of random amplified polymorphic DNA (RAPD) markers. They were used to assess genetic diversity in germplasm (Santalla et al. 1998) and in cultivars (Lakhanpaul et al. 2000) and to map resistance to the most important diseases and pests such as mungbean yellow mosaic disease (Selvi et al. 2006) and bruchid beetles (Chen et al. 2007). The more reproducible AFLP markers allowed refining diversity studies (Singh et al. 2013) and trait mapping (Chaitieng et al. 2002; Srinives et al. 2010), but none of the markers developed with this technology was reported to be used in breeding. Then, large numbers of SSR markers were assembled for mungbean from sequence data of various *Vigna* species (Somta et al. 2009), or were generated for mungbean genomic sequences (Tangphatsornruang et al. 2009) or transcriptome sequencing data (Gupta et al. 2014; Chen et al. 2015a). A genetic map resolving the 11 linkage groups of mungbean was constructed with 150 SSR markers (Kajonphol et al. 2017), and QTLs for resistances to *Cercospora* leaf spot (Chankaew et al. 2011), powdery mildew (Kasettranan et al. 2010), nutritional traits such as phytic acid content (Sompong et al. 2012), and domestication-related traits (Isemura et al. 2012) were mapped. SSR markers also were used to assess genetic diversity in a large germplasm collection to establish a mini-core collection (Schafleitner et al. 2015). Large-scale single-nucleotide polymorphic (SNP) marker detection was started in mungbean by transcriptome sequencing (Moe et al. 2011) and by comparing reads obtained by Illumina HiSeq sequencing of the genomes two mungbean cultivars (Van et al. 2013). Soon after, the whole genome sequence of mungbean cultivar VC1973A became available (Kang et al. 2014), paving the path for genotyping by sequencing approaches on this crop (Kang et al. 2014; Schafleitner et al. 2016). Current re-sequencing projects producing huge numbers of markers are likely to provide insight into genome re-arrangements in mungbean.

8.3.1 The First Molecular Markers for Mungbean Breeding: Markers Associated with Bruchid Resistance

The first application of molecular marker in mungbean was a study targeting bruchid-resistant loci. Promising bruchid resistance was discovered in wild mungbean *Vigna radiata* ssp. *sublobata* TC1966 (Fujii et al. 1989). At that time, the markers of choice were RFLPs. It was thought that RFPL may be a suitable marker system especially for crops with relatively small genome size such as mungbean to map genes and guide chromosome walking for gene cloning (Steinmetz et al. 1981). Young et al. (1992) analyzed 58 F_2 progeny derived from *V. radiata* ssp. *sublobata* TC1966 and a susceptible *V. radiata* line with 153 RFLP markers and succeeded to define a RFLP marker 3.6 cM distant from the bruchid resistance locus. One F_2 individual was identified carrying the bruchid resistance gene within a tightly linked double crossover. Such an individual would be highly valuable in developing resistant mungbean lines with reduced linkage drag. Later, Miyagi et al. (2004) succeeded to convert RFLP probes associated with bruchid resistance to PCR-based markers. They screened mungbean BAC libraries from resistant and susceptible lines with RFLP probes associated with bruchid resistance and identified SSR motives and sequence tagged sites (STS) on these BACs. This experiment yielded PCR-based markers STSbr1 and STSbr2 that co-segregated with an RFLP marker associated with bruchid resistance (Miyagi et al. 2004). STSbr1 was validated on Indian genotypes to be associated with bruchid resistance (Sarkar et al. 2011), while STSbr2 was associated with one of two bruchid-resistant loci in *V. radiata* V2709 (Sun et al. 2008).

In the 1990s, the low-cost and easy-to-use RAPD markers were adopted for mungbean. Conversion of these markers to more robust SCAR markers made this marker type more useful. Chen et al. (2007) identified ten RAPD markers to be associated with bruchid resistance in a bulked segregant analysis in recombinant inbred lines (RILs) derived from a cross between the bruchid-resistant *V. radiata* ssp. *sublobata* line TC1966 and the mungbean yellow mosaic disease-resistant *V. radiata* elite cultivar NM92 bruchid. Three pools were established: One pool consisted of 22 bruchid-resistant F_{12} RILs (0% infestation), and two pools were made of 20 susceptible RILs with 80–90% damage and 18 susceptible RILs of 90–100% damage. From the ten RAPD markers found to be associated with bruchid resistance in this experiment, the four most closely linked ones were cloned, sequenced, and transformed to SCAR and cleaved amplified polymorphism (CAP) markers. The CAP fragment derived from RAPD marker OPW02a4 was mapped to a location around position 6 mega base on chromosome 5, after the mungbean reference sequence (Kang et al. 2014) became available. QTL analysis using a mix of RAPD, SCAR, CAP, AFLP, and SSR markers (in total 489 markers) in the same population mapped bruchid resistance to linkage groups 7 and 9. Linkage group 9 was tagged with marker DMB158, which later was mapped to chromosome 5 (Schafleitner et al. 2016). QTL mapping using more than 9000 SNPs in population *V. radiata* ssp. *sublobata* TC1966 x *V. radiata* NM94 and more than 6000 SNPs in the cross of the two *V. radiata* lines V2802 x NM94 corroborated the presence of a bruchid resistance locus on chromosome 5 in both resistant lines, TC1966 and V2802. The markers are currently used in pyramiding bruchid resistance with disease resistances and good agronomic performance in mungbean breeding lines (Ramakrishnan Nair, personal communication).

The example bruchid resistance shows that molecular markers are a suitable tool for mungbean crop improvement. However, specifically for the trait bruchid resistance, evolution of marker technologies and access to large numbers of markers did not significantly improve the localization of the major resistance gene. The first RFLP markers associated with this resistance were mapped to a similar location like SSR and SNP markers in more recent experiments. The SSR marker DMB158 mapped nearest to the bruchid resistance locus (Chotechung et al. 2011; Chen et al. 2013). Fine-mapping the bruchid resistance locus confirmed the localization of this marker on chromosome 5 and resulted in two candidate genes VrPGIP1 and VrPGIP2 conferring the resistance (Chotechung et al. 2016; Kaewwongwal et al. 2017). The SNP markers linked to the resistance gene are not better than the SSR marker. This demonstrates that the trait, the biological material, and the assay conditions are by far more important for mapping traits than the marker system. Major resistance genes can be tagged also with simple methods, as long as the phenotyping data are sound. Including more markers and using more modern marker technologies do not necessarily improve the mapping result.

Bruchid resistance markers obtained by genotyping by sequencing were mapped to a reference sequence. The order of these markers on genetic maps was different to the order suggested by mapping the SNPs to the reference sequence (Schafleitner et al. 2016). This may be due to translocations, which caused differences in marker order in experimental populations compared to the reference sequence. Structural variations are important sources for phenotypic diversity. They are defined as genomic variations that involve segments of DNA larger than 1 kb in length and consist of insertions, deletions, inversions, translocations, and copy number variations (Feuk et al. 2006). Genotyping with markers do not capture all structural variations (Springer et al. 2011). Whole genome re-sequencing is likely to improve the knowledge about structural variations.

8.3.2 Markers for Diversity Analysis in Mungbean

Mungbean is an autogamous (cryptogamous) species. Current cultivars have a narrow genetic base, because only a limited number of genotypes were used for breeding (Kumar et al. 2003). Breeding improved varieties therefore needs access to new genetic diversity. Due to the replacement or disappearance of wild relatives and local cultivars, alleles that could be of high interest for future breeding are continuously lost. Therefore, germplasm collections of mungbean landraces and wild relatives are an important reservoir to source new genetic diversity for breeding. The genetic diversity and population structure of germplasm accessions in gene banks need to be characterized to improve management of the collections by identifying redundant accessions, produce germplasm subsets with certain properties, and identify genotypes of interest for breeding (de Vicente et al. 2006).

Before the advent of molecular markers, diversity analysis depended on geographic information on the site of origin of an accession, pedigree data, morphologic or agronomic traits, or on biochemical data. Geographic origin together with morphological traits, discrete ones like bean color, or continuous ones like plant height or seed size, has been used to classify mungbean germplasm and analyze the diversity of collections (Bisht et al. 1998). The joint analysis of discrete and continuous variables has higher potential than analysis of either discrete or continuous data alone (Gonçalves et al. 2008). However, morphological data are susceptible to environmental variability. Measuring morphological traits in large collections usually is done over several seasons, bearing the risk that environmental variability is causing variation of traits and subsequent errors in diversity analyses. In addition, morphological differences usually are determined by a small number of genes and may not be representative for the genetic diversity of the entire genome (Carroll 2008). In contrast, DNA markers are likely to reveal most accurately the genetic relationship among genotypes (reviewed by Crawford 1990). Marker genotypes are environment independent, and they are stable over different developmental stages of the plants. Small samples of plant tissue are sufficient for genotyping, and it is not necessary to grow plants to maturity, as it is required for morphological characterization, making genotyping a cheap option for diversity characterization. DNA markers likely provide information on homologous loci among genotypes, while morphological characteristics may be under the control of multiple genes, masking allelic relationships. DNA markers also are by far more abundant than morphological markers, increasing the power to discriminate between genotypes. Finally, scoring DNA markers is generally easier that measuring morphological parameters. Modern marker technologies also are amenable to automatization, further facilitating the approach.

Several marker technologies have been used to characterize mungbean germplasm. Santalla et al. (1998) have used RAPD markers to analyze genetic diversity in a small panel consisting of mungbean germplasm and three individuals of other *Vigna* species. Sixty random decamer primers were tested, and 28 pairs revealed being informative. The resulting phylogenetic tree showed three main clusters, which included *V. radiata* landraces, *Vigna mungo*, and *Vigna luteola*, respectively. Studies on the genetic diversity of Indian mungbean cultivars were also performed with RAPD markers (Lakhanpaul et al. 2000; Datta et al. 2012). These studies had in common that a relatively large number of primer combinations had to be tested and only about the half of the combinations yielded useful RAPD patterns for diversity analysis. RAPD markers can be readily applied on any organism, without previous sequence information. RAPD markers are generally abundant and evenly distributed over the genome. The main weakness of RAPD markers is their low reproducibility (Schierwater and Ender 1993). Hence, these markers are difficult to be used across laboratories and experiments. The scoring of the bands can be complex and is subject to different interpretation when analyzed by different persons. High-quality DNA is critical for these assays, adding costs to the experiment. All together, these properties make RAPD markers a poor tool to analyze genetic diversity in large genebank collections.

Chattopadhyay et al. (2005) applied a combination of RAPD and inter-simple sequence repeats (ISSR) markers to study genetic diversity in selected mungbean genotypes. ISSRs are regions in the genome flanked by microsatellite sequences. These regions are amplified in PCRs using a primer that contains a microsatellite motif at the 3′ end (Gupta et al. 1994). ISSR markers do not need any previous sequence information, are easy to use, and cause low costs. But ISSRs, like RAPD, may be affected by low reproducibility, and the obtained multiple bands may be derived from non-homologous loci and difficult to analyze.

A few reports describe the use of AFLP markers in mungbean diversity analysis (Bhat et al. 2005; Singh et al. 2013). Singh et al. (2013) compared phylograms obtained with ISSR and AFLP markers. They found that AFLP markers were more efficient than the ISSR in assessing genetic diversity, as they yielded more polymorphic markers than ISSR. The comparison of the Jaccard similarity matrices obtained with both marker systems showed only low correlation, and the clustering of genotypes within groups was not similar when AFLP- and ISSR-derived dendrograms were compared. It was hypothesized that the two marker technologies targeted different genomic regions and yielded different numbers of markers, which led to the different phylogenetic clustering of the accessions when the two methods were used. Advice for designing an ISSR experiment and recommendations on using ISSR markers in genetic variation studies has been disclosed in

Ng and Tan (2015). But in general, the easy-to-use SSR and SNP markers have widely replaced other marker systems, including ISSR.

Most diversity studies in mungbean have been accomplished with microsatellite (SSR) markers. These co-dominant markers are abundant in the genome, are easy and cheap to use, and are amenable to multiplexing and automatization (Hayden et al. 2008). Originally, SSR markers were developed from repeat-enriched libraries (Edwards et al. 1996), a labor intense technology. But with readily available sequence information from transcriptomes and genomes, microsatellites became much easier to access (Chen et al. 2015a). Specialized software tools to mine DNA sequences for microsatellite motifs and design primers to amplify microsatellite loci are widely available (da Maia et al. 2008; Wang and Wang 2016). Still, microsatellite markers need to be well chosen to obtain allelic bands in genotyping. Backward and forward mutations (homoplasy) may occur at microsatellite loci and cause underestimation of the genetic diversity (Spooner et al. 2005). In mungbean, SSR markers have been developed using the 5′-anchored polymerase chain reaction technique (Kumar et al. 2002), from genome shotgun sequences (Tangphatsornruang et al. 2009), transcriptome sequences of *V. radiata* (Chen et al. 2015a; Gupta et al. 2014), or have been transferred from other *Vigna* species (Isemura et al. 2012).

The first comprehensive study on mungbean diversity used a set of 19 SSR markers derived from adzuki bean (*Vigna angularis*) on 615 cultivated and wild mungbean accessions (Sangiri et al. 2008). The marker set was selected based on the marker location in the adzuki bean genome to contain at least one marker per linkage group. More alleles were detected in wild than in cultivated accessions, illustrating the lower diversity in the cultivated germplasm set. The study revealed that Australia and New Guinea represent a distinct center of diversity for wild mungbean, while cultivated mungbean has greatest diversity in South Asia. Soon after, the diversity and population structure of mungbean were analyzed with 15 different SSR markers in 692 mungbean accessions held by the National Agrobiodiversity Center of the Rural Development Administration, Korea. Mungbean germplasm obtained from 27 countries was grouped into seven phylogenetic clades and into two distinct genetic groups (Gwag et al. 2010). In total, 157 mungbean germplasm accessions were genotyped with EST-SSRs (Chen et al. 2015b).

A combination of morphological data and microsatellite markers was used to define a 300 accession mini-core collection that represents a large proportion of the overall diversity of the whole World Vegetable Center mungbean collection of more than 6700 accessions (Schafleitner et al. 2015). In the first step, geographic stratification was performed, and by cluster analysis of eight phenotypic descriptors, a phylogenetic tree was produced. From this tree, 20% of the accessions were randomly selected from each cluster as a core collection containing about 1400 genotypes. The core collection was subsequently genotyped with 20 microsatellite markers, and a mini-core set was selected to represent all detected 122 alleles (Schafleitner et al. 2015). The collection was small enough to be submitted to multilocation trials in various regions in Asia and Africa to discover new traits for mungbean breeding, and it is expected that it is large enough to map traits in genome-wide association studies.

Other marker types such as single-strand confirmation polymorphism, cleaved amplified polymorphic sequence, and SCAR markers that were used for diversity analysis in other crops (Spooner et al. 2005) were not reported for similar works in mungbean, while single-nucleotide polymorphic markers (SNPs) were applied for analyzing several germplasm collections.

The first SNPs for mungbean were reported for pairs of mungbean lines by Moe et al. (2011) from transcriptome sequences, followed by Van et al. (2013) from shotgun Illumina sequences. Availability of the mungbean whole genome sequence (Kang et al. 2014) strongly improved the access to SNPs for this species. Re-sequencing of selected lines yielded large numbers of SNPs (Liu et al. 2016), and genotyping by sequencing (Elshire et al. 2011) was applied to mungbean populations (Kang et al. 2014; Schafleitner et al. 2016). Genotyping of germplasm accessions of mungbean with SNPs was done on the USDA mungbean collection, the Australian mungbean mungbean diversity panel (Noble et al. 2018), on the World Vegetable Center mini-core (Breria et al. 2019).

The SNP-based diversity analysis of 94 cultivated mungbean genotypes from the USDA collection originating from 27 countries was done using a small set of SNP markers (Islam and Blair 2018). From a total of 42 known SNPs (Van et al. 2013), 18 were successfully converted to polymorphic KASP markers. The population could be divided in two subpopulations and one admixture group.

The Australian diversity panel was submitted to GBS. The germplasm set consisted of 466 cultivated and 16 wild accessions. In total, more than 22,000 polymorphic genome-wide SNPs were identified and used to analyze the genetic diversity, population structure, and linkage disequilibrium (Noble et al. 2018). As expected, polymorphism was lower in the cultivated than in the wild accessions. Linkage disequilibrium decay amounted to about 100 kb in cultivated lines and about 60 kb in wild mungbean. Structure analysis identified four distinct subgroups, which broadly corresponded to geographic origin and seed characteristics (Noble et al. 2018).

Genotyping using GBS of the World Vegetable Center mini-core produced more than 24,000 markers for a germplasm panel consisting of *V. radiata* and *V. mungo* and 8000 polymorphic markers for 296 *V. radiata* accessions. From this set, 5447 polymorphic SNPs were used for germplasm characterization and structure analysis, identifying two major populations, one of them falling into three subpopulations, in the World Vegetable Center germplasm set. The mini-core and the genotyping data are currently used to map a number of morpho-agronomic traits.

8.3.3 Molecular Marker for Cultivar Identification and Hybridity Tests

Molecular fingerprinting of varieties and determining purity of seed is a component of quality seed production. Testing seed purity with molecular markers is common for many crops and is considered to be quicker and more cost-effective than grow-out tests (Yashitola et al. 2002). This, however, may not be true for all cases. For example, much of the mungbean seed production and much of its growing area are located in developing countries where wages are low and where there is little access to infrastructure for low-cost genotyping. Therefore, grow-out tests may be still cheaper than genotyping for seed quality assessment. Ali et al. (2010) reported seed quality assessment of Bangladeshi mungbean varieties based on quantifying other seed than mungbean and inert matter in seed lots, seed moisture content, 1000 seed weight and germination tests. Molecular markers have been applied to produce molecular fingerprints of varieties (Tantasawat et al. 2010; Lestari et al. 2014; Reflinur et al. 2017), but reports on systematic use of markers for seed quality monitoring for mungbean are not available.

Monitoring the success of crosses by hybridity tests with molecular markers is a common practice (Solanki et al. 2010). In mungbean, SSR markers are being used to monitor crosses between mungbean germplasm and breeding lines, as well as in wide crosses between cultivated mungbean and wild relatives (Ramakrishnan Nair, personal communication). One or a few polymorphic SSR markers that are generally easy to define and cheap to apply are sufficient for this task.

8.3.4 Developing Markers Linked to Traits of Interest

Disease-resistant cultivars are the cheapest, simplest, and most environmentally safe way to manage disease. Likewise, improving abiotic stress tolerance of crops can stabilize yields and prevent crop failure. Disease resistance and abiotic stress tolerance are often sourced from landraces and wild relatives. Introgression of biotic and abiotic stress-tolerant traits from unadapted material into elite cultivars is a frequent breeding task. As outlined above, using molecular markers can improve introgression of these traits into elite breeding material.

Developing markers associated with traits of interest include the following steps:

(1) Establish the genetic resources for trait mapping, for example, a mapping population or a germplasm panel segregating for the trait of interest
(2) Phenotype the population and generate trait value data, e.g., on resistance or susceptibility to a pathogen or pest, or on tolerance to an abiotic stress factor
(3) Develop markers and genotype the experimental population
(4) Associate phenotypes to specific marker genotypes using appropriate statistical methods
(5) Validate the candidate markers in different genetic backgrounds and produce user-friendly markers for marker-assisted selection.

Once the genetic resources are phenotyped, a marker system has to be chosen to genotype the population or germplasm panel. Today, the most popular markers are SNPs. Genotyping by sequencing has been successfully used in mungbean to generate a large number of SNP markers for bi-parental populations and germplasm panels (Schafleitner et al. 2016; Noble et al. 2018). Ongoing whole genome re-sequencing efforts benefit from the available whole genome reference sequence of mungbean (Kang et al. 2014) and are likely to provide large number of SNPs for this species.

Several methods are available to associate phenotypic traits with genotypes. Bulked segregant analysis uses bulked DNA samples generated from individuals of a segregating population from a single cross (Michelmore et al. 1991). Each bulk contains DNA from individuals that are identical for a particular trait such as disease and pest resistance or susceptibility, but are arbitrary at all unlinked regions. The two bulks are therefore genetically dissimilar in the selected region but seemingly heterozygous at all other regions. The bulks are screened for genetic differences using suitable markers to identify loci that have contrasting alleles at homozygote state in the two bulks. Bulked segregant analysis is a rapid and simple method to determine association of markers to single gene or oligogenic traits, but is generally not suitable for multigenic traits. On mungbean, bulked segregant analysis has been used to map bruchid resistance (Cheng et al. 2005, 2007; Sun et al. 2008), Mungbean yellow mosaic disease resistance (Selvi et al. 2006; Dhole and Reddy 2013; Karthikeyan et al. 2012), and iron deficiency tolerance (Toojinda et al. 2001). Markers associated with the respective traits were obtained, but no application of these markers in breeding was reported.

Quantitative traits typically are tagged by QTL analysis (Liu 2017). For this task, first, the molecular markers are mapped, on either genetic or physical maps. Then, associations between the trait(s) of interest and the marker genotypes are tested using statistical methods. A large number of reports describe QTL studies on disease resistances, quality, and domestication traits in mungbean. The first reported QTL study on mungbean found associations between seed weight and RFLP marker genotypes (Fatokun et al. 1992). Humphry et al. (2005) investigated the relationships between hard-seededness and seed weight to support breeding of hard- and large-seeded genotypes. A large number of QTL analyses followed, targeting a wide range of morphological, agronomical, and nutritional traits. Disease and pest resistances were probably the traits that were most frequently targeted by such analyses. Results of many of these studies are summarized in the chapter "Genomic Approaches to Biotic Stresses" by Laosatit et al. in this book.

Genome-wide association studies (GWAS) are a quantitative method to test whether a genomic variant (marker genotype) is associated with a trait of interest using a germplasm panel as experimental population. It assumes that a specific property such as disease resistance, abiotic stress tolerance, or a nutritional trait shared by a subset of the germplasm panel is reflected by a specific marker genotype also shared by these individuals. The markers have to be in linkage disequilibrium with the genes conferring the trait. In comparison to QTL studies on bi-parental populations, GWAS have the advantage to work on germplasm panels and do not need specific mapping populations. Therefore, GWAS can analyze the function of all alleles and haplotypes present in the germplasm set under investigation, while QTL studies on bi-parental populations only take into account the alleles present in the mapping parents. The resolution of GWAS is generally higher than that of QTL analyses in bi-parental populations. Resolution depends on the number of recombination events that separate the investigated genotypes from each other. Bi-parental populations generally have undergone only a low number of recombination events until analysis, while germplasm panels have a long history of evaluation and therefore individuals are usually separated from each other by many recombination events. One of the major drawbacks of GWAS compared to QTL mapping in bi-parental populations is that population structure influences the outcome of the study, but inclusion of population structure into the GWAS model tries to mitigate this effect. Furthermore, GWAS generally requires a larger number of markers than mapping in bi-parental populations. The marker number needed depends on the linkage disequilibrium decay distance in the germplasm panel and is specific for the species and kind of population under investigation. However, modern genotyping technologies provide easy access to large numbers of molecular markers helping to overcome this drawback.

Up to date, only a few genome-wide association studies were undertaken in mungbean. This is probably due to the lack of high-quality phenotyping data for densely genotyped germplasm panels. A pilot GWAS on seed color in the Australian diversity panel revealed five genomic regions associated with this trait (Noble et al. 2018). Ongoing phenotyping efforts for mungbean diversity panels will likely lead to a broader application of GWAS to identify marker trait associations. As the resolution of GWAS often goes down to the gene level (Liu and Yan 2019), these studies not only give markers associated with a trait of interest. They are likely to improve the knowledge on the genetic basis of traits including the causative genes or alleles and their interactions (Hansen 2006).

The bottleneck for successful GWAS are, as mentioned above, high-quality phenotypic data, the complexity of the trait of interest, and the size of the germplasm panel used for determining marker trait associations. Even in associations that are statistically highly significant, false-positive associations may still occur. The large number of statistical inferences, inaccurate genotyping, and too small population sizes make the results prone to errors (Liu and Yan 2019). Like in QTL studies in bi-parental populations, the loci found to be associated in these studies need to be thoroughly validated before drawing conclusions on gene functions or using any of the markers apparently linked to a trait of interest in breeding. Candidate genes and alleles that are found association also in a different population can be assumed to be more likely linked with the trait of interest. In addition, candidate gene knock-out or overexpression studies are suitable methods for validation.

Both QTL analysis and GWAS are appropriate to tag loci conferring a trait with markers, but both approaches are poor tools to analyze complex traits, where a large number of loci contribute to a trait, such as yield and many abiotic stress tolerances (Heffner et al. 2009). Genomic selection became a powerful tool to use molecular markers for selection without associating specific markers to traits. Instead, phenotyping data and high-density genotyping information are used to calculate genomic breeding values for individuals (Heffner et al. 2009). Like that, 1000s of loci along the whole genome are included in the analysis, which reflects the contribution of 1000s of genes to complex traits. The method has been applied in animal breeding with great success (Hayes et al. 2009), is becoming more and more used in plant breeding (Voss-Fels et al. 2018), and has potential also in legume breeding (Mousavi-Derazmahalleh et al. 2019).

8.3.5 Marker-Assisted Selection

In section "Molecular Markers in Plant Breeding," some advantages were highlighted for selecting based on molecular markers that are tightly linked to traits instead of using trait values directly. The main advantages of marker-assisted selection are:

(1) MAS makes selection for traits that are difficult to measure easier
(2) It allows for selection of traits that are expressed during late developmental stages already at the seedling stage
(3) It eliminates environmental variability from the selection
(4) It makes selection of disease-resistant or abiotic stress-tolerant individuals independent of the presence of the biotic or abiotic stimuli (pathogens, pests, vectors, heat, etc.) required for selection
(5) It helps maintaining recessive alleles during backcrossing
(6) It facilitates pyramiding multiple traits, especially pyramiding multiple loci for the same trait.

Marker-assisted selection is designed to maintain introgressed loci in the population, while marker-assisted backcrossing (MABC) helps introgressing loci generally from unadapted material into an elite background. The introgressed fragment, in addition to the target gene, may contain genes that reduce the agronomic performance of the line. This effect is called linkage drag. Therefore, fragments that are as small as possible and contain as little genetic material from the donor line in addition to the target gene are preferred. Marker-assisted selection with markers can help to reduce the linkage drag and accelerate the reestablishment of the recurrent parent. Overall, the efficiency of MABC depends on the kind of the introgressed gene, the recurrent and the donor parent and the population size (Frisch and Melchinger 2005).

MABC combines foreground selection with markers associated with the trait of interest with background selection with markers that pinpoint offspring with maximal recovery of the recurrent parent genotype. The foreground selection monitors presence of the introgressed fragment in the progeny. Marker-assisted foreground selection with co-dominant markers such as SSRs or SNPs that are tightly associated with the trait of interest is particularly practical for traits that are not expressed at the heterozygote stage or are difficult to score. Introgression means that a double recombination occurring on both sides of a target locus has to occur. This event can best be monitored with a marker pair tightly flanking the target gene, and not with a single linked marker. Literature recommends markers to flank the target gene in a distance of maximally 5 cM, but new marker technologies that allow for much greater marker densities enable the choice of more tightly linked marker, to further reduce linkage drag.

Recombinant selection in foreground selection involves identifying backcross progeny with recombination events as near as possible to the target locus, to reduce the size of the donor chromosome segment containing the target locus and reduce linkage drag. As selection is applied on the target locus, there will be less recombination around the donor fragment than for unlinked regions (Hospital et al. Hospital 2001). As double crossover events occurring on both sides of the introgressed fragment are rare, the donor segment can remain very large, even with many backcross generations. The population size for backcrossing has to be adjusted to the crossover probability. The probability of a double crossover can be calculated from the product of the probabilities of a single recombination on both sides of the introgressed fragment. But for close markers, the probability of double crossovers is much lower than the probabilities for single crossover combinations (Young and Tanksley 1989). Producing a very large number of backcross plants would be necessary to achieve recombination on both sides of the gene in one cross. However, instead of working with a very large population, it is advantageous to select in the first backcross generation a single recombinant on one side and then selecting the recombinant on the other side in a second backcross generation (Young and Tanksley 1989). In summary, the distances between the flanking markers and the target gene, the population size during backcrossing, and the number of backcrosses are critical for reducing linkage drag (Hospital 2001).

Background selection in MABC involves selecting backcross progeny with the greatest proportion of the recovered recurrent parent genome using markers that are unlinked to the target locus and can be used to select against the donor genome. Background selection aims at accelerating the recovery of the recurrent genome. Without markers, the reconstitution of the recurrent phenotype, at least to 97%, can be accomplished within four backcross generations (Frisch et al. 1999), but selection for the introgressed trait affects the recovery and increases the required backcrosses by at least one cycle (Frisch et al. 1999). It was proposed to start with a large backcross population to increase the chance to identify an individual that has recovered the recurrent parent genome to an extent as large as possible, and to reduce the population size for the next generations. Simulation studies estimating the trade-offs between population size, MABC efficiency and costs are available and suggest steps to optimize MABC (Ribaut et al. 2002).

For mungbean, marker-assisted backcrossing efforts were not yet reported, but availability of markers associated with important traits makes it likely that this technology will be used also in mungbean improvement.

8.3.6 Pyramiding Multiple Traits in Breeding Lines

Pyramiding is the process of combining several genes together in a single genotype. Conventional breeding also applies gene pyramiding, but usually it is laborious and time consuming to check the results of this approach by phenotypic tests. For example, to improve agronomic properties, breeders combine multiple disease resistances in elite lines. Checking resistance to multiple diseases is laborious and requires multiple testing environments. The efficiency of this process can be enhanced by marker-assisted selection. To increase the durability of disease resistances, breeders pyramid various resistance genes from different sources (Hanson et al. 2016). Using conventional phenotypic selection, identification of stacked resistance genes is only possible when pathogen races that can detect specific resistance genes are available. In contrast, molecular markers greatly facilitate gene pyramiding, as they can be designed to be specific for each single resistance gene. Early selection by molecular markers also helps to keep the breeding populations small during gene pyramiding. However, in mungbean, no marker-assisted gene pyramiding efforts were reported so far.

8.3.7 Genomic Selection

Access to marker resources open up new methods for selecting favorable genotypes, if sufficient phenotypic data of the organism under investigation are available. As outlined in section "Developing markers linked to traits of interest," genomic selection is taking up momentum in crop breeding. It is thought that genomic selection is particularly advantageous for selecting favorable genotypes for complex, multigenic traits. However, the technology requires datasets from different environments and over a number of generations. These sets are not yet available for mungbean, so it will still take some time that this technology can be applied on this crop.

8.3.8 Constraints to Successful Marker-Assisted Selection

Great investments in marker-assisted selection, primary in the private sector, have resulted in several improved varieties for a range of crops including cereals, oil seed crops, cotton, legumes, and vegetables. Naturally, for minor crops such as mungbean, there has been much less investment in generating breeding tools, including genomic resources for breeding. But the available whole genome sequence for mungbean, germplasm panels displaying the diversity of the crop and coordinated breeding activities such as the Australian Centre for International Agricultural Research (ACIAR)-funded International Mungbean Breeding Network (https://www.aciar.gov.au/project/CIM/2014/079) make marker-assisted breeding also accessible for mungbean. Especially, disease and pest resistances are likely to be tackled by marker-assisted breeding in the very near future. Marker-assisted breeding for complex traits such as abiotic stress tolerance in mungbean probably will take longer, as it requires putting in place the phenotypic datasets to make use of molecular breeding for complex traits.

Cost savings compared to classical breeding are often mentioned as advantage of marker-assisted selection. Nevertheless, for some breeding programs, the investment required for marker-assisted selection may still be an issue. There are some early studies reporting several cases where marker-assisted selection was less cost-effective than phenotypic selection (Bohn et al. 2001; Dreher et al. 2003). In the meanwhile, the costs for genotyping have dropped, but the investment for molecular breeding may still be relatively high for small programs working on minor crops. Especially in developing countries, breeder may not have easy access to cost-effective genotyping infrastructure, and low labor costs may make field evaluations cheaper than laboratory work that requires relatively expensive consumables. However, the accelerated release of an improved crop variety may translate into more rapid profits. Therefore, if the additional income generated by improved varieties along the mungbean value chain over time is considered, calculations analyzing the costs and benefit of marker-assisted selection in plant breeding will most probably show that this technique is meaningful, also on mungbean.

Lack of access to molecular markers does not limit marker-assisted breeding in mungbean anymore, as technologies to obtain large numbers of markers at low cost are available. High-quality phenotypic data are being produced and expertise to combine phenotypic and genotypic datasets is available in mungbean breeding teams. Therefore, the first improved mungbean varieties produced by marker-assisted selection are in sight.

References

Ali MZ, Khan M, Rahaman AK, Ahmed M, Ahsan AF (2010) Study on seed quality and performance of some mungbean varieties in Bangladesh. Int J Exp Agric 2:10–15

Barata C, Carena MJ (2006) Classification of North Dakota maize inbred lines into heterotic groups based on molecular and testcross data. Euphytica 151:339–349

Bhagyawant SS (2016) RAPD-SCAR markers: an interface tool for authentication of traits. J Biosci Med 4:1–9

Bhat KV, Lakhanpaul S, Chadha S (2005) Amplified fragment length polymorphism (AFLP) analysis of genetic diversity in Indian mungbean [*Vigna radiata* (L.) Wilczek] cultivars. Indian J Biotechnol. http://nopr.niscair.res.in/handle/123456789/5634

Bisht IS, Mahajan RK, Patel DP (1998) The use of characterisation data to establish the Indian mungbean core collection and assessment of genetic diversity. Genet Resour Crop Evol 45:127–133

Bohn M, Groh S, Khairallah MM, Hoisington DA, Utz HF, Melchinger AE (2001) Re-evaluation of the prospects of marker assisted selection for improving insect resistance against *Diatraea* spp. in tropical maize by cross validation and independent validation. Theor Appl Genet 103:1059–1067

Bonierbale MW, Plaisted RL, Tanksley SD (1988) RFLP maps based on a common set of clones reveal modes of chromosomal evolution in potato and tomato. Genetics 120:1095–1103

Breria CM, Hsieh CH, Yen J-Y, Nair R, Lin C-Y, Huang S-M, Noble TJ, Schafleitner R (2019) Population structure of the world vegetable center mungbean mini core collection and genome-wide association mapping of loci associated with variation of seed coat luster. Trop Plant Biol

Carroll SB (2008) Evo-devo and an expanding evolutionary synthesis: a genetic theory of morphological evolution. Cell 134:25–36

Chaitieng B, Kaga A, Han OK, Wang XW, Wongkaew S, Laosuwan P, Tomooka N, Vaughan DA (2002) Mapping a new source of resistance to powdery mildew in mungbean. Plant Breed 121:521–525

Chankaew S, Somta P, Sorajjapinun W, Srinives P (2011) Quantitative trait loci mapping of *Cercospora* leaf spot resistance in mungbean, *Vigna radiata* (L.) Wilczek. Mol Breed 28:255–264

Chattopadhyay K, Ali MN, Sarkar HK, Mandai N, Bhattacharyya S (2005) Diversity analysis by RAPD and ISSR markers among the selected mungbean [*Vigna radiata* (L.) Wilczek] genotypes. Indian J Gent Plant Breed 65:173–175

Chen HM, Liu CA, Kuo CG, Chien CM, Sun HC, Huang CC, Lin YC, Ku HM (2007) Development of a molecular marker for a bruchid (*Callosobruchus chinensis* L.) resistance gene in mungbean. Euphytica 157:113–122

Chen HM, Ku HM, Schafleitner R, Bains TS, Kuo CG, Liu CA, Nair RM (2013) The major quantitative trait locus for mungbean yellow mosaic Indian virus resistance is tightly linked in repulsion phase to the major bruchid resistance locus in a cross between mungbean [*Vigna radiata* (L.) Wilczek] and its wild relative *Vigna radiata* ssp. *sublobata*. Euphytica 192:205–216

Chen H, Qiao L, Wang L, Wang S, Blair MW, Cheng X (2015a) Assessment of genetic diversity and population structure of mung bean (*Vigna radiata*) germplasm using EST-based and genomic SSR markers. Gene 566(2):175–183

Chen H, Wang L, Wang S, Liu C, Blair MW, Cheng X (2015a). Transcriptome sequencing of mung bean (*Vigna radiata* L.) genes and the identification of EST-SSR markers. PLoS One 10(4):e0120273

Cheng X, Wang S, Wu S, Zhou J, Wang S, Yang Y (2005) Tagging and utilization of bruchid resistance gene using PCR markers in mungbean. Zhongguo nongye kexue 38:1534–1539

Chotechung S, Somta P, Chankaew S, Srinives P, Somta P (2011) Identification of DNA markers associated with bruchid resistance in mungbean. Khon Khan Agri J 39:221–226

Chotechung S, Somta P, Chen J, Yimram T, Chen X, Srinives P (2016) A gene encoding a polygalacturonase–inhibiting protein (PGIP) is a candidate gene for bruchid (*Coleoptera*: *bruchidae*) resistance in mungbean (*Vigna radiata*). Theor Appl Genet 129:1673–1683

Crawford DJ (1990) Plant molecular systematics: macromolecular approaches. Wiley, New York, USA

Da Maia LC, Palmieri DA, De Souza VQ, Kopp MM, de Carvalho FI, Costa de Oliveira A (2008) SSR locator: tool for simple sequence repeat discovery integrated with primer design and PCR simulation. Int J Plant Genomics s.2008:412696

Datta S, Gangwar S, Kumar S, Gupta S, Rai R, Kaashyap M, Singh P, Chaturvedi SK, Singh BB, Nadarajan N (2012) Genetic diversity in selected Indian mungbean [*Vigna radiata* (L.) Wilczek] cultivars using RAPD markers. Am J Plant Sci 8:1085–1091

de Vicente MC, Guzman FA, Engels J, Rao VA (2006) Genetic characterization and its use in decision-making for the conservation of crop germplasm. The role of biotechnology in exploring and protecting agricultural genetic resources, pp 2129–138

Dhole VJ, Reddy KS (2013) Development of a SCAR marker linked with a MYMV resistance gene in mungbean (*Vigna radiata* L. Wilczek). Plant Breed 132:127–132

Dreher K, Khairallah M, Ribaut JM, Morris M (2003) Money matters (I): costs of field and laboratory procedures associated with conventional and marker-assisted maize breeding at CIMMYT. Mol Breed 11:221–234

Edwards KJ, Barker JH, Daly A, Jones C, Karp A (1996) Microsatellite libraries enriched for several microsatellite sequences in plants. Biotechniques 20:758–760

Elbasyoni IS, Lorenz AJ, Guttieri M, Frels K, Baenziger PS, Poland J, Akhunov E (2018) A comparison between genotyping-by-sequencing and array-based scoring of SNPs for genomic prediction accuracy in winter wheat. Plant Sci 270:123–130

Elshire RJ, Glaubitz JC, Sun Q, Poland JA, Kawamoto K, Buckler ES, Mitchell SE (2011) A robust, simple genotyping-by-sequencing (GBS) approach for high diversity species. PLoS One 6(5):e19379

Fan JB, Oliphant A, Shen R, Kermani BG, Garcia F, Gunderson KL, Hansen M, Steemers F, Butler SL, Deloukas P, Galver L (2003) Highly parallel SNP genotyping. In: Cold spring harbor symposia on quantitative biology, vol 68. Cold Spring Harbor Laboratory Press, pp 69–78

Fatokun CA, Menancio-Hautea DI, Danesh D, Young ND (1992) Evidence for orthologous seed weight genes in cowpea and mung bean based on RFLP mapping. Genetics 132:841–846

Feuk L, Marshall CR, Wintle RF, Scherer SW (2006) Structural variants: changing the landscape of chromosomes and design of disease studies. Human Mol Genet 15(suppl_1):R57–R66

Frisch M, Bohn M, Melchinger AE (1999) Comparison of selection strategies for marker-assisted backcrossing of a gene. Crop Sci 39(5):1295–1301

Frisch M, Melchinger A (2005) Selection theory for marker-assisted backcrossing. Genetics 170(2):909–917

Fujii K, Ishimoto M, Kitamura K (1989) Patterns of resistance to bean weevils (*Bruchidae*) in *Vigna radiata-mungo-sublobata* complex inform the breeding of new resistant varieties. Appl Entomol Zool 25:126–132

Gonçalves LS, Rodrigues R, Amaral Júnior AD, Karasawa M, Sudré CP (2008) Comparison of multivariate

statistical algorithms to cluster tomato heirloom accessions. Genet Mol Res 7:1289–1297
Gupta SK, Bansal R, Gopalakrishna T (2014) Development and characterization of genic SSR markers for mungbean (*Vigna radiata* (L.) Wilczek). Euphytica 195(2):245–258
Gupta M, Chyi YS, Romero-Severson J, Owen JL (1994) Amplification of DNA markers from evolutionarily diverse genomes using single primers of simple-sequence repeats. Theor Appl Genet 89:998–1006
Gwag JG, Dixit A, Park YJ, Ma KH, Kwon SJ, Cho GT, Lee GA, Lee SY, Kang HK, Lee SH (2010) Assessment of genetic diversity and population structure in mungbean. Genes Genomics 32:299–308
Hamada H, Kakunaga T (1982) Potential Z-DNA forming sequences are highly dispersed in the human genome. Nature 298(5872):396–398
Hansen TF (2006) The evolution of genetic architecture. Annu Rev Ecol Evol Syst 37:123–157
Hanson P, Lu SF, Wang JF, Chen W, Kenyon L, Tan CW, Tee KL, Wang YY, Hsu YC, Schafleitner R, Ledesma D (2016) Conventional and molecular marker-assisted selection and pyramiding of genes for multiple disease resistance in tomato. Scientia Hort 201:346–354
Hayden MJ, Nguyen TM, Waterman A, McMichael GL, Chalmers KJ (2008) Application of multiplex-ready PCR for fluorescence-based SSR genotyping in barley and wheat. Mol Breed 21:271–281
Hayes BJ, Bowman PJ, Chamberlain AJ, Goddard ME (2009) Invited review: genomic selection in dairy cattle: progress and challenges. J Dairy Sci 92:433–443
Heffner EL, Sorrells ME, Jannink JL (2009) Genomic selection for crop improvement. Crop Sci 49(1):1–2
Hospital F (2001) Size of donor chromosome segments around introgressed loci and reduction of linkage drag in marker-assisted backcross programs. Genetics 158 (3):1363
Humphry ME, Lambrides CJ, Chapman SC, Aitken EA, Imrie BC, Lawn RJ, McIntyre CL, Liu CJ (2005) Relationships between hard-seededness and seed weight in mungbean (*Vigna radiata*) assessed by QTL analysis. Plant Breed 124(3):292–298
Isemura T, Kaga A, Tabata S, Somta P, Srinives P, Shimizu T, Jo U, Vaughan DA, Tomooka N (2012) Construction of a genetic linkage map and genetic analysis of domestication related traits in mungbean (*Vigna radiata*). PLoS One 7:e41304
Islam AS, Blair MW (2018) Molecular characterization of mungbean germplasm from the USDA core collection using newly developed KASP-based SNP markers. Crop Sci 58(4):1659–1670
Jaccoud D, Peng K, Feinstein D, Kilian A (2001) Diversity arrays: a solid state technology for sequence information independent genotyping. Nucleic Acids Res 29:e25
Kaewwongwal A, Chen J, Somta P, Kongjaimun A, Yimram T, Chen X, Srinives P (2017) Novel alleles of two tightly linked genes encoding polygalacturonase–inhibiting proteins (VrPGIP1 and VrPGIP2) associated with the Br locus that confer bruchid (*Callosobruchus* spp.) resistance to mungbean (*Vigna radiata*) accession V2709. Front Plant Sci 8:1692
Kajonphol T, Sangsiri C, Somta P, Toojinda T, Srinives P (2017) SSR map construction and quantitative trait loci (QTL) identification of major agronomic traits in mungbean (*Vigna radiata* (L.) Wilczek). SABRAO J Breed Genet 44(1):71–86
Kang YJ, Kim SK, Kim MY, Lestari P, Kim KH, Ha BK, Jun TH, Hwang WJ, Lee T, Lee J, Shim S (2014) Genome sequence of mungbean and insights into evolution within *Vigna* species. Nat Commun 11:6443
Karthikeyan A, Sudha M, Senthil N, Pandiyan M, Raveendran M, Nagrajan P (2012) Screening and identification of random amplified polymorphic DNA (RAPD) markers linked to mungbean yellow mosaic virus (MYMV) resistance in mungbean (*Vigna radiata* (L.) Wilczek). Arch Phytopath Plant Prot 45:712–716
Kasettranan W, Somta P, Srinives P (2010) Mapping of quantitative trait loci controlling powdery mildew resistance in mungbean (*Vigna radiata* (L.) Wilczek). J Crop Sci Biotechnol 13(3):155–161
Kole C, Panigrahi J (2001) *Vigna glabrescens* is a natural allopolyploid of *V. radiata* and *V. umbellata*: evidences from seed protein electrophoresis. Crop Res 22:121–128
Kumar SV, Tan SG, Quah SC, Yusoff K (2002) Isolation of microsatellite markers in mungbean, *Vigna radiata*. Mol Ecol Notes 2:96–98
Kumar S, Gupta S, Chandra S, Singh BB (2003) How wide is the genetic base of pulse crops? In: National symposium on crop diversification and natural resources management, Kanpur, 20–22 Dec 2003
Lakhanpaul S, Chadha S, Bhat KV (2000) Random amplified polymorphic DNA (RAPD) analysis in Indian mung bean (*Vigna radiata* (L.) Wilczek) cultivars. Genetica 109(3):227–34
Lestari P, Kim SK, Kang YJ, Dewi N, Lee SH (2014) Genetic diversity of mungbean (*Vigna radiata* L.) germplasm in Indonesia. Plant Genet Res 12(S1):S91–S94
Liew M, Pryor R, Palais R, Meadows C, Erali M, Lyon E, Wittwer C (2004) Genotyping of single- nucleotide polymorphisms by high-resolution melting of small amplicons. Clin Chem 50(7):1156–1164
Liu BH (2017) Statistical genomics: linkage, mapping, and QTL analysis. CRC press
Liu HJ, Yan J (2019) Crop genome-wide association study: a harvest of biological relevance. Plant J 97 (1):8–18
Liu MS, Kuo TC, Ko CY, Wu DC, Li KY, Lin WJ, Lin CP, Wang YW, Schafleitner R, Lo HF, Chen CY (2016) Genomic and transcriptomic comparison of nucleotide variations for insights into bruchid

resistance of mungbean (*Vigna radiata* [L.] R. Wilczek). BMC Plant Biol 16:46

Mahmoud SH, Gatehouse JA, Boulter D (1984) Inheritance and mapping of isoenzymes in pea (*Pisum sativum* L.). Theor Appl Genet 68(6):559–66

Menancio-Hautea D, Fatokun CA, Kumar L, Danesh D, Young ND (1993) Comparative genome analysis of mungbean (*Vigna radiata* L. Wilczek) and cowpea (*V. unguiculata* L. Walpers) using RFLP mapping data. Theor Appl Genet 86:797–810

Mendel G, Corcos AF, Monaghan FV (1993) Gregor Mendel's experiments on plant hybrids: a guided study. Rutgers University Press

Michelmore RW, Paran I, Kesseli RV (1991) Identification of markers linked to disease-resistance genes by bulked segregant analysis: a rapid method to detect markers in specific genomic regions by using segregating populations. Proc Natl Acad Sci 88(21):9828–9832

Miyagi M, Humphry M, Ma ZY, Lambrides CJ, Bateson M, Liu CJ (2004) Construction of bacterial artificial chromosome libraries and their application in developing PCR-based markers closely linked to a major locus conditioning bruchid resistance in mungbean (*Vigna radiata* L. Wilczek). Theor Appl Genet 110:151–156

Moe KT, Chung JW, Cho YI, Moon JK, Ku JH, Jung JK, Lee J, Park YJ (2011) Sequence information on simple sequence repeats and single nucleotide polymorphisms through transcriptome analysis of mungbean. J Int Plant Biol 53:63–73

Mohanty JB, Naik BS, Panigrahi J, Kole, C (2011) Identification of mungbean cultivars based on electrophoretic patterns of seed proteins. Crop Res 21 (2):134–138

Moose SP, Mumm RH (2008) Molecular plant breeding as the foundation for 21st century crop improvement. Plant Physiol 147:969–977

Mousavi-Derazmahalleh M, Bayer PE, Hane JK, Valliyodan B, Nguyen HT, Nelson MN, Erskine W, Varshney RK, Papa R, Edwards D (2019) Adapting legume crops to climate change using genomic approaches. Plant, Cell Environ 42(1):6–19

Ng WL, Tan SG (2015) Inter-simple sequence repeat (ISSR) markers: are we doing it right. ASM Sci J 9 (1):30–39

Noble TJ, Tao Y, Mace ES, Williams B, Jordan DR, Douglas CA, Mundree SG (2018) Characterization of linkage disequilibrium and population structure in a mungbean diversity panel. Front Plant Sci 8:2102

Paran I, Michelmore RW (1993) Development of reliable PCR-based markers linked to downy mildew resistance genes in lettuce. Theor Appl Genet 85(8):985–993

Pattnaik J, Kole C (2002) Detection of a protein marker for screening of MYMV resistance mungbean genotype. Indian J Genet Plant Breed 62:77–78

Reflinur R, Lestari P, Lee SH (2017) The potential use of SSR markers to support the morphological identification of Indonesian mungbean varieties. Indonesian J Agric Sci 17(2):65–74

Ribaut JM, Jiang C, Hoisington D (2002) Simulation experiments on efficiencies of gene introgression by backcrossing. Crop Sci 42(2):557–565

Sangiri C, Kaga A, Tomooka N, Vaughan D, Srinives P (2008) Genetic diversity of the mungbean (*Vigna radiata*, Leguminosae) genepool on the basis of microsatellite analysis. Aust J Bot 55:837–847

Santalla M, Power JB, Davey MR (1998) Genetic diversity in mung bean germplasm revealed by RAPD markers. Plant Breed 117:473–478

Sarkar S, Ghosh S, Chatterjee M, Das P, Lahari T, Maji A, Mondal N, Pradhan KK, Bhattacharyya S (2011) Molecular markers linked with bruchid resistance in *Vigna radiata* var. *sublobata* and their validation. J Plant Biochem Biotech 20(2):155–160

Schafleitner R, Nair RM, Rathore A, Wang YW, Lin CY, Chu SH, Lin PY, Chang JC, Ebert AW (2015) The AVRDC–The World Vegetable Center mungbean (*Vigna radiata*) core and mini core collections. BMC Genom 16(1):344

Schafleitner R, Huang SM, Chu SH, Yen JY, Lin CY, Yan MR, Krishnan B, Liu MS, Lo HF, Chen CY, Long-fang OC (2016) Identification of single nucleotide polymorphism markers associated with resistance to bruchids (*Callosobruchus* spp.) in wild mungbean (*Vigna radiata* var. *sublobata*) and cultivated *V. radiata* through genotyping by sequencing and quantitative trait locus analysis. BMC Plant Biol 16 (1):159

Schierwater B, Ender A (1993) Different thermostable DNA polymerases may amplify different RAPD products. Nucleic Acids Res 21:4647

Selvi R, Muthiah AR, Manivannan N, Raveendran TS, Manickam A, Samiyappan R (2006) Tagging of RAPD marker for MYMV resistance in mungbean (*Vigna radiata* (L.) Wilczek). Asian J Plant Sci 5 (2):277–280

Semagn K, Babu R, Hearne S, Olsen M (2014) Single nucleotide polymorphism genotyping using Kompetitive Allele Specific PCR (KASP): overview of the technology and its application in crop improvement. Mol Breed 33(1):1–4

Sharp PJ, Chao S, Desai S, Gale MD (1989) The isolation, characterization and application in the *Triticeae* of a set of wheat RFLP probes identifying each homoeologous chromosome arm. Theor Appl Genet 78(3):342–348

Singh R, van Heusden AW, Yadav RC (2013) A comparative genetic diversity analysis in mungbean (*Vigna radiata* L.) using inter-simple sequence repeat (ISSR) and amplified fragment length polymorphism (AFLP). Afr J Biotechnol 12:6574–6582

Solanki RK, Singh S, Kumar J (2010) Molecular marker assisted testing of hybridity of F_1 plants in lentil. Food Leg 23:21–24

Sompong U, Somta P, Raboy V, Srinives P (2012) Mapping of quantitative trait loci for phytic acid and phosphorus contents in seed and seedling of mungbean (*Vigna radiata* (L.) Wilczek). Breed Sci 62 (1):87–92

Somta P, Seehalak W, Srinives P (2009) Development, characterization and cross-species amplification of mungbean (*Vigna radiata*) genic microsatellite markers. Conserv Genet 10(6):1939

Spooner D, van Treuren R, de Vicente MC (2005) Molecular markers for genebank management. IPGRI technical Bulletin No. 10. International Plant Genetic Resources Institute, Rome, Italy

Springer NM, Ying K, Fu Y, Ji T, Yeh CT, Jia Y, Wu W, Richmond T, Kitzman J, Rosenbaum H, Iniguez AL (2011) Maize inbreds exhibit high levels of copy number variation (CNV) and presence/absence variation (PAV) in genome content. PLoS Genet 5(11): e1000734

Srinives P, Kitsanachandee R, Chalee T, Sommanas W, Chanprame S (2010) Inheritance of resistance to iron deficiency and identification of AFLP markers associated with the resistance in mungbean (*Vigna radiata* (L.) Wilczek). Plant Soil 335:423–437

Steinmetz M, Minard K, Harvath S, McNicholas J, Srenlinger J, Wake C, Long E, Mach B, Hood L (1981) A molecular map of the immune response region of the major histocompatibility complex of the mouse. Nature 300:35–42

Sun L, Cheng XZ, Wang SH, Wang LX, Liu CY, Li ME, Ning XU (2008) Heredity analysis and gene mapping of bruchid resistance of a mungbean cultivar V2709. Agric Sci China 7(6):672–677

Storm N, Darnhofer-Patel B (2003) MALDI-TOF mass spectrometry-based SNP genotyping. In: Single nucleotide polymorphisms. Springer, Totowa, NJ, pp 241–262

Tangphatsornruang S, Somta P, Uthaipaisanwong P, Chanprasert J, Sangsrakru D, Seehalak W, Sommanas W, Tragoonrung S, Srinives P (2009) Characterization of microsatellites and gene contents from genome shotgun sequences of mungbean (*Vigna radiata* (L.) Wilczek). BMC Plant Biol 9(1):137

Tantasawat P, Trongchuen J, Prajongjai T, Thongpae T, Petkhum C, Seehalak W, Machikowa T (2010) Variety identification and genetic relationships of mungbean and blackgram in Thailand based on morphological characters and ISSR analysis. Afr J Biotechnol 9(27):4452–4464

Tapp I, Malmberg L, Rennel E, Wik M, Syvanen AC (2000) Homogeneous scoring of single- nucleotide polymorphisms: comparison of the 5′-nuclease TaqMan® Assay and molecular beacon probes. Biotechniques 28(4):732–739

Thiel T, Kota R, Grosse I, Stein N, Graner A (2004) SNP2CAPS: a SNP and INDEL analysis tool for CAPS marker development. Nucleic Acids Res 32(1):5

Toojinda T, Chunwongse J, Tragoonrung S, Vanavichit A, Srinives P, Sommanus W (2001) Identification of AFLP marker linked to genes controlling iron deficiency tolerance in mungbean by bulked segregant analysis. In: 12. Genetics: Gene Revolution Era, Bangkok (Thailand), 28–30 Mar 2001

Van Inghelandt D, Melchinger AE, Lebreton C, Stich B (2010) Population structure and genetic diversity in a commercial maize breeding program assessed with SSR and SNP markers. Theor Appl Genet 120 (7):1289–1299

Van K, Kang YJ, Han KS, Lee YH, Gwag JG, Moon JK, Lee SH (2013) Genome-wide SNP discovery in mungbean by Illumina HiSeq. Theor Appl Genet 126(8):2017–2027

Varshney RK, Pandey MK, Bohra A, Singh VK, Thudi M, Saxena RK (2018) Toward the sequence-based breeding in legumes in the post-genome sequencing era. Theor Applied Genet 17:1–20

Vos P, Hogers R, Bleeker M, Reijans M, Lee TV, Hornes M, Friters A, Pot J, Paleman J, Kuiper M, Zabeau M (1995) AFLP: a new technique for DNA fingerprinting. Nucleic Acid Res 23(21):4407–4414

Voss-Fels KP, Cooper M, Hayes BJ (2018) Accelerating crop genetic gains with genomic selection. Theor Appl Genet 19:1–8

Wang X, Wang L (2016) GMATA: an integrated software package for genome-scale SSR mining, marker development and viewing. Front Plant Sci 7:1350

Williams RC (1989) Restriction fragment length polymorphism (RFLP). Am J Phys Anthropol 32 (S10):159–184

Yashitola J, Thirumurugan T, Sundaram RM, Naseerullah MK, Ramesha MS, Sarma NP, Sonti RV (2002) Assessment of purity of rice hybrids using microsatellite and STS markers. Crop Sci 42(4):1369–1373

Young ND, Danesh D, Menancio-Hautea D, Kumar L (1993) Mapping oligogenic resistance to powdery mildew in mungbean with RFLPs. Theor Appl Genet 87(1–2):243–249

Young ND, Kumar L, Menancio-Hautea D, Danesh D, Talekar NS, Shanmugasundarum S, Kim DH (1992) RFLP mapping of a major bruchid resistance gene in mungbean (*Vigna radiata*, L. Wilczek). Theor Appl Genet 84(7–8):839–844

Young ND, Tanksley SD (1989) RFLP analysis of the size of chromosomal segments retained around the tm-2 locus of tomato during backcross breeding. Theor Appl Genet 77:353–359

Zagorski N (2006) Profile of Alec J. Jeffreys. Proceedings of the National Academy of Sciences of the United States of America, vol 103(24): 8918–8920

9 Mungbean Genome and Synteny with Other Genomes

Yang Jae Kang and Jungmin Ha

Abstract

Mungbean (*Vigna radiata* (L.) R. Wilczek var. *radiata*) is an important source of protein and carbohydrate in Asia. Despite its importance in the diet of the world's populace, genetic and genomic information of mungbean had been scarce compared to other legumes such as soybean or chickpea. The publication of mungbean reference genome sequence in 2014 has allowed diverse genetic and genomic studies in mungbean and its close relative legume species. The genome of adzuki bean (*Vigna angularis*) has been sequenced and assembled using the synteny relationship with mungbean genome and genetic information in soybean has been transferred to mungbean genome through translational genomics approach. Within a relatively short period time, a large amount of genetic and genomic resources has been accumulated in mungbean. To make use of the information and develop genomics-based breeding programs, it is essential to construct a data hub to share the information. Strong genomic resources and databases will accelerate mungeban breeding programs in the near future.

Y. J. Kang
Division of Applied Life Science Department, Gyeongsang National University, PMBBRC, Jinju 52828, Republic of Korea

Y. J. Kang
Division of Life Science Department, Gyeongsang National University, Jinju 52828, Republic of Korea

J. Ha
Department of Plant Science and Research Institute of Agriculture and Life Sciences, Seoul National University, Seoul 08826, Republic of Korea

J. Ha (✉)
Plant Genomics and Breeding Institute, Seoul National University, Seoul 08826, Republic of Korea
e-mail: saga16@snu.ac.kr

9.1 Introduction

Mungbean (*Vigna radiata* (L.) R. Wilczek var. *radiata*) is an important source of starch, protein, vitamin C, and isoflavonoid (vitexin), and plays a key role in heat alleviation (Cao et al. 2011). Despite its nutritional importance, until recently, mungbean was considered an orphan crop, with scarce genetic resources for marker-assisted breeding. The mungbean genome project boosted genetic and genomic resources and facilitated the development of breeding resources such as molecular markers, genetic maps, gene catalogs, DNA methylation, and genomic resources (Kang et al. 2014, 2017). The mungbean genome project also helped in the identification of genomic traces of domestication and stretches of genomic DNA sequences (contigs) for translational comparative genomics with a well-studied legume species, soybean (*Glycine max*). With the state-of-the-art sequencing technologies, mungbean, a once orphan species, has become one of

R. M. Nair et al. (eds.), *The Mungbean Genome*, Compendium of Plant Genomes,
https://doi.org/10.1007/978-3-030-20008-4_9

the reference legume species with rich genomic resources. After publication of the mungbean genome and resequencing efforts, genome sequences of other *Vigna* species, including adzuki bean (*Vigna angularis*) and cowpea (*Vigna unguiculata*), have been published (Kang et al. 2015; Muñoz-Amatriaín et al. 2017). *Vigna* species are now considered an important model for legume genomics. Moreover, advances in long-read sequencing technology continue to enhance the completeness of reference genomes, which facilitates the elucidation of unsequenced genomics regions and integration of singleton contigs into pseudomolecules (Sakai et al. 2015). In this chapter, we provide an overview of the current status of molecular breeding efforts with the utilization of mungbean genomic information, and summarize current genomic resources available for mungbean breeding. We also review the status of the reference genome update and curations of reference gene contents using additional data and algorithms. Additionally, we summarize data retrieved from whole genome resequencing projects of a collection of wild and cultivated mungbean genotypes for marker development. Moreover, we discuss the practical applications of translational comparative genomics of legume species ranging from soybean to mungbean using collinearity information of genes between different species.

9.2 Status of Mungbean Genome Assembly

The reference genome of mungbean contains 11 pseudochromosomes (Table 9.1). The total length of the whole genome sequence of mungbean is 463 Mb, including pseudomolecules (333 Mb). Genetic markers narrow down the quantitative trait locus (QTL) region to a gene responsible for a specific trait. Considering that fine mapping efforts using genetic markers rely highly on the gene content in continuous genomic sequences, it is necessary to minimize the number of unmapped scaffolds. The current mungbean genome assembly possesses approximately 130 Mb of unmapped scaffolds, thus highlighting the need for long-read sequencing technology and a high-resolution scaffolding strategy, such as Hi-C scaffolding (Burton et al. 2013). Moreover, the ratio of the number of transcripts to genes in mungbean is relatively low compared with that in the *Arabidopsis thaliana* gene model (1.7) (Cheng et al. 2017). This further emphasizes the need for RNA sequencing (RNAseq) using recent sequencing platforms such as PacBio Iso-Seq. Fortunately, the research group that published the first draft of the mungbean genome continues to update the mungbean genome sequence using the third-generation genome sequencer PacBio. The new mungbean genome assembly shows an N50

Table 9.1 Current status of mungbean reference genome

Parameter	Value
Total length of whole genome sequence	463,085,359 bp
Total length of pseudomolecules	333,308,464 bp
N50 value of whole genome sequence	25,360,630
N50 value of pseudochromosomes	37,180,910
Number of pseudomolecules	11
Number of unmapped scaffolds	2486
Number of Ns	33,558,739
Number of genes	22,368
Number of transcripts	23,181

value of 5 Mb, before the construction of pseudochromosomes, with only 0.7% Ns. Sequencing with third-generation technology has reduced the total number of scaffolds from ~2400 to 330; this would minimize the number of unmapped scaffolds.

9.3 Improvement of the Mungbean Gene Model

In the new mungbean genome assembly, the gene model has been updated for the number of genes (30,882) and mRNAs (30,958) using previously generated RNAseq data from various tissues, including leaf, root, flower, and pod, which were used for the first version of the gene model prediction. Although the number of genes has increased because of gap ('N') filling by long reads, the number of mRNAs is still not sufficient to increase the ratio of the number of mRNAs to genes (Cheng et al. 2017). Additional mRNAs can be identified if more RNAseq data are collected from several different tissues of plants grown under diverse conditions.

After publication of the mungbean draft genome, RNAseq analysis of mungbean has yielded 200 Gb of data. These RNAseq experiments have been performed mostly using Illumina platforms, and data have been deposited in the Sequence Read Archive (SRA) database of the National Center for Biotechnology Information (NCBI) (Table 9.2). These RNAseq analyses have been conducted with the aim to understand leaf development (Jiao et al. 2019), desiccation tolerance (Tian et al. 2016), adventitious rooting (Li et al. 2015), bruchid resistance (Liu et al. 2016), and development of expressed sequence

Table 9.2 Mungbean RNAseq data deposited in the NCBI sequence read archive (SRA) database

SRA_Study	Run	Mbp	Tissue	Platform
SRP043316	SRR1407784	5021	Root, stem, leaf	Illumina HiSeq 2000
SRP043316	SRR1412914	4930	Root, stem, leaf	Illumina HiSeq 2000
SRP049538	SRR1640452	2773	Root, stem, leaf	Illumina Genome Analyzer IIx
SRP049538	SRR1642852	2668	Root, stem, leaf	Illumina Genome Analyzer IIx
SRP049538	SRR1642854	2663	Root, stem, leaf	Illumina Genome Analyzer IIx
SRP055970	SRR2177452	8430	Flower	Illumina HiSeq 2000
SRP055970	SRR2177454	7703	Pod	Illumina HiSeq 2000
SRP077637	SRR3735179	14,562	Seed	Illumina HiSeq 2500
SRP077637	SRR3735193	15,340	Seed	Illumina HiSeq 2500
SRP077637	SRR3735547	14,985	Seed	Illumina HiSeq 2500
SRP077637	SRR3735572	15,436	Seed	Illumina HiSeq 2500
SRP077637	SRR3735589	14,526	Seed	Illumina HiSeq 2500
SRP077637	SRR3735674	14,552	Seed	Illumina HiSeq 2500
SRP077637	SRR3735739	16,193	Seed	Illumina HiSeq 2500
SRP077637	SRR3735764	17,140	Seed	Illumina HiSeq 2500
SRP110723	SRR5768043	5615	Shoot apices	HiSeq X Ten
SRP110723	SRR5768042	6171	Shoot apices	HiSeq X Ten
SRP110723	SRR5768041	7039	Shoot apices	HiSeq X Ten
SRP110723	SRR5768040	6129	Shoot apices	HiSeq X Ten
SRP110742	SRR5769684	5190	7 mm bud	HiSeq X Ten
SRP110742	SRR5769683	5960	7 mm bud	HiSeq X Ten
SRP047207	SRR1653637	7020	Hypocotyl	Illumina HiSeq 2000

tag-simple sequence repeat (EST-SSR) markers (Chen et al. 2015, 2016). Incorporation of these RNAseq data in the new mungbean gene model would enrich the mRNA content. Once the near-complete mRNA reference set is constructed, a transcriptome atlas of mungbean can be built on the reference, which would assist mungbean researchers and breeders to understand the behavior of candidate genes under specific conditions.

9.4 DNA Methylation Profile of the Mungbean Genome

The majority of DNA methylation occurs on cytosine residues and affects gene transcription and consequently phenotype. In soybean, a previous study showed that epigenetic profiles are inherited as genetic variation, suggesting that phenotypic polymorphisms can be explained by epigenetic variation (Schmitz et al. 2013). Seed development is controlled not only by the expression of a combination of several genes but also by the dynamics of DNA methylation (An et al. 2017). Moreover, DNA methylation has shaped the genetic/genomic selection of soybean during domestication (Shen et al. 2018).

In mungbean, the genome-wide DNA methylation profile has been investigated using bisulfite sequencing (BSseq) in two cultivars, Sunhwanogdu (VC1973A) and Kyung-Ki Jaerae #5 (V2974) (Kang et al. 2017). DNA methylation patterns of cytosine residues in three genomic sequence contexts, including CG, CHG, and CHH, revealed that CHH methylation (mCHH) in mungbean is higher than in other plant species such as Arabidopsis and soybean. Interestingly, increased mCHH sites are also observed in common bean (*Phaseolus vulgaris*), implicating that a large CHH methylation event occurred in the genome of the common ancestor of *Phaseolus* and *Vigna* species.

Comparison of DNA methylation patterns between the mungbean cultivars VC1973A and V2974 revealed polymorphic sites of DNA methylation. Among the single nucleotide polymorphisms (SNPs) between VC1973A and V2974, SNPs that cause nucleotide changes from/to CG, CHG, and CHH showed different methylation patterns; SNPs resulting in changes from/to CHH showed interchange of methylation and un-methylation of the cytosine residue (Kang et al. 2017). Thus, these SNPs, with both DNA and DNA methylation polymorphisms, may affect plant phenotype. Hence, for breeding purposes, SNPs in CG, CHG, and CHH contexts should be considered equally as important as SNPs in coding sequences.

9.5 Genomics of Mungbean Germplasm Collection

Genomic diversity among germplasm is essential for breeding, as it facilitates the selection of parental lines with maximum allelic diversity. Although genomics technologies have been developed at an exponential rate, we still cannot guarantee the phenotypic results based on allelic variation among genotypes and environmental conditions during plant growth. However, we can obtain the whole genome sequence using next-generation sequencing (NGS) technology, and select parental lines for the development of breeding populations with diverse alleles. These breeding populations can be used for genome-wide association study (GWAS) or QTL analysis for the identification of loci associated with specific phenotypes. To secure a wide range of alleles, collections of mungbean genotypes from all over the world are being developed and reported. A total of 83 mungbean accessions were collected from Indonesia and evaluated using 29 polymorphic SSR markers (Lestari et al. 2014). Similarly, 94 mungbean cultivars from the USDA germplasm center originating in 27 countries were reported and evaluated using 20 SNP markers (Islam and Blair 2018). Additional 415 mungbean cultivars, including 189 wild and 11 intermediate accessions, were collected from 29 countries and evaluated using 19 SSR markers (Sangiri et al. 2008). Moreover, AVRDC-The World Vegetable Center maintains 6700 mungbean accessions; it also developed 1481 core collections that are accessible to breeders, and 289 accessions of the mini core set were evaluated using 20 SSR markers (Schafleitner et al. 2015).

These collections of mungbean accessions would gradually be genotyped using resequencing approaches or other cheaper genotyping platforms, such as genotype-by-sequencing (GBS) and SNP chip; the status of mungbean resequencing will be revisited in another chapter of this book. The resulting genome sequences would be assembled into reference genome sequences, ultimately constructing the mungbean allele database, which will assist breeders in the selection of parental lines with maximum allelic variation at loci studied in model plants such as Arabidopsis and soybean.

9.6 Listing Functional Loci in the Mungbean Genome by Translational Genomics

To utilize allele information obtained from resequencing projects, previous knowledge of genes/loci is highly valuable. Comparative genomics plays an important role in knowledge transfer and enables the collection of information on loci across different species. The legume family has very rich resources for comparative genomics, as it is one of most intensively sequenced plant families. To date, nine legume species have been sequenced including soybean, medicago, lotus, ground nut, pigeon pea, chickpea, cowpea, mungbean, and adzuki bean (Sato et al. 2008; Schmutz et al. 2010; Young et al. 2011; Varshney et al. 2012, 2013; Kang et al. 2014, 2015; Bertioli et al. 2015; Muñoz-Amatriaín et al. 2017). Soybean, the model legume species, has a nearly complete reference genome sequence and rich genomic resources such as genotyping platforms, breeding populations for genetic/functional marker identification, and known genes/QTLs underlying agricultural traits. To advance from conventional to genomics-based breeding in mungbean, transferring the accumulated knowledge from soybean to mungbean is key.

The transfer of knowledge from soybean to other legume species can be achieved using synteny-based comparative genomics (Paterson et al. 2010). Synteny indicates genetic colinearity, where a stretch of genes is highly conserved between genomic regions of different species, indicating a shared common ancestor as well as a strong genomic relationship among species. Analysis of the relationship between genomes of soybean and other legumes provides general information for QTLs in newly sequenced legume genomes. Functional conservation of loci in syntenic regions has been reported in studies on soybean allotetraploidy. Comparison of *Rxp* loci, controlling bacterial leaf pustule resistance, between soybean and *Medicago truncatula* and among soybean allopolyploidy genomes based on synteny showed that the multiple *Rxp* QTLs are syntenic to each other in different soybean linkage groups (Do Kim et al. 2009). Mungbean and soybean genomes exhibit high synteny. The seed size/germination and bruchid resistance QTL region in mungbean genome are syntenic with the soybean genomic region containing SSR markers linked to seed weight and nematode resistance (Kang et al. 2014).

In addition, well-studied gene families, such as nucleotide-binding site leucine-rich repeat (NBS-LRR) and WRKY transcription factor families involved in disease resistance, can be used for the annotation of loci. The *NBS-LRR* genes are known to contribute to the perception of pathogens by plants (DeYoung and Innes 2006). The number of *NBS-LRR* genes shows a significant correlation with the number of disease resistance QTLs in the soybean genome; moreover, *NBS-LRR* genes and disease resistance QTLs exist in close proximity to each other (Kang et al. 2012). These data suggest that diversification of *NBS-LRR* genes is needed for breeding disease-resistant soybean genotypes. WRKY transcription factors are key contributors to disease resistance, and their level of activity is a primary target of genomics-based breeding for trait improvement (Srivastava et al. 2018).

9.7 Perspectives

Within a relatively short period of time, a large amount of genomic resources of mungbean have been accumulated, and data from ongoing projects are currently pending submission in sequence databases. To develop genomics-based breeding pipelines, improvement of the reference genome and gene model, continuous collection and resequencing of germplasm, and omics approaches to broaden observable data space are needed. To achieve this goal, it is essential to construct a data hub where mungbean breeders can: (1) share phenotype/genotype information of germplasm, (2) estimate breeding values calibrated with genomic/environmental factors, and (3) annotate genomic regions by traits using translational genomics. The data hub must have user friendly interfaces, easy and fair sharing policies, sustainable computation resources for hosting, and manpower for maintenance. Strong genomic resources and databases will lead to the acceleration of mungbean breeding programs in the near future.

References

An YC, Goettel W, Han Q et al (2017) Dynamic changes of genome-wide dna methylation during soybean seed development. Sci Rep 7:12263

Bertioli DJ, Cannon SB, Froenicke L et al (2015) The genome sequences of *Arachis duranensis* and *Arachis ipaensis*, the diploid ancestors of cultivated peanut. Nat Genet 47:438

Burton JN, Adey A, Patwardhan RP et al (2013) Chromosome-scale scaffolding of de novo genome assemblies based on chromatin interactions. Nat Biotechnol 31:1119

Cao D, Li H, Yi J et al (2011) Antioxidant properties of the mung bean flavonoids on alleviating heat stress. PLoS ONE 6:e21071

Chen H, Wang L, Wang S et al (2015) Transcriptome sequencing of mung bean (*Vigna radiata* L.) genes and the identification of EST-SSR markers. PLoS One 10:e0120273

Chen J, Somta P, Chen X et al (2016) Gene mapping of a mutant mungbean (*Vigna radiata* L.) using new molecular markers suggests a gene encoding a yuc4-like protein regulates the chasmogamous flower trait. Front Plant Sci 7:830

Cheng C, Krishnakumar V, Chan AP et al (2017) Araport11: a complete reannotation of the *Arabidopsis thaliana* reference genome. Plant J 89:789–804

DeYoung BJ, Innes RW (2006) Plant NBS-LRR proteins in pathogen sensing and host defense. Nat Immunol 7:1243

Do Kim K, Shin JH, Van K et al (2009) Dynamic rearrangements determine genome organization and useful traits in soybean. Plant Physiol 151:1066–1076

Islam A, Blair MW (2018) Molecular Characterization of mung bean Germplasm from the USDA Core Collection Using Newly Developed KASP-based SNP Markers. Crop Sci

Jiao K, Li X, Guo Y et al (2019) Regulation of compound leaf development in mungbean (*Vigna radiata* L.) by CUP-SHAPED COTYLEDON/NO APICAL MERIS TEM (CUC/NAM) gene. Planta 1–10

Kang YJ, Bae A, Shim S et al (2017) Genome-wide DNA methylation profile in mungbean. Sci Rep 7:40503

Kang YJ, Kim KH, Shim S et al (2012) Genome-wide mapping of NBS-LRR genes and their association with disease resistance in soybean. BMC Plant Biol 12:139

Kang YJ, Kim SK, Kim MY et al (2014) Genome sequence of mungbean and insights into evolution within Vigna species. Nat Commun 5:5443

Kang YJ, Satyawan D, Shim S et al (2015) Draft genome sequence of adzuki bean. Vigna angularis. Sci Rep 5:8069. https://doi.org/10.1038/srep08069

Lestari P, Kim SK, Kang YJ et al (2014) Genetic diversity of mungbean (*Vigna radiata* L.) germplasm in Indonesia. Plant Genet Resour 12:S91–S94

Li S-W, Shi R-F, Leng Y (2015) De novo characterization of the mung bean transcriptome and transcriptomic analysis of adventitious rooting in seedlings using RNA-Seq. PLoS ONE 10:e0132969

Liu M-S, Kuo TC-Y, Ko C-Y et al (2016) Genomic and transcriptomic comparison of nucleotide variations for insights into bruchid resistance of mungbean (*Vigna radiata* [L.] R. Wilczek). BMC Plant Biol 16:46

Muñoz-Amatriaín M, Mirebrahim H, Xu P et al (2017) Genome resources for climate-resilient cowpea, an essential crop for food security. Plant J 89:1042–1054

Paterson AH, Freeling M, Tang H, Wang X (2010) Insights from the comparison of plant genome sequences. Annu Rev Plant Biol 61:349–372

Sakai H, Naito K, Ogiso-Tanaka E et al (2015) The power of single molecule real-time sequencing technology in the de novo assembly of a eukaryotic genome. Sci Rep 5:16780. https://doi.org/10.1038/srep16780

Sangiri C, Kaga A, Tomooka N et al (2008) Genetic diversity of the mungbean (*Vigna radiata*, Leguminosae) genepool on the basis of microsatellite analysis. Aust J Bot 55:837–847

Sato S, Nakamura Y, Kaneko T et al (2008) Genome structure of the legume, *Lotus japonicus*. DNA Res 15:227–239

Schafleitner R, Nair RM, Rathore A et al (2015) The AVRDC–The World Vegetable Center mungbean (*Vigna radiata*) core and mini core collections. BMC Genom 16:344

Schmitz RJ, He Y, Valdés-López O et al (2013) Epigenome-wide inheritance of cytosine methylation variants in a recombinant inbred population. Genome Res

Schmutz J, Cannon SB, Schlueter J et al (2010) Genome sequence of the palaeopolyploid soybean. Nature 463:178–183. https://doi.org/10.1038/nature08670

Shen Y, Zhang J, Liu Y et al (2018) DNA methylation footprints during soybean domestication and improvement. Genome Biol 19:128

Srivastava R, Kumar S, Kobayashi Y et al (2018) Comparative genome-wide analysis of WRKY transcription factors in two Asian legume crops: Adzuki bean and Mung bean. Sci Rep 8:16971

Tian X, Li S, Liu Y, Liu X (2016) Transcriptomic profiling reveals metabolic and regulatory pathways in the desiccation tolerance of Mungbean (*Vigna radiata* [L.] R. Wilczek). Front Plant Sci 7:1921

Varshney RK, Chen W, Li Y et al (2012) Draft genome sequence of pigeonpea (*Cajanus cajan*), an orphan legume crop of resource-poor farmers. Nat Biotechnol 30:83

Varshney RK, Song C, Saxena RK et al (2013) Draft genome sequence of chickpea (*Cicer arietinum*) provides a resource for trait improvement. Nat Biotechnol 31:240

Young ND, Debellé F, Oldroyd GE et al (2011) The Medicago genome provides insight into the evolution of rhizobial symbioses. Nature 480:520

Genomic Approaches to Biotic Stresses

10

Kularb Laosatit, Prakit Somta, Xin Chen
and Peerasak Srinives

Abstract

Average seed yield of mungbean (*Vigna radiata*) grown in major growing countries is low, being only about one-third to one-fourth of its yield potential. A main factor causing the low yield in mungbean is biotic stresses (disease infection and insect infestation) that happen at all stages of plant growth and development and after harvest. Common and important diseases and insect pests of mungbean include powdery mildew, Cercospora leaf spot, yellow mosaic virus, bruchids and pod sucking bugs. Employing host plant resistance is the best way to manage the diseases and insect pests. However, progress in the development of new mungbean cultivar(s) with the biotic resistance is slow due to bottleneck in evaluation for the resistance which is environmental-dependent or time-consuming, although germplasm with immune or highly or moderately resistance for these biotic stresses is available and the genetics of the resistance appears to be simple. Genomic approaches, especially gene mapping and marker-assisted selection, are promising in the acceleration of cultivar development for biotic stress resistance in mungbean. Mungbean is a slow runner in genomics research, although it is among the forefront crops targeted for genome analysis at the beginning of the crop genomics era; e.g., powdery mildew resistance and bruchid resistance in mungbean are among the plant diseases and insects being investigated nearly 30 years ago. However, the recent release of a reference genome sequence of mungbean and current advanced sequencing technology has enabled fast and efficient DNA marker development, fine-mapping and identification of candidate gene(s) for the biotic resistance in mungbean possible. This chapter covers past, present and future research on molecular and genomics approaches to biotic stresses for mungbean genetic improvement.

10.1 Introduction

Mungbean (*Vigna radiata* (L.) R. Wilczek var. *radiata*) is an important legume crop of Asia with the production area of more than 6 million ha, playing its role as a cash crop for farmers and as a source of protein and carbohydrate for consumers. India is the largest producer of

K. Laosatit · P. Somta (✉) · P. Srinives
Department of Agronomy, Faculty of Agriculture at Kamphaeng Saen, Kasetsart University, Kamphaeng Saen Campus, Kamphaeng Saen, Nakhon Pathom 73140, Thailand
e-mail: agrpks@ku.ac.th

X. Chen
Institute of Industrial Crops, Jiangsu Academy of Agricultural Sciences, Nanjing 210014, Jiangsu, China

P. Srinives
The Royal Society of Thailand, Sanam Suea Pa, Dusit, Bangkok 10300, Thailand

R. M. Nair et al. (eds.), *The Mungbean Genome*, Compendium of Plant Genomes,
https://doi.org/10.1007/978-3-030-20008-4_10

mungbean followed by Myanmar and China. Mungbean is the third most important pulse crop in India, after chickpea (*Cicer arietinum* L.) and pigeon pea (*Cajanus cajan* L.), which it accounts for 3.4 million ha or 15% of the total legume area sown (Nair et al. 2012). Mungbean is principally grown after rice, wheat and maize. Trends on the consumption demand and production of the mungbean are increasing. This is reflected by the increasing production area of the mungbean in Australia which is a leading mungbean seed exporter. The production area in this country increased from about 1000 ha in the 1970s to 125,000 ha in 2015–2016 (Clarry 2016).

Abiotic and biotic stresses cause significant decline in mungbean yield. Among the biotic stresses, diseases and insect pests are major factors. The disease pathogens and insect pests can infect or infest mungbean plants at all growing stages, from emerging, seedling, vegetative and reproductive stages causing substantial damage leading to yield loss or complete failure of the crop (Pandey et al. 2018). Common diseases of mungbean include powdery mildew, Cercospora leaf spot, yellow mosaic virus, anthracnose, Fusarium wilt, root rot, charcoal rot/dry root rot, bacterial leaf blight. Fungal diseases alone accounted for up to 40–60% yield loss in mungbean (Pandey et al. 2018). However, currently, powdery mildew, Cercospora leaf spot and yellow mosaic virus diseases are the most common and economically important. Important insect pests of mungbean include whitefly (*Besimi tabaci* Gennadius), bean fly (stem borer) (*Ophiomyia phaseoli* Tryon and *Melanagromyza sojae* (Zehntner), armyworms (*Spodoptera exigua* Hübner and *Spodoptera litura* Fabricius), bean pod borers (*Helicoverpa armigera* Hübner and *Maruca vitrata* Fabricius) and bruchids (*Callosobruchus* spp.) are serious pests worldwide in the major production areas of mungbean. Breeding for resistance appears to be the best approach to reduce yield loss due to disease and insect pest infestation in mungbean. Conventional plant breeding despite its limitations has contributed to significant improvement in yield and disease and insect resistance in mungbean (Fernandez and Shanmugasundaram 1988). The conventional breeding is also important in conserving genetic resources and creating genetic variation by hybridization between contrasting genotypes and by induced mutations.

Molecular breeding through marker-assisted selection is a promising approach to improve speed, efficiency and precision of conventional plant breeding for the development of new crop cultivars with improved traits, including resistance to biotic and abiotic stresses (Collard and Mackill 2008; Das et al. 2017). Genetic linkage map and markers associated with quantitative trait locus (QTL) are indispensable for marker-assisted selection (Collard and Mackill 2008). Mungbean is a slow runner in genomics research, although it is among the forefront crops targeted for genome analysis nearly 30 years ago (Young et al. 1992, 1993). Until 2012, the first genetic linkage map of mungbean with 11 haploid chromosomes resolved was developed using DNA markers from various legume species (Isemura et al. 2012). Then, with the advance in DNA sequencing technology, a reference genome of mungbean became available in 2014 (Kang et al. 2014). This reference genome appears to be useful for gene/QTL mapping for biotic stress resistance in mungbean. In this chapter, we present a review of genome mapping of the resistance to biotic stresses including powdery mildew disease, Cercospora leaf spot disease, yellow mosaic virus disease, stored insects and pod sucking bugs in the mungbean.

10.2 Powdery Mildew Disease

Powdery mildew disease caused by the ascomycete fungus *Erysiphe polygoni* D.C. of the order *Erysiphales* is one of the most widespread and serious foliar diseases of mungbean. Besides, *E. polygoni*, *Sphaerotheca phaseoli* (Z. Y. Zhao) U. Braun & S. Takam also cause the powdery mildew in mungbean (Khajudparn et al. 2007). The disease occurs throughout the mungbean production areas of Asia. *E. polygoni* is an obligate fungus infecting mungbean leaves, prevailing in cool dry season and favored by cloudy weather. Low temperatures of 22–26 °C and relative humidity of 80–90% are suitable for disease development (Grewal 1978). A white-gray powdery mildew appears first in circular patches, but later spreads over the surface of the leaves, stem and pods. Yield loss due to the disease was reported to be 20–40% at the reproductive stages (Fernandez and Shanmugasundaram 1988; Khunti et al. 2002; Mandhare and Suryawanshi 2008), but the damage can be more serious when the disease starts at the seedling stages (Reddy et al. 1994).

Although resistance to powdery mildew in mungbean can be easily screened for in either field or greenhouse condition, an efficient method is necessary to identify the resistant genotype(s). A simple and reliable method for assessing resistance to the disease using excised leaves has been developed (Reddy et al. 1987). In this method, the third fully expanded trifoliate leaves (from the base) at 21–25-day-old plants are used. The leaves are excised a little above the pulvinus and cultured in trays filled with tap water under the temperature of 21 ± 1 °C and photoperiod of 12 h/day. After culturing for 9–12 days, the leaves with substantial root development are sprayed with conidial suspension of the mildew fungi. In the susceptible germplasm, the powdery mildew symptom starts at 8–10 days after inoculation and becomes very clear at 20 days after inoculation. This screening method gives the same result with field screening.

10.2.1 Sources of the Resistance to Powdery Mildew

Several sources of resistance to powdery mildew caused by *E. polygoni* in mungbean have been reported, including ML-3, ML-5, M0231, M0613, M1936, IC669-3, PLM187, PLM190, PLM224, PLM731, PLM741, PLM822, PLM857, PLM863, PLM873, PLM944, PLM1060, LM211, RUM-1, RUM-5, RUM-7, RUM-11, RUM-20, RUM-22, RUM-33, OB 24-1 (dull yellow), Mulmarada, TJM-3, TM-96-2, S-158-18 and ATF 3640 (Fernandcz and Shanmugasundaram 1988; Reddy et al. 1987, Hartman et al. 1993; Humphry et al. 2003; Reddy 2009). In contrast, only three germplasms, V4718 (PLM945), V4758 and V4785, have been identified for resistance to the disease caused by *S. phaseoli* (Khajudparn et al. 2007). All these resistant accessions are originated from India. Genetic control of the resistance in some germplasm has been reported as being controlled by one or two genes depending on sources of the resistance. The resistance in ML-3 and ML-5 is conditioned by a single dominant gene (AVRDC 1978). Similarly, the resistance in V4718, V4758, and V4785 is each governed by a single gene, but they are non-allelic (Khajudparn et al. 2007). Reddy et al. (1994) demonstrated that the resistance in Raipur Utera Mung (RUM) lines is controlled by two dominant genes, designated as *Pm1* and *Pm2*. Mungbean plant possessing both *Pm1* and *Pm2* is immune to the fungus that possessing *Pm1* is highly resistance, that possessing *Pm2* is moderately resistance, while lacking both *Pm1* and *Pm2* is susceptible to the fungus. In addition, Reddy (2009) found that the resistance in Mulmarada mungbean is controlled by a single dominant gene which is non-allelic to those in RUM lines, thus designated as *Pm3*. By using quantitative genetic analysis, Gawande and Patil (2003) reported that the resistance in TARM18 is governed by more than one gene with both additive and dominance gene actions. Sorajjapinun et al. (2005) and Kasettranan et al. (2009) demonstrated that the resistance in VC6468-11-1A, which derived its resistance from both ML-3 and ML-5, is predominantly controlled by additive gene action, and the resistance is highly affected by environments.

10.2.2 Genome Analysis of Powdery Mildew Resistance

Several mungbean varieties with the powdery mildew resistance have been developed and released to the farmers through conventional breeding programs. However, since the occurrence of the disease is seasonal dependent, the progress in breeding for the resistance is slow. Marker-assisted selection for the resistance can overcome this problem. In fact, although mungbean is an orphan crop, it has been well subjected to genome research and powdery mildew resistance in mungbean was among the plant disease being investigated at the beginning of the crop genomics era (Young et al. 1993). Quantitative trait locus/loci (QTL) controlling powdery mildew resistance has been conducted using different resistance sources (Table 10.1). An analysis in the partial resistance breeding line VC3890A that obtained its resistance from ML-3 and ML-5 was mapped using restriction fragment length polymorphism (RFLP) markers (Young et al. 1993). The resistance was conditioned by three QTLs on linkage groups (LG) 3, 7 and 8. These QTLs together accounted for a maximum of 58% of the disease variation. Two QTLs were associated with the resistance at 65 days after planting (DAP), while one QTL was associated with the resistance at 85 DAP. By using VC1210A that derived the resistance from ML-3, Chaitieng et al. (2002) employed a single marker analysis to reveal two RFLP probes associated with the resistance, but an interval mapping using 96 RFLP markers failed to identify the QTL. They subsequently used bulked segregant analysis (BSA) and AFLP markers to show that four AFLP bands were linked to the resistance. They then cloned these AFLP bands and used as probes for RFLP analysis in which five RFLP markers constituting a new LG were constructed and a major QTL, named *qPMR-2*, associated with the resistance on this LG accounting for 64.9% of the trait variation. Humphry et al. (2003) used ATF3640 which is highly resistance to powdery mildew as the gene source to identify QTL for the resistance in a recombinant inbred line population using RFLP markers. The population was assessed for the resistance in three environments. A single major QTL on LG K was consistently found controlling the resistance. This locus explained as high as 86% of the variation in disease reaction. RFLP marker VrCS65 was closely linked to the QTL with about 1.3 cM away. Comparative genome analysis revealed no co-localization between the QTLs identified in ATF 6340 and in VC3890A (Humphry et al. 2003).

Due to the fact that RFLP marker analysis is expensive, time-consuming and laborious, it is not suitable for genomic study and MAS, and thus, an alternative type of DNA markers is needed. Zhang et al. (2008) identified two simple sequence repeats (SSRs), VrC SSR1 and VrC SSR3, and two sequence-tagged site (STS), VrC STS1 and VrC STS2, markers co-segregating with the marker VrCS65 that is closely linked to a major gene for powdery mildew resistance in ATF 6340 (Zhang et al. 2008). These four markers were developed from bacterial artificial chromosome (BAC) clones showing positive relationship with the marker VrCS65. After this period, many SSR markers have been developed for mungbean (Gwag et al. 2006; Somta et al. 2008b, 2009; Seehalak et al. 2009; Tangphatsornruang et al. 2009) and other related legumes such as azuki bean (*Vigna angularis* (Ohwi) Ohwi and Ohashi) (Wang et al. 2004) and cowpea (*Vigna unguiculata* (L.) Walps.) (Gupta and Gopalakrishna 2010; Kongjaimun et al. 2012). Kasettranan et al. (2010) used partially resistance breeding line VC6468-11-1A that obtained the resistance from ML-3 and ML-5 to locate the QTLs for powdery mildew resistance in an RIL population using SSR markers from mungbean and azuki bean. The population was evaluated for the resistance under field and greenhouse conditions. One major QTL named *qPMR-2* and one minor QTL named *qPMR-1* were associated with the resistance in both environments. *qPMR-1* was located at about 5 cM marker interval between CEDG282 and CEDG191, while *qPMR-2* was localized between 0.6 cM marker interval of MB-SSR238 and CEDG166. *qPMR-1* and *qPMR-2* accounted for 20.1% and 57.8% of the total disease variation, respectively. Comparative linkage analysis revealed that *qPMR-2* for the resistance in VC6468-11-1A and the single major QTL identified for the resistance ATF3640 (Humphry et al. 2003) locate on the same linkage group. Chankaew et al. (2013) identified QTLs for highly resistance and complete resistance in V4718 and RUM-5, respectively. F_2 and BC_1F_1 mapping populations derived from these two resistance sources were assessed for the resistance for more than one year. One major QTL, *qPMV4718-3* on LG9, and two minor QTLs, *qPMV4718-1* and *qPMV4718-2* on LG4, were identified for the resistance in V4718, while two major QTLs, *qPMRUM5-2* on LG6 and *qPMRUM5-3* on LG9, and one minor QTL,

Table 10.1 Sources of resistance and QTLs for the resistance to powdery mildew, Cercospora leaf spot, mungbean yellow mosaic virus, bruchids and bean bug in mungbean

No.	Trait	Population	Resistance source	Marker type(s)	Number of QTLs	Reference
1	Powdery mildew	F_2 (TC1966 × VC3980A)	VC3980A	RFLP	3	Young et al. (1993)
2	Powdery mildew	F_2 (TC1966 × VC1210A)	VC1210A	RFLP, AFLP	1	Chaitieng et al. (2002)
3	Powdery mildew	RIL (Berken × ATF3640)	ATF3640	RFLP	1	Humphry et al. (2003)
4	Powdery mildew	RIL (KPS1 × VC6468-11-1A)	VC6468-11-1A	SSR	2	Kasettranan et al. (2010)
5	Powdery mildew	F_2 (KPS1 × V4718)	V4718	SSR	3	Chankaew et al. (2013)
6	Powdery mildew	F_2, BC_1F_1(CN60 × RUM5)	RUM5	SSR	3	Chankaew et al. (2013)
7	Powdery mildew	RIL (CN72 × V4718)	V4718	ISSR, ISSR-RGA	1	Poolsawat et al. (2017)
8	Cercospora leaf spot	F_2, BC_1F_1 (KPS1 × V4718)	V4718	SSR	1	Chankaew et al. (2011)
9	MYMV	RIL (NM92 × TC1966)	NM92	RAPD, AFLP, SCAR, CAP, SSR	3	Chen et al. (2013)
10	MYMV	RIL (KPS2 × NM10-12-1)	NM10-12-1	AFLP, SSR	5	Kitsanachandee et al. (2013)
11	MYMV	F_2, BC_1F_1 (BM1 × BM6)	BM6	RGA, SCAR, SSR	2	Alam et al. (2014)
12	Bruchid	F_2 (VC3890 × TC1966)	TC1966	RFLP	1	Young et al. (1992)
13	Bruchid	$BC_{20}F_2$ (Osaka-ryokuto × TC1966)	TC1966	RFLP, RAPD	1	Kaga and Ishimoto (1998)
14	Bruchid	RIL (Berken × ACC41)	ACC41	RFLP	1	Mei et al. (2009)
15	Bruchid	RIL (TC1966 × NM92)	TC1966	RAPD, AFLP, SCAR, CAP, SSR	1	Chen et al. (2013)
16	Bruchid	F_2 (Sunhwa × Jangan)	Jangan (V2709)	RAPD, CAP, SSR, STS	2	Hong et al. (2015)
17	Bruchid	RIL (Berken × ACC41)	ACC41	RFLP, SSR, EST-SSR, STS	1	Wang et al. (2016)
18	Bruchid	RIL (V2802 × NM94)	V2802	SNP	1	Schafleitner et al. (2016)
19	Bruchid	RIL (TC1966 × NM92)	TC1966	SNP	1	Schafleitner et al. (2016)
20	Bruchid	$BC_{11}F_2$ (KPS1 × V2802)	V2802	SSR, EST-SSR	1	Chotechung et al. (2016)
21	Bruchid	$BC_{11}F_2$ (KPS1 × V2709)	V2709	SSR, EST-SSR, STS, Indel	1	Kaewwongwal et al. (2017)
22	Bruchid	F_2 (Jilyu7 × V1128)	V1128	SSR, EST-SSR, STS, Indel	1	Liu et al. (2018)
23	Bean bug	$BC_{20}F_2$ (Osaka-ryokuto × TC1966)	TC1966	RFLP, RAPD	1	Kaga and Ishimoto (1998)
24	Bean bug	F_2 (Sunhwa × Jangan)	Jangan	RAPD, CAP, SSR, STS	1	Hong et al. (2015)

qPMRUM5-1 on LG4, were found for the resistance in RUM-5 (Fig. 10.1). *qPMV4718-1*, *qPMV4718-3*, *qPMRUM5-1* and *qPMRUM5-3* were consistently identified. The major QTL *qPMV4718-3* in V4718 and *qPMRUM5-3* in RUM-5 on LG9 appeared to be the same locus; similarly, the minor QTL *qPMV4718-1* in V4718 and *qPMRUM5-1* in RUM-5 on LG4 might also be the same locus. While *qPMRUM5-3* and *qPMRUM5-2* are possibly respectively the same loci as *Pm1* and *Pm2*, as reported by Reddy et al. (1994). Moreover, a comparative genome analysis demonstrated that the major QTLs *qPMV4718-3* and *qPMRUM5-3* for the resistance on LG9 are possibly the same locus with the QTL *qPMR-2* in VC6468-11-1A and the major QTL in ATF 3640 (Chankaew et al. 2013; Fig. 10.1). The analysis also demonstrated that the major QTL *qPMRUM5-2* for resistance in RUM5 and the major QTL *qPMR-1* for resistance in VC6468-11-1A are both on LG6 and possibly the same locus or closely linked (Fig. 10.2). However, in contrast to the resistance to *E. polygoni*, Poolsawat et al. (2017) identified a QTL controlling powdery mildew resistance caused by *P. phaseoli* in an RIL population obtaining the resistance from V4718 by using ISSR and ISSR-anchored resistance gene analog (ISSR-RGA) markers. A major QTL explaining up to 92.4% of the disease score variation was detected for the resistance, designated as *qPMC72V18-1*. *qPMC72V18-1* located between markers I42PL229 and I85420 which are about 13 cM apart. Since the number and effect of QTLs for the resistance to *E. polygoni* and those to *P. phaseoli* in V4718 are different, it is conceivable to conclude that the genetics of the resistance

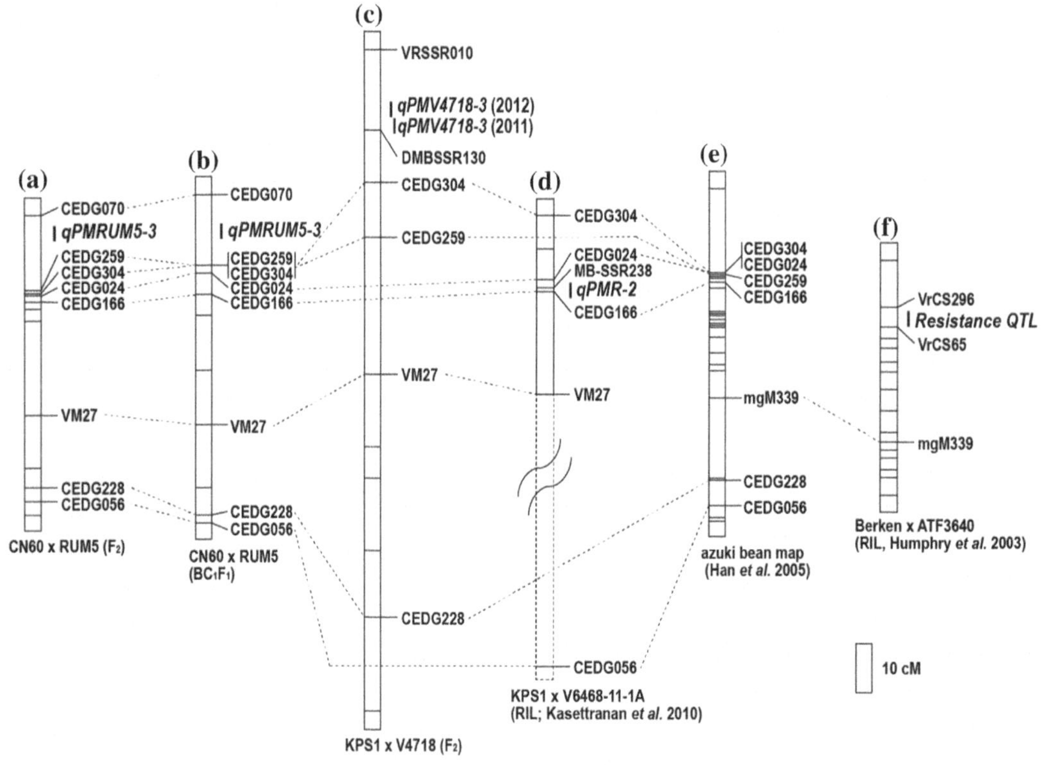

Fig. 10.1 Comparative map illustrating a conserved major QTL on LG9 for powdery mildew resistance detected in RUM5 (**a** and **b** Chankaew et al. 2013), V4718 (**c** Chankaew et al. 2013), VC6468-11-2A (**d** Kasettranan et al. 2010) and ATF340 (**f** Humphry et al. 2003). Map of azuki bean (**e** Han et al. 2005) is used to facilitate comparison between maps. Reprinted with the permission from Springer Nature. *Source* Chankaew et al. (2013)

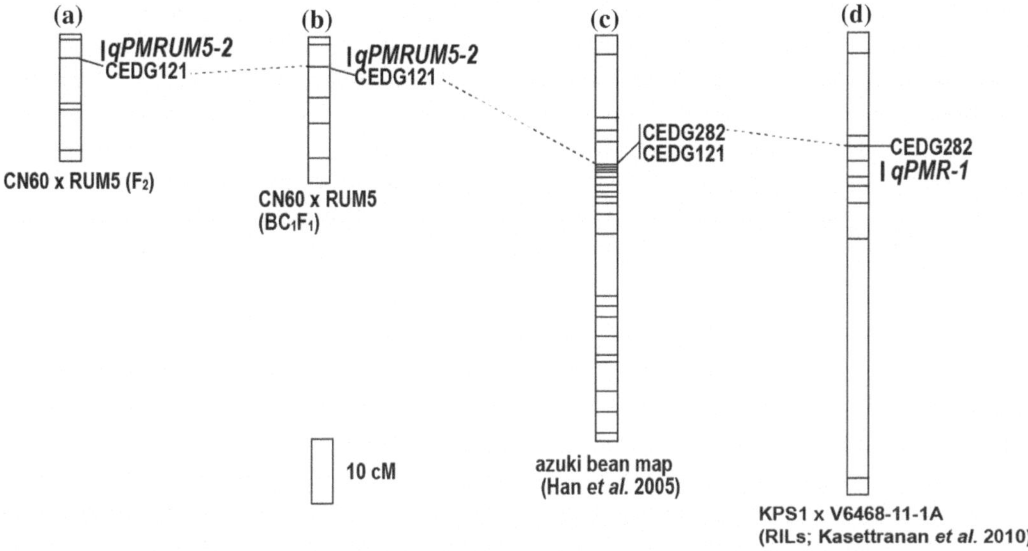

Fig. 10.2 Map illustrating a conserved minor QTL on linkage group 6 conferring powdery mildew resistance identified in RUM5 (**a** and **b** Chankaew et al. 2013) compared with that in VC6468-11-2A (**d** Kasettranan et al. 2010). Map of azuki bean (**c** Han et al. 2005) is used to facilitate comparison between different maps. Reprinted with the permission from Springer Nature. *Source* Chankaew et al. (2013)

to these two powdery mildew fungi in V4718 are different. Recently, markers I85420, I42PL229 and I42PL222 flanking the *qPMC72V18-1* were shown to be able to efficiently select the backcross progenies of V4718 possessing resistance to powdery mildew disease caused by *P. phaseoli* at the accuracy of 94.4 percent (Chathiranrat et al. 2018). Nonetheless, the fact that combination of two or more QTLs provides highly or completely resistance to *E. polygoini* (e.g., *qPMRUM5-2* and *qPMRUM5-3*) pyramiding such QTLs into a target genotype is necessary to develop cultivar(s) with effective and durable resistance. Markers tightly linked to such QTLs must be developed to facilitate the QTL pyramiding through MAS.

Plant *Mildew Locus O* (*MLO*) gene family codes for protein possessing seven transmembrane domains and one calmodulin-binding domain, topological reminiscent of metazoan and fungal G-protein-coupled receptors (GPCRs) (Devoto et al. 2003). The total number of *MLO* genes in plant genome varies depending on plant species (Deshmukh et al. 2014; Kusch et al. 2016). Specific homologs of the *MLO* gene family have been primarily involved in the interaction between plant and fungi causing the powdery mildew disease. Some *MLO* genes act as susceptibility factors (*S* gene) toward powdery mildew. Loss or reduction in the function of specific *MLO* genes results in resistance to powdery mildew in several plant species (Pavan et al. 2009). This type of resistance is known as *mlo*-based resistance. The *mlo*-based resistance has been shown to provide durable resistance in barley and pea. The first *MLO* gene required for powdery mildew pathogenesis was *HvMLO* in barley (Büschges et al. 1997). Then *MLO* susceptibility genes have been functionally characterized in several plant species including rice (*Oryza sativa* L.) (*OsMLO3*; Devoto et al. 2003), thale cress (*Arabidopsis thaliana* (L.) Heynh.) (*AtMLO2*, *AtMLO6* and *AtMLO12*; Consonni et al. 2006), tomato (*Solanum lycopersicum* L.) (*SlMLO1*; Bai et al. 2008), pea (*Pisum sativum* L.) (*PsMLO1*; Humphry et al. 2011; Pavan et al. 2011), wheat (*Triticum aestivum* L.) (*TaMLO_A1* and *TaMLO_B1*; Várallyay et al. 2012), pepper (*Capsicum annuum* L.) (*CaMLO2*; Zheng et al. 2013), tobacco (*Nicotiana tabacum* L.) (*NtMLO1*; Appiano et al. 2015), lotus (*Lotus japonicus* (Regel) K. Larsen) (*LjMLO1*; Humphry et al. 2011) and barrel clover (*Medicago truncatula* Gaertn.) (*MtMLO1*; Humphry et al. 2011). Recently, a genome-wide survey of the *MLO* family in eight legume species including mungbean has been conducted (Rispail and Rubiales 2016). Eighteen *MLO* genes were identified for mungbean. These *MLO* genes distribute onto all the 11 chromosomes of the mungbean reference genome (VC1973A) except chromosome 6 (Fig. 10.3). Interestingly, among all the *MLO* genes, *VrMLO15* and *VrMLO6* localizing on chromosome 4 are in the same genome region with the major QTL *qPMRUM5-1* in RUM5 and *qPMV4718-1* in V4718 for the powdery mildew resistance reported by Chankaew et al. (2013) (Fig. 10.4). In addition, *VrMLO12* locating on chromosome 9 is in the same genome region with the major QTL *qPMRUM5-3* in RUM5 and *qPMV4718-3* in V4718 for the resistance reported by Chankaew et al. (2013). This suggests that *VrMLO5*, *VrMLO6* and *VrMLO12* are possibly the *MLO* gene required for susceptibility to powdery mildew in mungbean. Thus, these MLO genes can be considered as candidate genes for fine-mapping of powdery mildew resistance in RUM5 and V4718.

10.3 Cercospora Leaf Spot Disease

Cercospora leaf spot (CLS) disease is among the most important biotic stresses of mungbean production in the humid tropics. The disease is caused by the fungus *Cercospora canescens* Illis and Martin. The disease is commonly widespread in Asia and devastating in the warm, wet growing season. *Cercospora* is recognized by the appearance of leaf spots that are circular to irregularly shape with grayish-white centers and reddish-brown to dark brown margins. The fungus initially causes spotting on the mungbean leaves. The spots increase in number and size during flowering, but the increment is most rapid at the pod-filling stage. In the susceptible varieties, infection expands rapidly resulting in a premature defoliation and reduction in the size of pods and seeds (Grewal et al. 1980). The disease can cause yield loss of up to 50% if there is no protection (AVRDC 1984; Iqbal et al. 1995).

10.3.1 Sources and Genetics of the Resistance to Cercospora Leaf Spot

Mungbean germplasm resistance to Cercospora leaf spot has been identified. Some of the resistant accessions include EC27087-2, EC27261, Blanco, STB#121, ML-1, ML-3, ML-5, ML-6, LM612, PLM88, PLM926, PLM944, V4718, Shanhua 1, LM013 yellow, Pusa 105, PDM 2, PDM 14, PDM 15 and PDM 113 (Fernandez and Shanmugasundaram 1988; Hartman et al. 1993; Thakur et al. 1977). Some of these accessions such as ML-3, ML-5 and V4718 are also resistant to powdery mildew disease. The resistance in ML-3 and ML-5, EC27087-2, EC27261 and ML-1 is governed by a single dominant gene (AVRDC 1974; Thakur et al. 1977). The resistance in V4718 is also likely controlled by a single dominant gene, although the segregation of the resistance is continuous (Chankaew et al. 2011). The monogenic resistance in these accessions is contradictory to the observation of Mishra et al. (1988) who reported that the resistance in Pusa 105, PDM 2, PDM 14, PDM 15 and PDM 113 is controlled by a single recessive gene. AVRDC (1980) and Leabwon and Oupadissakoon (1984) reported that the resistance is conditioned by quantitative genes.

10.3.2 Genome Analysis of Cercospora Leaf Spot Resistance

Similar to *E. polygoni* causing powdery mildew, *C. canescens* is an obligate fungus and the disease incidence is seasonal. Breeding progress against this disease is slow because the disease only occurs in rainy seasons. Mungbean breeders have tried to identify QTL and DNA markers associated with the resistance for MAS. However, up until now, there is only one report on the QTL for

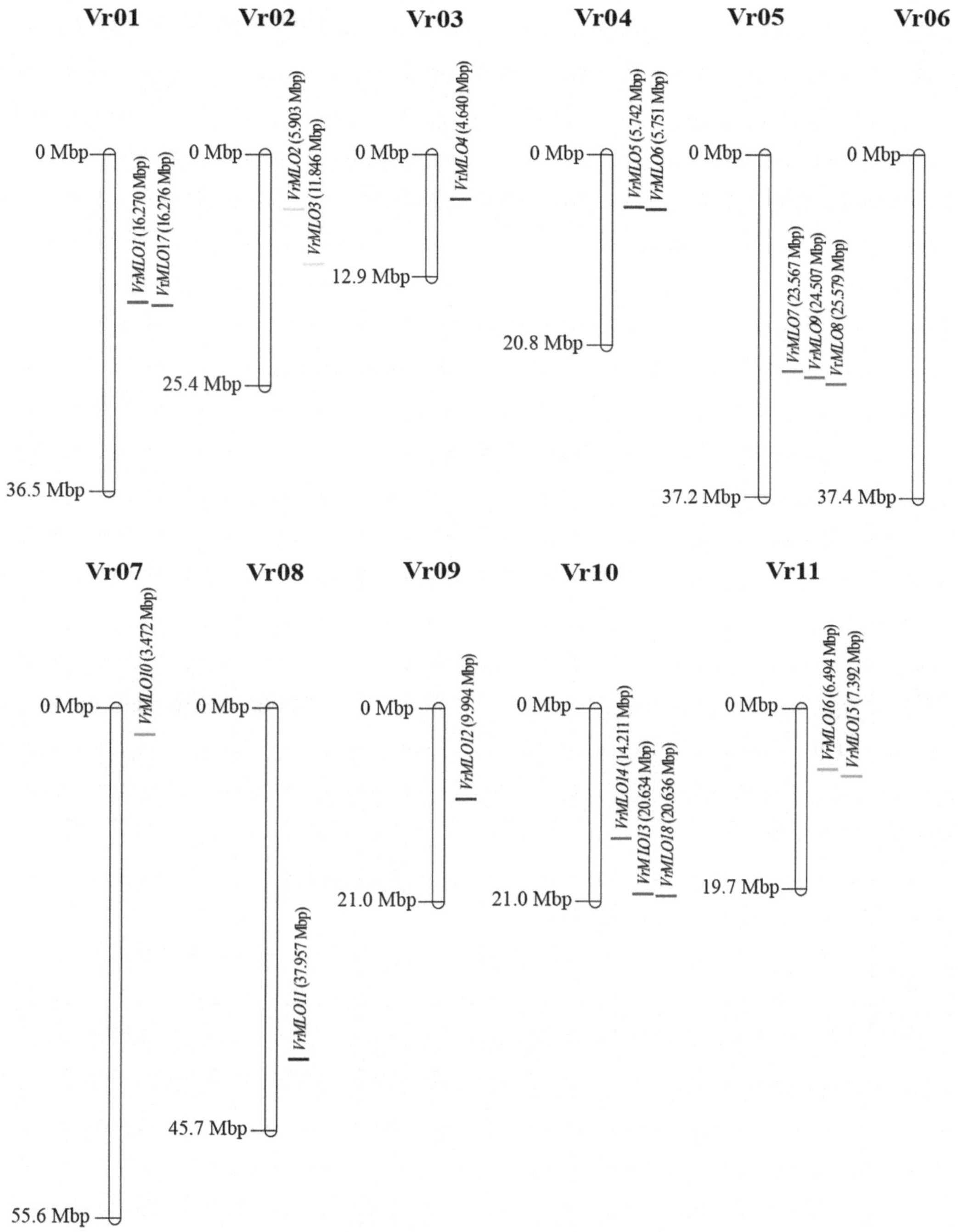

Fig. 10.3 Distribution of 18 mungbean *MLO* genes on mungbean reference genome, VC1973A (Kang et al. 2014). Location of each *MLO* gene is based on Rispail and Rubiales (2016)

the Cercospora leaf spot resistance in mungbean (Table 10.1). This is possibly due to the scarce of genomic tools and lack of reliable gene source conferring the resistance. Gene mapping for Cercospora leaf spot resistance was conducted on F_2 and BC_1F_1 populations developed from V4718

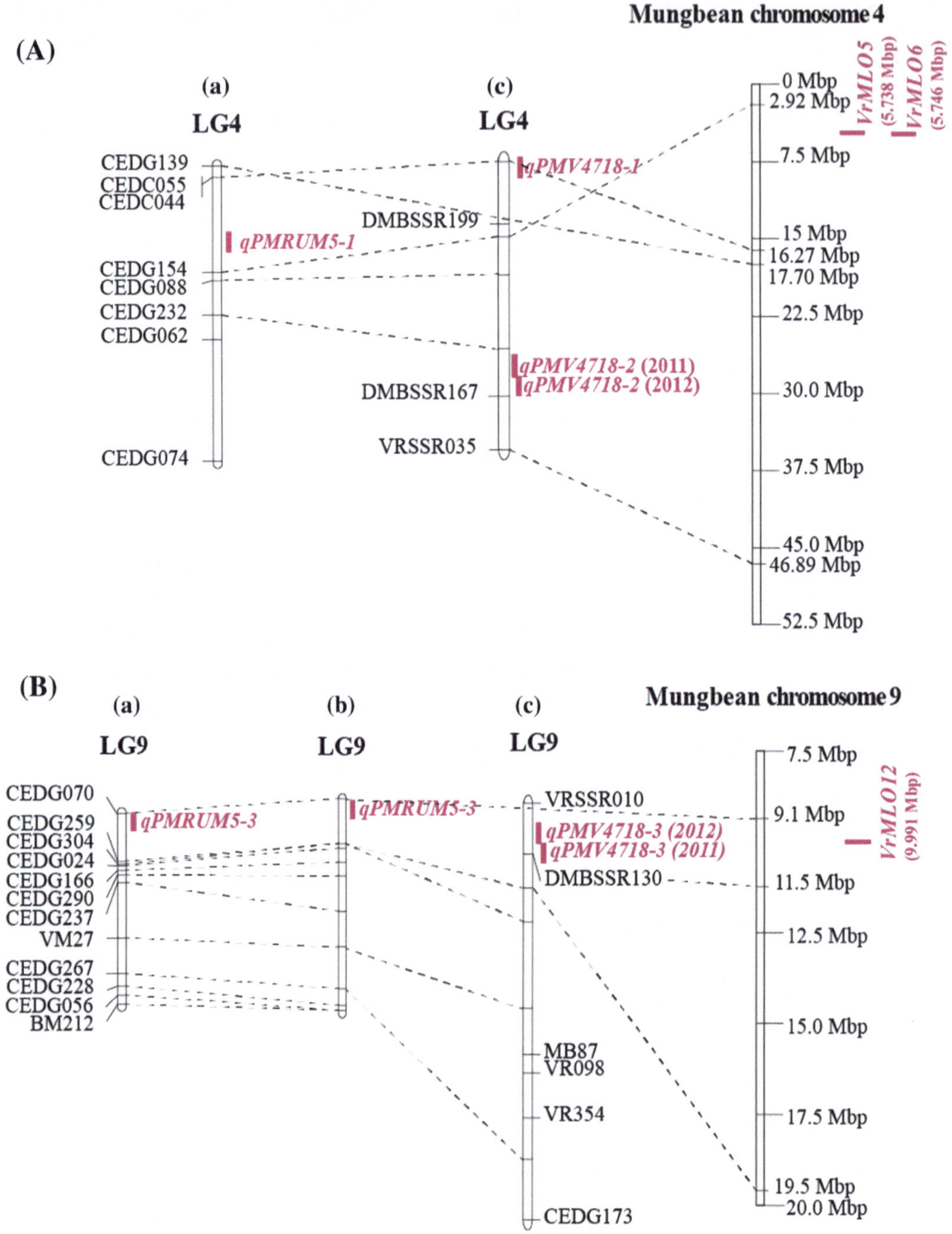

Fig. 10.4 Comparative map illustrating co-localization between quantitative trait locus (QTL) on linkage groups (LG) 4 (**A**) and 9 (**B**) conferring powdery mildew resistance in mungbeans RUM5 (*qPMRUM5-1* and *qPMRUM5-3*) and V4718 (*qPMV4718-1* and *qPMV4718-3*) (Chankaew et al. 2013) and *MLO* genes on mungbean chromosomes. Locations of *VrMLO* genes are based on Rispail and Rubiales (2016)

as the resistance source using SSR markers from various legume species (Chankaew et al. 2011). V4718 showed a stable resistance in Taiwan (Hartman et al. 1993), Thailand (Somta, unpublished data) and China (Xin, unpublished data). In both F_2 and BC_1F_1 populations, QTL analysis revealed a major QTL for the resistance, designated as *qCLS*. The QTL was located on LG3 between SSR markers CEDG117 and VR393 and accounted up to 81% of the disease score variation. *qCLS* shows no linkage with any QTLs for the powdery mildew resistance (Chankaew et al. 2013). This agreed with a previous report that genes conferring resistance to CLS and powdery mildew disease inherit independently (Thakur et al. 1977). There is a high genome conservation among legumes in the genus *Vigna* and among *Vigna* species, common bean (*Phaseolus vulgaris* L.) and soybean (*Glycine max* (L.) Merr.) (Boutin et al. 1995; Isemura et al. 2007; Kaga et al. 2011). For example, a QTL for seed weight is common in mungbean, cowpea and azuki bean (Isemura et al. 2007). Comparative genome analysis can be useful for identification of genes/QTLs controlling other traits in mungbean. A major QTL, *qCLS9.1*, for Cercospora leaf spot caused by *C. canescens* in cowpea appears to be the same with a major QTL for leaf spot disease caused by *Pseudocercospora griseola* (Sacc.) Crous and U. Braun in common bean (Duangsong et al. 2016). However, the major QTL, *qCLS9.1*, for Cercospora leaf spot in cowpea and the *qCLS* in mungbean appears to be different loci. The *qCLS9.1* locates on LG9, whereas *qCLS* resides on LG3. This suggests that the mechanism of the resistance to *C. canescens* in mungbean V4718 is different from that in cowpea. In soybean, there is an attempt to identify gene(s) controlling resistance to frogeye leaf spot disease caused by *Cercospora sojina* K. Hara (Pham et al. 2015). Since the Cercospora leaf spot in mungbean and frogeye leaf spot in soybean are both caused by *Cercospora* fungi, genes for the resistance in both legumes may be the same. Fine-mapping, gene expression and sequencing for the resistant gene in the resistance soybean accessions PI 594891 and PI 594774 revealed three candidate genes: *Glyma13g25320*, *Glyma13g25340* and *Glyma13g25350* (Pham et al. 2015). Among them, *Glyma13g25350* is the most probable candidate gene for *C. sojina* resistance because (i) it is derived from a mutation that resulted in an amino acid change for the protein function In PI 594774, but there is no such mutation in PI 594891, and (ii) PI 594774 and PI 594891 share two common single-nucleotide polymorphisms (SNPs) in the promoter region of this gene. Annotation suggests that *Glyma13g25350* is one of heterotrimeric G-proteins (www.soybase.org) which plays a central role in plant signal transduction involving in programmed cell death in plant immunity to pathogens (Zhang et al. 2012). It is worth investigating whether *Glyma13g25350* is involved in the resistance to *C. canescen* in V4718.

10.4 Yellow Mosaic Disease

Yellow mosaic disease in mungbean is caused by geminivirus (genus *Begomovirus*, family *Geminiviridae*), which has a bipartite genome (DNA A and DNA B). Although several begomoviruses cause yellow mosaic disease in mungbean, two of them are predominant, namely mungbean yellow mosaic virus (MYMV) and mungbean yellow mosaic India virus (MYMIV). MYMV is the major pathogen responsible for the yellow mosaic disease in western and southern India, Thailand and Indonesia, while MYMIV is the main pathogen for the yellow mosaic disease in central, eastern and northern India, Pakistan, Bangladesh, Nepal and Vietnam (Hussain et al. 2004; Malathi and John 2009; Ilyas et al. 2010). Nowadays, yellow mosaic disease is the most important and devastating disease in mungbean production in South Asia, especially India, and it is becoming a major threat in mungbean production in Myanmar and Thailand. Begomoviruses are transmitted by whitefly (*Bemisia tabaci*) (Gennadius) (Hemiptera: Aleyrodidae) and can infect mungbean at all growth stages. In susceptible cultivars, the disease can reduce seed yield up to 100% (Marimuthu et al. 1981) or even kill the infected plants at an early vegetative stage. Disease symptoms on the leaf start as small yellow specks along the veinlets and spread over the lamina (Karthikeyan et al. 2014). The symptoms vary from a few small yellow specks or spots on a few leaves, to yellowing or chlorosis of all leaves of the whole plant followed by necrosis. In highly susceptible plants, the symptoms include shortening of internodes, severe stunting of plants with no yield or with a few flowers and deformed pods (thin and curling upward) producing small, immature and shriveled seeds.

Chemical control of yellow mosaic disease by eliminating the virus vector, whitefly, is ineffective, expensive and detrimental to the environment. Deployment of genetically improved resistant mungbean cultivar(s) provides an efficient, cost-effective and environmentally friendly measure to manage the disease. Currently, enhancing resistance to yellow mosaic disease is a major goal in mungbean breeding programs in India, Pakistan, Bangladesh and Myanmar.

10.4.1 Sources of Resistance to Yellow Mosaic Disease

Although many germplasms have been identified as being tolerant or resistant to yellow mosaic disease, in most cases the causative begomovirus is not known and resistance evaluation does not take account of the strain diversity among the various strains of the virus and among biotypes of the whitefly transmitter. Mungbean lines, NM10-12-1, NM92 and NM94, have been reported to be resistant to yellow mosaic disease in Pakistan and India (Ali et al. 1997; Chen et al. 2013; Kitsanachandee et al. 2013). NM92 and NM94 are from the same cross combination carrying the resistance from 6601, a landrace mungbean of Pakistan (Ali et al. 1997). Possibly, NM10-12-1 obtains the resistance from 6601 too (Kitsanachandee et al. 2013). The development of NM92 and NM94 is a breakthrough in mungbean breeding for the yellow mosaic disease (Ali et al. 1997). Nonetheless, recent resistance evaluation of mungbean against yellow mosaic disease under field conditions in India where different begomovirus and white fly cryptic species are predominant revealed that mungbean line NM94 is tolerant to the disease in several locations but highly susceptible in some locations (Nair et al. 2017). The evaluation also revealed high level of tolerance in mungbean line ML-1628.

10.4.2 Genetics of the Resistance to Yellow Mosaic Disease

Inheritance studies of the yellow mosaic disease resistance revealed monogenics or digenics of the resistance, including single recessive gene (Khattak et al. 2000; Reddy 2009; Dhole and Reddy 2013; Cayalvizhi et al. 2017; Malik et al. 1986; Thakur et al. 1977), two recessive genes (Shukla and Pandya 1985; Ammavasai et al. 2004; Vinod and Kandali 2012; Alam et al. 2014a; Bhanu et al. 2019) and single dominant gene (Lekhi et al. 2018). The resistance in NM92 is controlled by a single recessive gene (Khattak et al. 2000), while that in NM94 (BARImung 6 in the original paper) is controlled by two recessive genes (Alam et al. 2014a). The difference in number of genes conferring the resistance in NM92 and NM94 is surprising. The pedigrees of NM92 and NM94 (BARImung 6) are nearly identical (NM36 × VC2768B and NM36 × VC2768A, respectively). Both of them acquired the resistance from NM36 which is developed from hybridization between 6601 and VC1973A (Ali et al. 1997). Nonetheless, the difference in number of genes conditions the resistance in various resources provides opportunity for pyramiding those genes into a target genotype to develop cultivar(s) with stable and durable resistance to yellow mosaic disease.

10.4.3 Genome Analysis of the Yellow Mosaic Disease

Monogenic and digenetic inheritance of the resistance to yellow mosaic disease in mungbean suggests that the development of new resistant cultivar(s) may not be difficult. The major difficulty is field screening for the resistance, which is usually hampered by non-uniform development of the disease due to the fluctuation of the whitefly population in different locations and season, and by the variation in strains of the viruses. MAS may overcome this difficulty. Efforts to identify DNA markers associated with yellow mosaic disease in mungbean have been done by bulked segregate analysis (BSA). Lambrides et al. (1999) identified resistance gene from NM92 in two RIL populations. Randomly amplified polymorphic DNA (RAPD) marker OPAJ20 was found linking to the gene. Selvi et al. (2006) reported that RAPD marker $OPS7_{900}$ is linked to MYMV resistance in mungbean ML-267. Later, sequence-characterized amplified region (SCAR) marker MYMVR-583 developed from a RAPD marker was found linking to the MYMV resistance in mungbean TM-99-37 (Dhole and Reddy 2013). Two SCAR markers CM9 and CM815 developed from RAPD are associated with MYMV resistance in mungbean KMG189 (Cayalvizhi et al. 2017). Linkage analysis revealed that the marker CM815 is 5.56 cM away from the MYMV resistance gene, while the marker CM9 co-segregates with the resistance gene. Efforts have been made to locate QTLs for yellow mosaic disease resistance in 10-12-1, NM92 and NM94 (BARImung6) to the genetic map (Table 10.1). A RIL population developed of a cross between NM92 and TC1966 (wild mungbean susceptible to yellow mosaic disease) was evaluated for MYMIV resistance in India for three years. QTL analysis by composite interval mapping (CIM) in this population identified four QTLs for the resistance (Chen et al. 2013). The most significant and largest effect QTL, $MYMIV^r9\_25$, was on LG9. It accounted for 59% of the disease variation. SSR markers are closely linked to the $MYMIV^r9\_25$, being about 1 cM away from the QTL. However, only QTL $MYMIV^r8\_48.8$ locating on LG8 was consistently identified across all the three years. Another RIL population derived from NM10-12-1 × Kamphaeng Saen 2 (susceptible to yellow mosaic disease) was evaluated for MYMIV resistance under field condition in India and Pakistan in 2012 and 2008, respectively. CIM in this population revealed three QTLs for the resistance in India (*qMYMIV1*, *qMYMIV2* and *qMYMIV3*) and two QTLs for the resistance in Pakistan (*qMYMIV4* and *qMYMIV5*) (Kitsanachandee et al. 2013). *qMYMIV1*, *qMYMIV3*, *qMYMIV4* and *qMYMIV5* are conferred by NM10-12-1, while *qMYMIV2* is by Kamphaeng Saen 2. Only QTL *qMYMIV4* accounted more than 20% of the disease variation and can be considered as a major QTL. This QTL is probably the same with *qMYMIV4* because it is located in the same genome region. SSR marker CEDG100 is located at the same position with the QTL *qMYMIV4*. In another study, F_2 and BC_1F_1 populations developed from across between BARImung6 (NM94) and BARImung1 (susceptible) were assessed for field resistance to MYMIV in two locations in Bangladesh. QTL analysis in these two populations gave the same results that two major QTLs, *qMYMIV2.1* on LG2 and *qMYMIV7.1* on LG7, confer the resistance in BARImung6 (Alam et al. 2014b). The marker interval CEDG275 and CEDG006 flanking the QTL *qMYMIV2.1* is only about 3 cM apart, while the marker CEDG041 is closely linked to the QTL *qMYMIV2.1*, being less than 1 cM from the QTL. These two QTLs each accounted for over 40% of the disease score variation. The result agrees with Mendelian genetic and quantitative genetic analyses that two recessive genes control the yellow mosaic disease in BARImung6 (Alam et al. 2014a).

Linkage map comparison revealed that the major QTL $MYMIV^r9\_25$ for the resistance in NM92, *qMYMIV4* (*qMYMIV1*) in NM10-12-1 and *qMYMIV2.1* in BARImung6 are located to the same linkage group and possibly the same locus (Kitsanachandee et al. 2013; Alam et al. 2014b). This is very likely because the resistance in NM92, BARImung6 and NM10-12-1 is all derived from the same resistance source, 6601 (Ali et al. 1997). If these QTLs are truly the same locus, fine-mapping and QTL cloning to identify the function of this resistance locus should be done because it showed stable resistance in different seasons, locations, environments and genetic backgrounds. However, the challenge in fine-mapping is that a large segregating population must be assessed for the resistance under field condition where the variation of the fields and whiteflies can be very high. Such variation can hinder the accuracy of the phenotypic data and thus the accuracy of the fine-mapping.

It appears from gene mapping for the yellow mosaic disease that the genetics of resistance to MYMIV and to MYMV are different. QTL analysis showed that at least two loci are involved in the resistance to MYMIV. Successful marker tagging for the resistance to MYMV by bulked segregant analysis suggested that a single gene is involved in the resistance, and this agrees with the results from a segregation analysis that MYMV resistance is monogenics (Selvi et al. 2006; Dhole and Reddy 2013; Cayalvizhi et al. 2017). Alam et al. (2014b) noted that SCAR marker MYMVR-583 which is closely linked to a single recessive gene governing MYMV resistance in mungbean TM-99-37 revealed polymorphism between the parents of F_2 and BC_1F_1 mapping populations, but this marker was not found associated with the QTLs identified for the MYMIV resistance. This supports the difference between genetic control of the resistance to MYMV and MYMIV.

Apart from the resistance to yellow mosaic disease in the primary gene pool of mungbean, scientists are also interested in the resistance in the secondary gene pool. Black gram (*Vigna mungo* (L.) Hepper) and rice bean (*Vigna umbellata* (Thunb.) Ohwi and Ohashi) are considered the secondary gene pool of mungbean. In general, level of resistance to yellow mosaic disease in black gram is higher than mungbean, while rice bean is immune to the disease (Gill et al. 1983; Singh et al. 1998). Attempts have been made to transfer the resistance from black gram to mungbean (Gill et al. 1983; Singh et al. 1998; Lekhi et al. 2018; Sudha et al. 2013b). High yield mungbean breeding lines with yellow mosaic disease resistance are developed from crosses between mungbean and black gram (Gill et al. 1983; Singh et al. 1998). Resistance gene analog (RGA) markers YR4 and CYR1 are found to be completely linked with the resistance to MYMIV in black gram (Maiti et al. 2011). The complete association suggests that CYR1 is a portion of the candidate disease resistance (*R*) gene (Maiti et al. 2011). The marker CYR1 is also associated with resistance in mungbean. Full sequence of the *R* gene CYR1 is successfully isolated from black gram (Maiti et al. 2012). CYR1 codes a protein sequence of 1176 amino acids with coiling structure at the N-terminus, central nucleotide-binding site (NBS) and C-terminal leucine-rich repeats (LRR) that belongs to non-TIR-NBS-LRR subfamily of plant *R* genes. In silico analysis of the NBS and LRR regions of CYR1 and MYMIV-coat protein (CP) showed that CYR1-LRR forms an active pocket and interacts with MYMIV-CP during docking. This suggests that CYR1 may act as a signaling molecule involving in recognizing the effector molecule of the patho-system to contribute to plant–MYMIV incompatible interactions (Maiti et al. 2012). CYR1 and YR4 are used in the QTL mapping for MYMIV resistance in NM10-12-1 (Kitsanachandee et al. 2013), and BARImung6 (Alam et al. 2014a). These markers showed no polymorphism between the parents of mapping population in both studies. BLASTN analysis of the complete sequence of *CYR1* gene and sequence of the SSR markers associated with the major QTLs for MYMIV resistance in NM10-12-1, NM92 and BARImung6 on the reference genome of mungbean and azuki bean (*Vigna angularis* (Ohwi) Ohwi and Ohashi) revealed that *CYR1* locates on different chromosomes with the QTLs for the MYMIV resistance in these mungbeans (Somta, unpublished data). This suggested that resistance to MYMIV in mungbean and black gram is different. Similarly, SSR marker CEDG180 which was found associated with a major gene controlling MYMIV resistance in black gram (Gupta et al. 2013) showed no polymorphism between mapping parents used by Alam et al. (2014b). Based on the linkage map of mungbean reported by Isemura et al. (2012), CEDG180 is located on LG10. The difference in resistance genes between mungbean and black gram provides an opportunity to develop durable resistant cultivar (s) by gene pyramiding. Gene mapping of the resistance in black gram would be greatly helpful for identifying the resistance QTLs for molecular breeding of the trait in black gram per se and mungbean. It should be noted that up until now, there is no report on gene or QTL mapping for yellow mosaic disease in black gram.

Although rice bean is highly resistance to yellow mosaic disease in general (Sudha et al. 2015), transferring yellow mosaic disease resistance gene(s) from rice bean to mungbean is only successful in a few cases (Sudha et al. 2013a; Bhanu et al. 2017) due to low cross-compatibility between the two species. As compared to black gram, rice bean is genetically distinct from mungbean (Tomooka et al. 2002; Sudha et al. 2015). Nonetheless, genetics of the resistance to yellow mosaic disease in rice bean has been reported. F_1, F_2 and F_3 populations were generated from the cross between MYMV-susceptible mungbean VRM(Gg)1 and rice bean TNAU RED and evaluated for MYMV resistance. Segregation analysis in these populations revealed that the resistance in TNAU RED is controlled by a single recessive gene (Sudha et al. 2013b). DNA marker analysis showed severe segregation distortion and chromosome elimination in the F_2 population and none of F_2 progenies possessed homozygous allele from rice bean (Sudha et al. 2013a). Such anomaly hampers genetic and gene mapping studies for the resistance. However, very recently, QTL mapping for the resistance in an F_9 RIL population of 108 lines derived from the cross between VRM (Gg)1 and TNAU RED was conducted by constructing SNP-based linkage map by genotyping by sequencing (Mathivathana et al. 2019). The population was evaluated for the MYMV resistance in two years, 2015 and 2016. QTL analysis revealed five QTLs: two (*qMYMV4_1* and *qMYMV5_1*) for 2015 and three (*qMYMV4_1*, *qMYMV6_1* and *qMYMV10_1*) for 2016. These QTLs had low phenotypic variation explained, for the resistance, being 10.11–20.04%. Only the major *qMYMV4_1* was consistently detected in both years. It was located in a 1.2-Mb (14,504,302–15,788,321) region on mungbean chromosome 4. Although there are as many as 83 annotated genes in this genome region, 18 out of these genes are suggestive as candidate genes at the *qMYMV4_1* based on their associated with disease resistance in various plant species (Mathivathana et al. 2019). Adding more markers can narrow down the *qMYMV4_1* region and reduce the number of candidate genes for this QTL. Nonetheless, the number of QTLs detected for the MYMV resistance in TNAU RED in this study does not agree with the result from segregation analysis that the resistance is controlled by a single recessive gene (Sudha et al. 2013b). Additional study is needed to confirm the contrast results.

Apart from black gram and rice bean, genome information of the yellow mosaic disease in soybean may be useful for genetic improvement of the resistance in mungbean. Genetic analysis and gene mapping for MYMIV resistance were conducted in RIL population derived from the cross JS335 PI171443 and F_2 population developed from the cross SL525 × NRC101 revealed that the resistance is controlled by a single recessive gene and that the resistance gene locates between SSR markers GMAC7L and Satt322 on LG C2 or chromosome 6 (Rani et al. 2017). Based on the reference genome sequence of soybean, these markers are 77.12 Kbp apart (position of 12,259,594–12,336,709 bp) and there are 12 genes in this region (www.phytozome.net). Two of these genes, *Glyma.06G151000* and *Glyma.06G150700*, each codes for probable serine/threonine-protein kinase that is known for its role in signal transduction for the disease and insect resistance in plants (Afzal et al. 2008). These genes can be used as candidate gene for MYMIV resistance in mungbean.

10.5 Bruchids

Bruchids or seed weevils (Coleoptera: Chrysomelidae: Bruchinae) are stored insect pests causing serious damage to legume seeds worldwide. About 20 bruchid species are recognized as devastating storage pests in different pulse crops (Southgate 1979). Among them, azuki bean weevil (*Callosobruchus chinensis* L.) and cowpea weevil (*Callosobruchus maculatus* Fab.) are the most serious species infesting mungbean in the tropics and subtropics. The origin of azuki bean weevil is in Asia, whereas that of cowpea weevil is in Africa. They are polyphagous insects infesting wide host range in the legume family. For example, hosts of cowpea weevil include cowpea (*Vigna unguiculata* (L.) Walps.), bambara groundnut (*Vigna subterranea* (L.) Verdc.), mungbean, black gram, azuki bean, faba bean (*Vicia faba* L.), chickpea (*Cicer arietinum* L.) and pigeon pea (*Cajanus cajan* L.) (Arora 1977; Anton et al. 1997; Desroches et al. 1995; Tuda et al. 2005; Srinives et al. 2007). At present, both azuki bean and cowpea weevils can be found in nearly all continents due to their obligatory larval, pupal and pre-emergence stages within the legume seeds which promote their spreading through international seed trade. Azuki bean and cowpea weevils can cause seed damage of mungbean under both field and seed storage conditions, but the greater losses occur in the storage stage which may lead to a total damage of the seed lot within 3–4 months (Srinives et al. 2007). The bruchid infestation to seed results in weight loss, low germination and changes of nutrition in seeds. The damaged seeds are not fit for human consumption or agricultural and commercial uses (Talekar 1988). The bruchids also cause allergy to some consumers. In addition, bruchid infestation may bring about a negative publicity and lost in consumer trust in the product brand. Several methods including cultural, physical and biochemical methods can be used to control the bruchid infestation (Mishra et al. 2018), but those methods are impractical to small-scale mungbean growers and traders and in most cases are not effective. In addition, the use of synthetic chemicals to control the bruchids is unsafe for users and consumers and is environmental unfriendly. Using host plant resistance is the best method to manage the bruchids in mungbean (Srinives et al. 2007). Although several insect pests are of economically important in mungbean, bruchids receive the most attention to mungbean scientists and breeders. This is due to its economic importance together with the facts that (i) level of bruchid resistance is very high or complete, (ii) inheritance of the resistance is simple, and (iii) the standard protocol for resistance evaluation is simple, easy and inexpensive.

10.5.1 Sources of Resistance to Bruchids

Resistance to bruchids in mungbean and some other *Vigna* species is none or rare (Srinives et al. 2007). No moderately or highly resistance to *C. maculatus* is found in cultivated and wild cowpea. No useful resistance to *C. chinensis* has been identified in cultivated and wild azuki beans. No resistance to *C. maculates* was discovered in cultivated black gram. In mungbean, some resistance accessions showing highly or completely resistance against *C. chinensis* and *C. maculates* have been identified in both wild and cultivated mungbeans. The first resistance source is identified in wild mungbean (*V. radiata* var. *sublobata* (Roxb.) Verdc.) accession TC1966 (Fujii and Miyazaki 1987; Fujii et al. 1989). TC1966 exhibits complete resistance to both *C. chinensis* and *C. maculatus*. Later, cultivated mungbean accessions V2709 and V2802 were found resistant to *C. chinensis* (Talekar and Lin 1992). Wild mungbean accessions ACC23 and ACC41 were reported to be resistant to *C. chinensis* and *C. maculatus* (Lambrides et al. 2000). Cultivated mungbean accessions V1128, V2709, V2802 and V2817 were resistant to *C. chinensis* and *C. maculatus* (Somta et al. 2007). Wild mungbean accession Sub2 and cultivated accessions IC333175, IC325770, IC329039, Dantan Sonamung, RS4, RMG11 and Khargone 1 were identified resistant against *C. chinensis* (Sarkar et al. 2011). These accessions are highly or completely resistant to *C. chinensis* and/or *C. maculatus*. TC1966, ACC41, V2709 and V2802 have been shown to be resistant to several bruchid biotypes. TC1966 and ACC41 are also resistant to graham bean weevil (*Callosobruchus analis* F.), *Callosobruchus phaseoli* (Gyll.), common bean weevil (*Scanthoscelides obtectus* Say.) and Mexican bean weevil (*Zabrotes subfasciatus* Boh.) (Fujii and Miyazaki 1987; Fujii et al.

1989; Lambrides and Imries 2000; Kashiwaba et al. 2003).

10.5.2 Biochemical Basis of the Bruchid Resistance

Secondary metabolites, storage proteins and enzyme inhibitors in legume seeds are major biochemical compounds causing resistance (antibiosis) to bruchid infestation. Feeding test using artificial seed diets revealed that the resistance in TC1966, V1128, V2709, V2802 and V2817 is due to the presence of a particular biochemical in seeds (Kitamura et al. 1990; Talekar and Lin 1992; Somta et al. 2008a). The chemical substance for resistance in TC1966 is water-soluble with high molecular weight, and heat- and protease-stable characteristics, suggesting that it may not be a protein but a polysaccharide (Kitamura et al. 1990). Two novel cyclopeptide alkaloids, vignatic acids A and B, were isolated from $BC_{20}F_4$ isogenic lines carrying the *Br* gene from TC1966 (Sugawara et al. 1996). Feeding test showed that the vignatic acid A provides resistance to *C. chinensis*. Vignatic acid A is composed of L-tyrosine, 3(S)-hydroxyl-L-leucine, L-phenylalanine and 2-hydroxyisocaproic acid, possessing a 14-membered ring (Sugawara et al. 1996). However, fine-mapping demonstrated that location of the *Va* gene encoding vignatic acids is not the same as the *Br* gene; the two genes are very near with the distance of 0.2 cM apart (Kaga and Ishimoto 1998). This indicates that vignatic acid is not a principal factor responsible for the resistance. A peptide compound "GIF-5" toxic to the bruchids was also identified from a similar material that is used for isolating vignatic acids (Kaga et al. 2000). A cysteine-rich protein (VrCRP or VrD1) of the plant defensin family has been isolated from mungbean isogenic line VC6089A carrying the resistance gene from TC1966 (Chen et al. 2002). Feeding test revealed that VrD1 is lethal to *C. chinensis* larvae. VrD1 protein is a specific α-amylase inhibitor that inhibited α-amylase of the bruchids (Liu et al. 2006). However, the DNA marker developed from *VrD1* was located on different linkage groups with the gene conferring the resistance (*Br*), indicating that VrD1 is not the product of the *Br* gene (Isemura et al. 2012). Proteome analysis in seeds of bruchid-susceptible and bruchid-resistant isogenic lines derived from TC1966 together with its susceptible parent Osaka-Ryokuto suggested that chitinase, alanine aminotransferase in membrane, and SalF1R, beta-1,3-glucanase, protein phosphatase 2A, peroxidase BP 2A, provicilin and canavalin play important roles in the resistance (Khan et al. 2003). However, it has been shown that chitinases and beta-1,3-glucanases from cowpea and common bean are ineffective against *C. maculatus* and *Z. subfasciatus* (Sales et al. 2000). Thus, the biochemical basis for the resistance due to the *Br* gene in mungbean is still not known. Additional research is necessary to elucidate the factor(s) responsible for bruchid resistance conferred by the *Br* gene.

10.5.3 Genetics of the Resistance to Bruchids

Seed resistance to *C. chinensis* and *C. maculatus* in mungbean is a monogenic trait. The resistance to *C. chinensis* in the wild mungbean TC1966 is controlled by a single dominance gene locus, designated *Br*, possibly in combination with minor gene(s) and the resistance is dependent on genotypes of the seeds (Kitamura et al. 1988; Fujii et al. 1989). The resistance to *C. chinensis* in wild mungbean ACC41 is controlled by a single dominant locus (Miyagi et al. 2004). Also, the resistance to *C. chinensis* in wild mungbean Sub2 is controlled by a single dominant gene with some modifiers (Sarkar et al. 2011). The resistance to *C. chinensis* and *C. maculatus* in cultivated mungbean V1128, V2709, V2802 and V2817 is also controlled by a single dominant gene with modifiers (Somta et al. 2007; Sun et al. 2008). An F_2 population generated from a cross between ACC41 and TC1966 showed no segregation for the resistance to *C. chinensis*, suggesting that ACC41 and TC1966 share the same resistance gene (Lambrides and Godwin 2007). Allelism test for the *C. chinensis* resistant in

TC1966, V2709 and V2802 suggested that they are controlled by the same or closely linked genes (Sun 2007). Therefore, *Br* appears to confer bruchid resistance in TC1966, ACC41, V2709 and V2802.

10.5.4 Genome Analysis of the Bruchid Resistance

Genetic improvement of mungbean for bruchid resistance using the *Br* gene should not be difficult; however, evaluating for the resistance is time-consuming. To test for the resistance, dried seeds of tested genotypes must be harvested (takes about 3 months) and tested for the resistance (about 2 months); thus, the overall process of testing takes at least 5-month time. In addition, bruchid population must be maintained all the time. Moreover, all of the resistance germplasm, especially the wild mungbean TC1966, ACC41 and Sub2, possess some undesirable agronomic traits such as small seed size, pod shattering, hard seededness, photoperiod sensitivity and determinate growth habit. These defective traits may accompany when transferring the *Br* gene to the susceptible cultivars (linkage drag) (Watanasit and Pichitporn 1996). These problems hamper progress in genetic improvement of the resistance, although some bruchid-resistant mungbean cultivars have been developed using the *Br* gene (Lee et al. 2000; Yao et al. 2015). MAS can accelerate developing the resistant plants and reducing the linkage drag. Similar to the case of gene mapping for powdery mildew resistance in mungbean, mapping for bruchid resistance is among the earliest work done for insect resistance in crop plants. Young et al. (1992) used RFLP markers to locate the *Br* locus conferring *C. chinensis* resistance in an F_2 population developed from the cross between TC1966 and VC3890 by considering the resistance to *C. chinensis* as either qualitative or quantitative trait. Linkage analysis and QTL analysis gave the same result that the *Br* gene was located between RFLP probe pA882 and pM151on LG VIII. Later, additional RFLP markers were added to the mungbean genetic map and LG VIII containing the *Br* gene was re-named LG9. The *Br* locus was labeled as the *Bruc* locus on this map (Menancio-Hautea et al. 1993). Linkage mapping for the resistance to *C. chinensis* in the RIL population derived from the cross Berken × ACC41 using RAPD and AFLP markers showed that the resistance is linked to the RFLP marker pR26 (Imrie and Lambrides 1998) which is closely linked to the *Br* gene in TC1966 (Young et al. 1992), suggesting that they may be the same *Br* gene conditioning bruchid resistance in TC1966 and ACC41. Kaga and Ishimoto (1998) conducted fine-mapping of the *Br* gene in $BC_{20}F_2$ population of 414 individuals developed from TC1966 using RAPD and RFLP markers. The *Br* gene was narrowed down to a region of 0.7 cM between the RFLP markers Bng110 and Bng143. Bng143 is 0.2 cM away from the *Br* gene. Moreover, one $BC_{20}F_2$ individual produced vignatic acids but was susceptible to the bruchid. These suggested that the vignatic acids were not the principal factors responsible for bruchid resistance. QTL analysis by evaluating *C. chinensis* resistance in seeds of a RIL population derived from Berken × ACC41 across four environments (locations and years) revealed that the resistance was controlled by a single major QTL consistently located between RFLP markers mgM213 and VrCS161 (Mei et al. 2009). The QTL analysis also revealed that the resistance QTL was associated with a seed-weight QTL. All these gene mapping studies used RFLP markers which are not suitable for genome analysis and marker-assisted selection. So efforts have been made to develop or identify more suitable markers for the resistance. BAC clones containing RFLP marker mgM213 closely associated with the QTL for *C. chinensis* resistance in ACC41 (1.3 cM away from the resistance gene) were identified and used to develop six sequence-tagged site (STS) and one SSR markers (Miyagi et al. 2004). Two of those markers, STSbr1 and STSbr2, co-segregate with the marker mgM213. STSbr2 is also associated with the *Br* gene in V2709 (*Br1* gene in the original paper; Sun et al. 2008) and the *C. chinensis* resistance in Sub2 (Sarkar et al. 2011). Chen et al. (2007) developed some cleaved amplified polymorphism (CAP) markers closely linked to the *Br* gene in TC1996 from RAPD markers, being less than 1 cM away from the *Br* gene. Some CAP markers showed tighter linkage with the *Br* gene than the original RAPD markers. Expressed sequence tag-simple sequence repeat (EST-SSR) marker DMB-SSR158 (DMB158 in the original report) was found co-segregated perfectly with the *Br* gene in V2802 (Chotechung et al. 2011). SSR and amplified fragment length (AFLP) markers were added to the linkage map used by Chen et al. (2007) for QTL analysis of *C. chinensis* resistance. The result showed that DMB-SSR158 is tightly linked (<0.1 cM) with the major QTL for the resistance (Chen et al. 2013). The RFLP linkage map of RIL population used to locate QTL for the bruchid resistance in ACC41 (Mei et al. 2009) was saturated with SSR markers from mungbean and other related legumes, resulting in mapping the *Br* gene between SSR markers BM202 and Vr2-627 (*Br1* gene in the original paper; Wang et al. 2016; Zhao et al. 2010). BM202 is only 0.7 cM from the *Br* gene. Although most of these genomics studies had concluded that a single gene control bruchid resistance in mungbean, QTL analysis in 420 F_2 plants of the cross between Jangan carrying the *Br* gene from V2709 and Sunhwa, using SSR, STS and CAP markers revealed that resistance to *C. chinensis* is controlled by two major QTLs that are closely linked (Hong et al. 2015). One QTL was between markers GBssr-MB87 (MB87 in the original paper) to COPU11, while another QTL was between markers RP to COPU06. However, these two QTLs have not been confirmed.

Apart from the localization of *Br* gene and QTL for the bruchid resistance and identification of marker associated with the *Br* gene, identification of candidate gene(s) for the resistance has been carried out. The recently available reference mungbean genome sequence of mungbean line VC1973A (Kang et al. 2014) is greatly helpful for marker development, fine-mapping and gene identification in mungbean. Genome of bruchid-resistant recombinant inbred line 59 (RIL59) derived from the cross TC1966 × NM92 and transcriptome of RIL59, TC1966, NM92 together with pairs of resistant and susceptible RILs which derived from the same F_2 progenies of TC1966 × NM92 was sequenced (Liu et al. 2016). Next-generation sequencing and de novo assembly of the genome of the RIL59 revealed 42,223 genes. The number of annotated genes in RIL59 is 14,512 genes higher than that in the reference mungbean genome VC1973A. Comparison of the transcriptomic sequences of bruchid-resistant and -susceptible parental lines and their offspring revealed 91 differentially expressed genes (DEGs). The comparison also revealed 408 nucleotide variations between bruchid-resistant and -susceptible lines in regions spanning 2 kb of the promoters of 68 DEGs in which 282 nucleotide variations were found in exons of 148 sequence-changed-protein (SCPs) genes. Sixty-seven bruchid resistance-associated genes, including DEGs and SCPs, were mapped to the mungbean chromosome 5 where markers associated with the *Br* gene OPW02a4 and DMB-SSR158 are located. Based on the sequencing results, some gene-specific markers associated with the *Br* genes are developed, including *g779p*, *g34480p* and *g34458p*. *g779p* and *g34480p* exhibited 93.4% accuracy in predicting the resistance/susceptible genotypes. However, the marker DMB-SSR158 exhibited the highest accuracy, 98.3%. Jeong et al. (2015) developed molecular markers from BAC library made from genomic DNA of mungbean cultivar Jangannogdu that receive the *Br* gene from V2709 and used the markers in map-based cloning for *C. chinensis* resistance using 450 near-isogenic lines of a cross between Jangannogdu and Seonhwa (susceptible cultivar). The results showed that a gene, named *VrBURP1*, is the gene responsible for the *C. chinensis* resistance in V2709. *VrBURP1* encodes a protein containing 457 amino acids (Fig. 10.5). VrBURP1 protein is composed of a signal peptide, repeated units and C-terminal BURP (BNM2, USP, RD22 and PG1β) domain. They also showed that *VrBURP1* exists in V2709 and

V2808 and TC1966 possess the same *VrBURP1* allele, but different from V2709 and that *VrBURP1* exists in the resistant but not in the susceptible mungbean. A gene coding for a protein containing BURP domain is also found associated with the *C. chinensis* resistance in TC1966 (Lin et al. 2016). Bruchid resistance near-isogenic line VC6089A possessing the *Br* gene from TC1966 and its susceptible parent VC1973A was subjected to transcriptome sequencing and proteomic analysis. These analyses together identified three differentially expressed genes/differential proteins, including resistant-specific protein (*g39185*), gag/pol polyprotein (*g34458*) and aspartic proteinase (*g5551*). A real-time PCR analysis confirmed that these genes were highly expressed in resistant mungbeans (VC6089A, TC1966 and RIL59) and less expressed in susceptible mungbeans (VC1973A and NM92). These genes locate on chromosomes 5, 1, and 7, respectively. *g39185* and *g34458* genes encode a protein containing a BURP domain that covers a signal peptide, repeat regions and a C-terminal BURP domain. g39185 BURP protein comprises 402 amino acids (Fig. 10.5). The N-terminal BURP protein of g34458 contains 433 amino acids (Fig. 10.5). *g39185* locates at the position 5236,101 on the mungbean chromosome 5 near the *Br* gene (Lin et al. 2016; Liu et al. 2016) which corresponds to the annotated gene *LOC106759697* encoding a PG in the GenBank database. These results suggest that *g39185* is possibly the *Br* gene in TC1966. Sequence alignment between VrBURP1 and g39185 and N-terminal BURP protein of g34458 demonstrated that these proteins are very similar but still different (Fig. 10.5). This suggests that the *Br* genes in V2709 and TC1966 are different. BURP domain-containing proteins have only been found in plants. Expression patterns of BURP domain-containing proteins are diverse and most of their functions are still unknown, although they are reported to play important roles in maintaining normal plant metabolism or development (Shao et al. 2011; Tang et al. 2014; Li et al. 2016). The association between *VrBURP1* and *g39185* and the bruchid resistance in mungbean is interesting and worth further investigation.

Chotechung et al. (2016) performed high-resolution mapping for the *Br* gene in mungbean V2802. They used mungbean reference genome sequence data to identify locations of RFLP markers Bng110 and Bng143 flanking to the *Br* gene in TC1966 as reported by Kaga and Ishimoto (1998) and developed new SSR markers between these two markers for QTL analysis of the resistance to *C. chinensis* and *C. maculatus* in 418 $BC_{11}F_2$ plants derived from a cross between V2802 and Kamphaeng Saen 1 (KPS1). They found that Bng143 and Bng110 are 1.72 Mbp apart on the mungbean chromosome 5. QTL analysis revealed a single major QTL, *qBr*, locating between markers VrBr-SSR013 and DMB-SSR158 (Fig. 10.6). The *qBr* explained about 93% of the total resistance variation. When the resistance was considered as a qualitative trait, DMB-SSR158 gene co-segregated perfectly with the resistance. There are two annotated genes, *Vradi05g03940* and *Vradi05g03950*, between the markers VrBr-SSR013 and DMB-SSR158. In fact, VrBr-SSR013 and DMB-SSR158 are part of *Vradi05g03940* and *Vradi05g03950*, respectively (Fig. 10.6). *Vradi05g03940* (*VrPGIP1*) and *Vradi05g03950* (*VrPGIP2*) encode a polygalacturonase inhibitor (polygalacturonase-inhibiting protein; PGIP). Comparison of VrPGIP2 coding sequences between bruchid-resistant mungbeans V2802, V1128, V2817 and TC1966 and bruchid-susceptible mungbeans KPS1, Sulu-1, CM, together with an unknown accession mungbean lines revealed six single-nucleotide polymorphisms (SNPs) between the resistant and susceptible mungbeans (Fig. 10.7) (Chotechung et al. 2016). Three SNPs cause amino acid changes in the *VrPGIP2* sequence in the resistant mungbean (Fig. 10.8). The nucleotide sequence comparison also revealed that V2802, V1128, V2817 and TC1966 have the same *VrPGIP2* allele (Fig. 10.7). Based on these findings, they concluded that *VrPGIP2* is very likely the gene at the *Br* locus. The same approaches were used to investigate the candidate genes for the *Br* gene in 355 $BC_{11}F_2$ plants derived from a cross between

```
g39185              MEFQCLALFFSLIVILMAAQAALPPEVYWERMLPNTPIPKVIRQFSELDGGQEIASKDEF  60
N-ter-BURP-g34458   MEFQCLALFFSLIVILMAAQASLPSEVYWERXLPNTPIPKVIRQFSKQDGGKDIASKDEF  60
VrBURP1             MEFQCLALFFSLIVILMAAQASLPSEVYWKRKLPNTPIPKVIRQFSKQDGGKDIASKDEF  60
                    *********************:** ****:* **************: ***::*******

g39185              LLFGSGDKKNKDKLFHFGCGDKKNELQDDVKDISPEDENALLFYTKYANKKNELQDDVQD  120
N-ter-BURP-g34458   LLFGSGDKKNKDKLLRFGCGDKENKLQDDVQDISPEDENLLLIYDRYAK--NKLQDDLQD  118
VrBURP1             LLFGSGDKKNKDKLLRFGCGDKENKLQDDVQDISPEDENLLLIYDRYAK--NKLQDDVQD  118
                    **************::******:*:*****:******** **:* :**:  *:****:**

g39185              ISPEDENALLFYTKYANKKNELQDDVQDISPEDENALLLYTKYAN---------------  165
N-ter-BURP-g34458   ISPEDENFLLFYDRYAK--NKLQDDVQDISPEDENLLLFYDRYAKNKLQDNVQDISPEDE  176
VrBURP1             ISPEDENLLLFYDRYAK--NKLQDDVQDISPEDENLLLFYDRYAKNKLQDDVQDISPD-E  175
                    ******* **** :**:  *:************** **:* :**:

g39185              ---------------------------------------------KKNGVVLRRIGPIVN  180
N-ter-BURP-g34458   NLLDYEKSKLQDDVQDISPEDEN------------------LSGYKKNGVVLRGIGPIAN  218
VrBURP1             NLLDYQENKLQDDVQHISPEDENLLDYEKNKLQDDVQDISPLSGYKKNGVVLRGIGPIAN  235
                                                                ******** ****.*

g39185              HHHHDHLKPSNFFSEERLRRGAKLDVLFRKRNFSTPLLTREIAEHLPFSSEKINEILEIL  240
N-ter-BURP-g34458   HHHHDHLKPSSYFSEEGLRRGAKLVMLFHKRKFSTPLLTREIAEHLPFSSEKINEILEIL  278
VrBURP1             HHHHDHLKPSSYFSEEGLRRGAKLVMLFHKRKFSTPLLTREIAEHLPFSSEKINEILEIL  295
                    **********.:**** ******* :**:**:****************************

g39185              AVKPDSKDAKNVEETLNHCEKPALKGEEKQCATSVESMVDFVTSKLGNNARVTSTELEIG  300
N-ter-BURP-g34458   AVKPDSKNAKNVEKTLNNCEEPALKGEEKHCATSVESMVDFVTSKLGNNARVTSTELEIE  338
VrBURP1             AVKPDSKNAKNVEKTLNNCEEPALKGEEKHCATSVESMVDFVTSKLGNNARVTSTELEIE  355
                    *******:*****:***:**:********:*****************************

g39185              SKFQKFIVKDGVKILAEEKIIACHPMSYPYVVFYCHKMANSTAHFLPLEGEDGTRVKAVA  360
N-ter-BURP-g34458   SKFQKFIVKDGVKILAEEEIIACHPMSYPYVVFYCHKMSNSTAHVVPLEGEDGTRVKAIV  398
VrBURP1             SKFQKFIVKDGVKILAEEEIIACHPMSYPYVVFYCHKMSNSTAHVVPLEGEDGTRVKAIV  415
                    *****************:*******************:*****.:************:.

g39185              ICHKDTSQWDPHHVAFQVLKVKPGTSSACHFFPEGHLVWYAK    402
N-ter-BURP-g34458   ICHKDTSQWDPDHVAFQVLKVKPGTSPVCHFFPND-------    433
VrBURP1             ICHKDTSQWDPDHVAFQVLKVKPGTSPVCHFFPNGHLLWYAK    457
                    ***********.**************  .*****:.
```

Fig. 10.5 Alignment of amino acid sequences of VrBURP1, g39185 and N-terminal BURP protein of g34458 (N-ter-BURP-g34458). Bold amino acids represent BURP domain. VrBURP1 sequence is from mungbean Jangannogdu that harbors the *Br* gene from V2709 (Jeong et al. 2015). g39185 and N-terminal BURP protein of g34458 sequences are from VC6089A that obtains the *Br* gene from TC1966 (Lin et al. 2016)

V2709 and KPS1 (Kaewwongwal et al. 2017). The results showed that a single major QTL, *qBr5.1*, for the bruchid resistance in V2709 locate in a confident region between markers VRID5 and VrBr-SSR037 corresponding to the positions 5,410,272 to 5,647,621 on the mungbean chromosome 5. *qBr5.1* accounted for about 93% of the bruchid resistance variation. Eight annotated genes including *VrPGIP1* and *VrPGIP2* are in these regions (Fig. 10.6). Sequence analysis revealed that V2709 possesses different *VrPGIP1* and *VrPGIP2*, compared to V2802. Amino acid mutations in VrPGIP2 (Fig. 10.8) and VrPGIP1 (Fig. 10.9) of V2709 occur in the regions that may affect interactions between PGIP and polygalacturonase (Kaewwongwal et al. 2017). These results suggest that the tightly linked markers *VrPGIP1* and *VrPGIP2* may be the genes at the *Br* locus in mungbean V2709. Many SSR, STS and InDel markers locating very close to the *Br* gene were developed (Chotechung et al. 2016; Kaewwongwal et al. 2017). Recently, Liu et al. (2018) mapped *Br* gene for *C. chinensis* resistance in V1128 using an F_2 population. Gene mapping by considering the resistance as a qualitative trait and quantitative trait gave the same result that the *Br* gene (*Br3* in the original paper) was located between SSR markers DMB-SSR158 and a cluster of markers (VrID5, VrBr-SSR032 and VrBr-SSR033) (Fig. 10.6). The QTL region corresponded to the position 5,310,107 to 5,597,891 (288 kb) on the chromosome 5. There were ten annotated genes including *VrPGIP1* and *VrPGIP2* in this region (Fig. 10.6). These further indicate the association between *VrPGIP1* and *VrPGIP2* and the bruchid resistance/*Br* gene. Therefore, the results from fine-mapping for the bruchid resistance in V1128, V2709 and V2808 suggest that *VrPGIP2* alone or *VrPGIP2* and *VrPGIP1* are the candidate gene(s) at the *Br* locus in these resistant mungbeans. Enzyme inhibitor(s) is a major class of plant biochemical that plays pivotal roles in plant defense against phytophagous insects (Lawrence and Koundal 2002), including bruchids (Ishimoto and Kitamura 1989; Shade et al. 1994; Ishimoto et al. 2006). Polygalacturonase (PG) is an important digestive enzymes found in the midgut of *C. maculatus* (Pedra et al. 2003; Pauchet et al. 2010; Nogueira et al. 2012). PG hydrolyzes α-1,4-glycosidic bonds between galacturonic acid units and acts preferentially on pectic acid. PGs and pectinesterases degrade pectin into oligosaccharides that can be absorbed in the insect gut. VrPGIP2 may provide resistance to the bruchids by inhibiting PG in the bruchid midgut to digest seed starch and thus causing growth and development retardation of the bruchids.

One of the problems in identifying the gene responsible for bruchid resistance in mungbean is the genome re-arrangement around the QTL region containing the *Br* gene or the QTL for the resistance. In a recent study, QTL mapping for *C.*

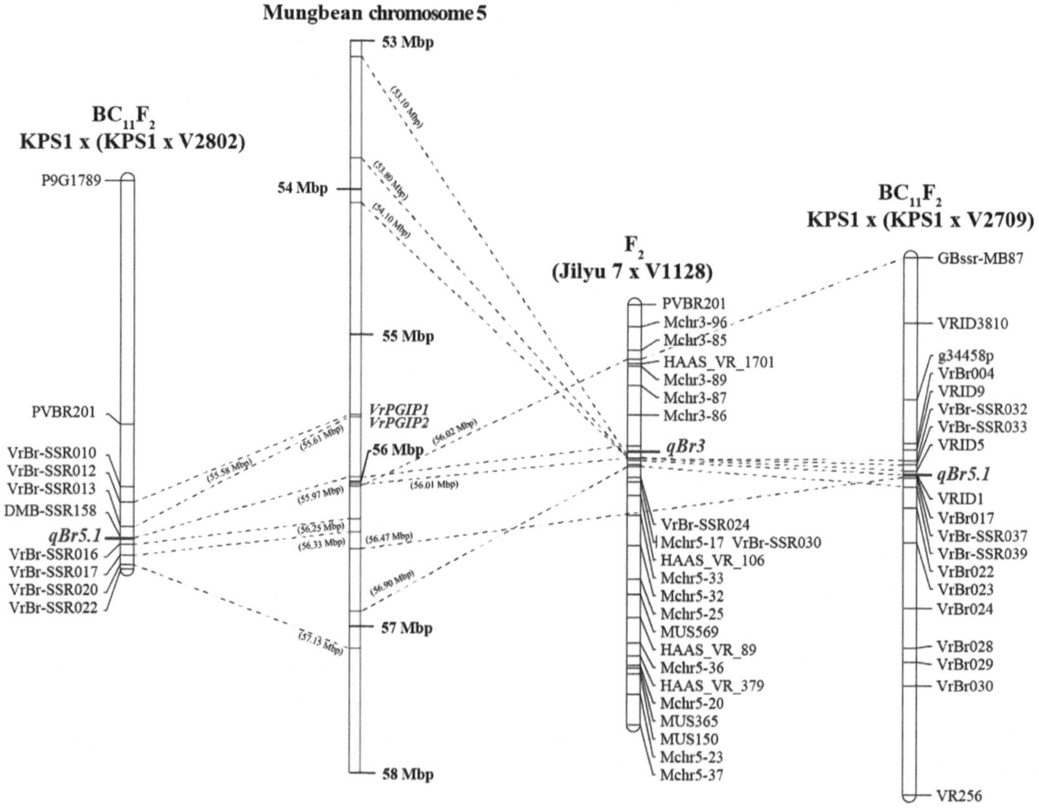

Fig. 10.6 Location of the *Br* locus detected in the $BC_{11}F_2$ [KPS1 × (KPS1 × V2802)] (Chotechung et al. 2016), $BC_{11}F_2$ [KPS1 × (KPS1 × V2709)] (Kaewwongwal et al. 2017) and F_2 (Jilyu7 × V1128) (Liu et al. 2018) populations and its positions and corresponding genes on chromosome 5 of the mungbean reference genome

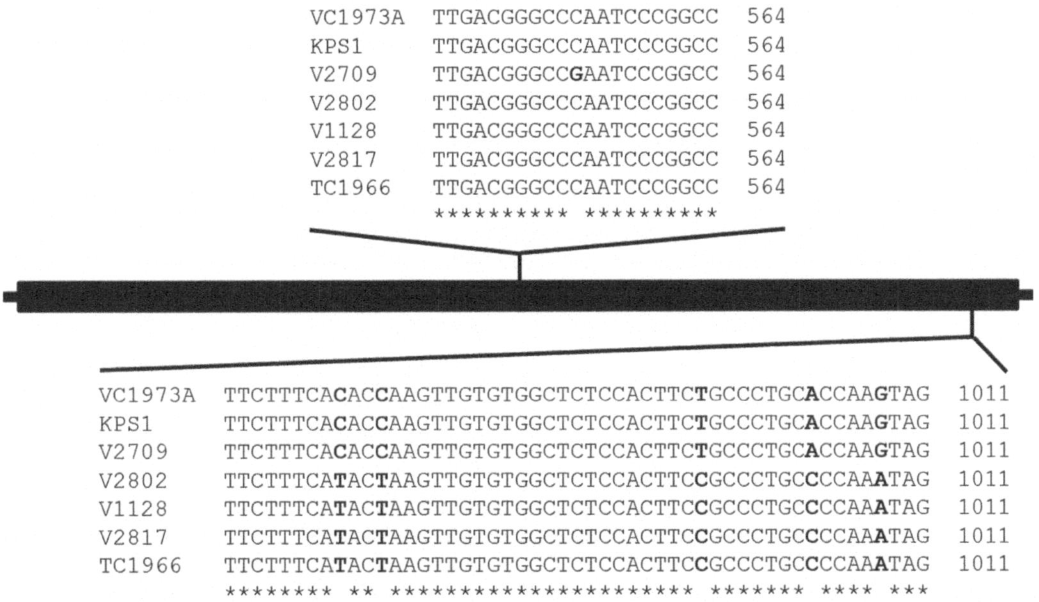

Fig. 10.7 Single-nucleotide polymorphism in the *VrPGIP2* (LOC106760237) sequences in mungbean accession VC1973A (reference sequence), KPS1, V2709, V2802, V1128, V2817 and TC1966. VC1973A and KPS1 are susceptible to the bruchids, while V1128, V2709, V2802, V2817 and TC1966 are resistant to the bruchids. Polymorphic sites are presented in bold

chinensis resistance in two RIL populations (one with 61 F_{12} lines from the cross TC1966 × NM92 and another with 150 F_7 lines from the cross V2802 × NM94) using SNP markers from genotyping by sequencing (Schafleitner et al. 2016). In both populations, the linkage map constructed by ordering the markers based on their positions on the reference genome sequence of VC1973A and the one constructed based on recombination frequency are strikingly different, especially the chromosome 5 where the *Br* locus locating. In the latter mapping strategy, chromosome 5 of the map TC1966 × NM92 was split into three LGs and the QTL region for the bruchid resistance contained markers from chromosomes 5, 4 and 3. Similarly, chromosome 5 of the map V2802 × NM94 split into two LGs and the QTL region for the resistance harbored markers from chromosomes 5, 4 and 3. More importantly, the orders of the markers from the chromosome 5 on the linkage map were highly different from the reference genome. These implied genome reorganizations in the reference genome VC1973A. When some GBS-based SNP markers around the QTL regions were converted into PCR-based SNP markers (CAPS and dCAPS markers) and used together with some other PCR-based markers for QTL mapping of the resistance, the results showed that the QTL is located on the chromosome 5 and a gene between positions 5,178,332 and 6,066,948 on this chromosome confers the resistance. SNPs associated with amino acid changes were found in *Vradi05g03780*, *Vradi05g03980* and *Vradi05g04130*. SNPs in *Vradi05g04130* were detected in both TC1966 and V2802. This gene probably codes for LRR receptor-like serine/threonine-protein kinase and can be considered as another candidate gene for the *Br* locus. In addition, the study by Schafleitner et al. (2016) also validated the association of the markers DMB-SSR158, *g779p*, *g34480p* and *g34458p* with the bruchid resistance.

Similar to the case of yellow mosaic disease, mungbean scientists are also interested in the bruchid resistance from the secondary gene pool. Black gram and rice bean are more resistance to bruchids than mungbean (Tomooka et al. 2000).

```
VC1973A   MRRLLMMLGLYVLVLSPFPVALSERCHPQDKKVLLRFKKELNNPYFLASWDPKEDCCDWY   60
KPS1      MRRLLMMLGLYVLVLSPFPVALSERCHPQDKKVLLRFKKELNNPYFLASWDPKEDCCDWY   60
V2709     MRRLLMMLGLCVQVLSLFPVALSERCHPQDKKVLLRFKKELNNPYFLASWDPKEDCCDWY   60
V2802     MRRLLMMLGLYVLVLSPFPVALSERCHPQDKKVLLRFKKELNNPYFLASWDPKEDCCDWY   60
V1128     MRRLLMMLGLYVLVLSPFPVALSERCHPQDKKVLLRFKKELNNPYFLASWDPKEDCCDWY   60
V2817     MRRLLMMLGLYVLVLSPFPVALSERCHPQDKKVLLRFKKELNNPYFLASWDPKEDCCDWY   60
TC1966    MRRLLMMLGLYVLVLSPFPVALSERCHPQDKKVLLRFKKELNNPYFLASWDPKEDCCDWY   60
          ********** * *** *******************************************

VC1973A   CVKCDDRNNRIYTIFLKSDEPDPNVTGQIPPSVGDLPYLRGLNFHNLPNLVGPIQPTLAK   120
KPS1      CVKCDDRNNRIYTIFLKSDEPDPNVTGQIPPSVGDLPYLRGLNFHNLPNLVGPIQPTLAK   120
V2709     CIKCDDRNNRIYTIFLRSDEPDPNVTGQIPPSVGDLPYLRGLSLHNLPNLVGPIQPTLAK   120
V2802     CVKCDDRNNRIYTIFLKSDEPDPNVTGQIPPSVGDLPYLRGLNFHNLPNLVGPIQPTLAK   120
V1128     CVKCDDRNNRIYTIFLKSDEPDPNVTGQIPPSVGDLPYLRGLNFHNLPNLVGPIQPTLAK   120
V2817     CVKCDDRNNRIYTIFLKSDEPDPNVTGQIPPSVGDLPYLRGLNFHNLPNLVGPIQPTLAK   120
VC1973A   CVKCDDRNNRIYTIFLKSDEPDPNVTGQIPPSVGDLPYLRGLNFHNLPNLVGPIQPTLAK   120
          *:**************:*************************.:****************

VC1973A   LHKLEGIFFSNTGISGPIPEFLAHIKTLQYIELSSNRLSGPIPSSLSQLPNLISLQLDRN   180
KPS1      LHKLEGIFFSNTGISGPIPEFLAHIKTLQYIELSSNRLSGPIPSSLSQLPNLISLQLDRN   180
V2709     LHKLEGIFISNTGISGPIPEFLAQIKTLQYIELSSNRLSGPIPSSLSQLPNLISLQLDRN   180
V2802     LHKLEGIFFSNTGISGPIPEFLAHIKTLQYIELSSNRLSGPIPSSLSQLPNLISLQLDRN   180
V1128     LHKLEGIFFSNTGISGPIPEFLAHIKTLQYIELSSNRLSGPIPSSLSQLPNLISLQLDRN   180
V2817     LHKLEGIFFSNTGISGPIPEFLAHIKTLQYIELSSNRLSGPIPSSLSQLPNLISLQLDRN   180
TC1966    LHKLEGIFFSNTGISGPIPEFLAHIKTLQYIELSSNRLSGPIPSSLSQLPNLISLQLDRN   180
          ********:**************:************************************

VC1973A   KLTGTIPASFGSFKKPGPDLILSHNQLSGPIPTSLGNLDPDRIDLSGNNFVGDASFLIGS   240
KPS1      KLTGTIPASFGSFKKPGPDLILSHNQLSGPIPTSLGNLDPDRIDLSGNNFVGDASFLIGS   240
V2709     KLTGPIPASFGSFKKPGPDLILSHNKLSGPIPASLGNLDPDRIDLSRNNFVGDASFLFGS   240
V2802     KLTGTIPASFGSFKKPGPDLILSHNQLSGPIPTSLGNLDPDRIDLSGNNFVGDASFLIGS   240
V1128     KLTGTIPASFGSFKKPGPDLILSHNQLSGPIPTSLGNLDPDRIDLSGNNFVGDASFLIGS   240
V2817     KLTGTIPASFGSFKKPGPDLILSHNQLSGPIPTSLGNLDPDRIDLSGNNFVGDASFLIGS   240
TC1966    KLTGTIPASFGSFKKPGPDLILSHNQLSGPIPTSLGNLDPDRIDLSGNNFVGDASFLIGS   240
          ****.********************:******:************* **********:**

VC1973A   KKKIRVFDLSRNAFSFDLSRLTFPKETLTWLDISHNNIYGSIPTALTKVRYFQQLNFSFN   300
KPS1      KKKIRVFDLSRNAFSFDLSRLTFPKETLTWLDISHNNIYGSIPTALTKVRYFQQLNFSFN   300
V2709     KKKTQILDLSRNAFSFDLSRLTFPKETLTWLDISHNKIYGSIPMALTKVRYFQQLNFSFN   300
V2802     KKKIRVFDLSRNAFSFDLSRLTFPKETLTWLDISHNNIYGSIPTALTKVRYFQQLNFSFN   300
V1128     KKKIRVFDLSRNAFSFDLSRLTFPKETLTWLDISHNNIYGSIPTALTKVRYFQQLNFSFN   300
V2817     KKKIRVFDLSRNAFSFDLSRLTFPKETLTWLDISHNNIYGSIPTALTKVRYFQQLNFSFN   300
TC1966    KKKIRVFDLSRNAFSFDLSRLTFPKETLTWLDISHNNIYGSIPTALTKVRYFQQLNFSFN   300
          *** :::*****************************:****** ****************

VC1973A   PGLKGQIPQGGGLQTFDKYSYFHTNLCGSPLPPCTK   336
KPS1      PGLKGQIPQGGGLQTFDKYSYFHTNLCGSPLPPCTK   336
V2709     PGLKGQIPQGGALQTFDKYSYFHTNLCGSPLPPCTK   336
V2802     PGLKGQIPQGGGLQTFDKYSYFHTNLCGSPLPPCTK   336
V1128     PGLKGQIPQGGGLQTFDKYSYFHTNLCGSPLPPCTK   336
V2817     PGLKGQIPQGGGLQTFDKYSYFHTNLCGSPLPPCTK   336
TC1966    PGLKGQIPQGGGLQTFDKYSYFHTNLCGSPLPPCTK   336
          ***********.************************
```

Fig. 10.8 Alignment of the protein sequences encoded by *VrPGIP2* (LOC106760236) from the accession VC1973A (reference sequence), KPS1, V2709, V2802, V1128, V2817 and TC1966. VC1973A and KPS1 are susceptible to the bruchids, while V2709, V2802, V1128, V2817 and TC1966 are resistant to the bruchids. Polymorphic sites are presented in bold

```
GBXO01011683    MRSLLMMLGLCVLVLSPFPVAVSERCHPQDKKVLLRFKKELNNPYFLASWDPKEDCCDWY  60
Sulu_1          MRSLLMMLGLCVLVLSPFPVAVSERCHPQDKKVLLRFKKELNNPYFLASWDPKEDCCDWY  60
CM              MRSLLMMLGLCVLVLSPFPVAVSERCHPQDKKVLLRFKKELNNPYFLASWDPKEDCCDWY  60
KPS1            MRSLLMMLGLCVLVLSPFPVAVSERCHPQDKKVLLRFKKELNNPYFLASWDPKEDCCDWY  60
V2709           MRSLLMMLGLCVLVLSPFPVAVSERCHPQDKKVLLRFKKELNNPYFLASWDPKEDCCDWY  60
V2802           MRSLLMMLGLCVLVLSPFPVAVSERCHPQDKKVLLRFKKELNNPYFLASWDPKEDCCDWY  60
V1128           MRSLLMMLGLCVLVLSPFPVAVSERCHPQDKKVLLRFKKELNNPYFLASWDPKEDCCDWY  60
V2802           MRSLLMMLGLCVLVLSPFPVAVSERCHPQDKKVLLRFKKELNNPYFLASWDPKEDCCDWY  60
TC1966          MRSLLMMLGLCVLVLSPFPVAVSERCHPQDKKVLLRFKKELNNPYFLASWDPKEDCCDWY  60
                ************************************************************

GBXO01011683    CVKCDDRNNRIYEVFLVSSFPDPNVTGQIPPSVGDLPYLEYLDFHKLPNLVGPIPTTITK  120
Sulu_1          CVKCDDRNNRIYEVFLVSSFPDPNVTGQIPPSVGDLPYLEYLDFHKLPNLVGPIPTTITK  120
CM              CVKCDDRNNRIYEVFLVSSFPDPNVTGQIPPSVGDLPYLEYLDFHKLPNLVGPIPTTITK  120
KPS1            CVKCDDRNNRIYEVFLVSSFPDPNVTGQIPPSVGDLPYLEYLDFHKLPNLVGPIPTTITK  120
V2709           CVKCDDRNNRIYEVFLVSSFPDPNVTGQIPPSVGDLPYLEYLDFHKLPNLVGPIPTTITK  120
V2802           CVKCDDRNNRIYEVFLVSSFPDPNVTGQIPPSVGDLPYLEYLDFHKLPNLVGPIPTTITK  120
V1128           CVKCDDRNNRIYEVFLVSSFPDPNVTGQIPPSVGDLPYLEYLDFHKLPNLVGPIPTTITK  120
V2817           CVKCDDRNNRIYEVFLVSSFPDPNVTGQIPPSVGDLPYLEYLDFHKLPNLVGPIPTTITK  120
TC1966          CVKCDDRNNRIYEVFLVSSFPDPNVTGQIPPSVGDLPYLEYLDFHKLPNLVGPIPTTITK  120
                ************************************************************

GBXO01011683    LTKLKSLSITYTGLSGSIPEFLGQIKTLEYIALSFNSLSGSIPSSLSQLPNLGSLQLDRN  180
Sulu_1          LTKLKSLSITYTGLSGSIPEFLGQIKTLEYIALSFNSLSGSIPSSLSQLPNLGSLQLDRN  180
CM              LTKLKSLSITYTGLSGSIPEFLGQIKTLEYIALSFNSLSGSIPSSLSQLPNLGSLQLDRN  180
KPS1            LTKLKSLSITYTGLSGSIPEFLGQIKTLEYIALSFNSLSGSIPSSLSQLPNLGSLQLDRN  180
V2709           LTKLKSLSITYTGLSGSIPEFLGQIKTLEYIALSFNSLSGSIPSSLSQLPNLGSLQLDRN  180
V2802           LTKLKSLSITYTGLSGSIPEFLGQIKTLEYIALSFNSLSGSIPSSLSQLPNLGSLQLDRN  180
V1128           LTKLKSLSITYTGLSGSIPEFLGQIKTLEYIALSFNSLSGSIPSSLSQLPNLGSLQLDRN  180
V2817           LTKLKSLSITYTGLSGSIPEFLGQIKTLEYIALSFNSLSGSIPSSLSQLPNLGSLQLDRN  180
TC1966          LTKLKSLSITYTGLSGSIPEFLGQIKTLEYIALSFNSLSGSIPSSLSQLPNLGSLQLDRN  180
                ************************************************************

GBXO01011683    KLTGPIPASFGSFKKPGPDLKLSHNQLSGTLPPSLGNLDPDNIDLSRNNLVGDASFLFGS  240
Sulu_1          KLTGPIPASFGSFKKPGPDLKLSHNQLSGTLPPSLGNLDPDNIDLSRNNLVGDASFLFGS  240
CM              KLTGPIPASFGSFKKPGPDLKLSHNQLSGTLPPSLGNLDPDNIDLSRNNLVGDASFLFGS  240
KPS1            KLTGPIPASFGSFKKPGPDLKLSHNQLSGTLPPSLGNLDPDNIDLSRNNLVGDASFLFGS  240
V2709           KLTGRIPASFGSFKKPGPDLKLSHNQLSGTLPPSLGNLDPDNIDLSRNNLVGDASFLFGS  240
V2802           KLTGPIPASFGSFKKPGPDLKLSHNQLSGTLPPSLGNLDPDNIDLSRNNLVGDASFLFGS  240
V1128           KLTGPIPASFGSFKKPGPDLKLSHNQLSGTLPPSLGNLDPDNIDLSRNNLVGDASFLFGS  240
V2817           KLTGPIPASFGSFKKPGPDLKLSHNQLSGTLPPSLGNLDPDNIDLSRNNLVGDASFLFGS  240
TC1966          KLTGPIPASFGSFKKPGPDLKLSHNQLSGTLPPSLGNLDPDNIDLSRNNLVGDASFLFGS  240
                **** *******************************************************

GBXO01011683    KKKTQVLDLSRNAFSFDLSPVTFPKKTLIWFDINHNKIYGSLPVALTKVENLQQLNVSYN  300
Sulu_1          KKKTQVLDLSRNAFSFDLSPVTFPKKTLIWFDINHNKIYGSLPVALTKVENLQQLNVSYN  300
CM              KKKTQVLDLSRNAFSFDLSPVTFPKKTLIWFDINHNKIYGSLPVALTKVENLQQLNVSYN  300
KPS1            KKKTQVLDLSRNAFSFDLSPVTFPKKTLIWFDINHNKIYGSLPVALTKVENLQQLNVSYN  300
V2709           KKKTQVLDLSRNAFSFDLSPVTFPKKTLIWFDINHNKIYGSLPVALTKVENLQQLNVSYN  300
V2802           KKKTQVLDLSRNAFSFDLSPVTFPKKTLIWFDINHNKIYGSLPVALTKVENLQQLNVSYN  300
V1128           KKKTQVLDLSRNAFSFDLSPVTFPKKTLIWFDINHNKIYGSLPVALTKVENLQQLNVSYN  300
V2817           KKKTQVLDLSRNAFSFDLSPVTFPKKTLIWFDINHNKIYGSLPVALTKVENLQQLNVSYN  300
TC1966          KKKTQVLDLSRNAFSFDLSPVTFPKKTLIWFDINHNKIYGSLPVALTKVENLQQLNVSYN  300
                ************************************************************

GBXO01011683    PQLKGQIPQGGELHRFDKYAFFHTKLCGSPLLPCTK  336
Sulu_1          PQLKGQIPQGGELHRFDKYAFFHTKLCGSPLLPCTK  336
CM              PQLKGQIPQGGELHRFDKYAFFHTKLCGSPLLPCTK  336
KPS1            PQLKGQIPQGGELHRFDKYAFFHTKLCGSPLLPCTK  336
V2709           PQLKGQIPQGGELHRFDKYAFFHTKLCGSPLLPCTK  336
V2802           PQLKGQIPQGGELHRFDKYSFFHTKLCGSPLPPCPK  336
V1128           PQLKGQIPQGGELHRFDKYSFFHTKLCGSPLPPCPK  336
V2817           PQLKGQIPQGGELHRFDKYSFFHTKLCGSPLPPCPK  336
TC1966          PQLKGQIPQGGELHRFDKYSFFHTKLCGSPLPPCPK  336
                *******************:*********** **.*
```

Fig. 10.9 Alignment of the deduced amino acid sequences encoded by *VrPGIP1* in Sulu_1, CM, GBXO01011683, KPS1, V1128, V2709, V2802, V2817 and TC1966. Sulu_1, CM, GBXO01011683 and KPS1 are susceptible to the bruchids, while V1128, V2709, V2802, V2817 and TC19966 are resistant to the bruchids. Polymorphic sites are presented in bold

By employing the same RIL population derived from interspecific hybridization between mungbean and rice bean (VRM(Gg)1 × TNAU RED) used for QTL mapping for MYMV resistance reported by (Mathivathana et al. 2019), Mariyammal et al. (2019) reported identification of *C. maculatus* resistance in this population grown in two years by evaluating percentage of damaged seeds, developmental period and percent of adult emergence in two. In the population, some RILs possessed seeds showing typical characteristics of mungbean but resistance to *C. maculatus* similar to the rice bean. In each year, the number of QTLs detected for each trait was one or two. At some QTLs, alleles from susceptible also conferred the resistance. However, only two QTLs, *qSD05* for percentage of damaged seeds and *qAE08* for percent of adult emergence, were consistently detected in both years. *QSD05* was mapped between the positions 17,429,208 and 19,012,443 (1.58 Mb) on mungbean chromosome 5, whereas *qAE08* was mapped between the positions 541,709,231 and 44,902,000 (3.19 Mb) on mungbean chromosome 8. These QTL accounted for about 35% and 10% of the total variation trait variation, respectively. Based on the functions of the annotated genes in the 1.58-Mb genomic region for the *QSD05*, 17 candidate genes were proposed as candidate genes for the resistance (Mariyammal et al. 2019). Among those 17 genes, *Vradi05g10460* predicted to have protein kinase superfamily protein/concanavalin A-like lectin/- glucanase/protein phosphorylation function is interesting. Artificial seeds containing Concanavalin A isolated from jack bean (*Canavalia ensiformis* (L.) DC) seeds are highly toxic to *C. maculatus* (Oliveira et al. 1999). Artificial seeds possessing a lectin isolated from seeds of leguminous plant *Griffonia simplicifolia* are detrimental to *C. maculatus* (Zhu et al. 1996). *Vradi05g10460* can be a major focused candidate gene for bruchid resistance from rice bean for breeding for the resistance in mungbean.

10.6 Bean Bug

Bean bug or pod sucking bug (*Riptortus pedestris* Fab.) (Heteroptera: Alydidae) (also known as *Riptortus clavatus*) is an insect pest of grain legumes including mungbean. This insect is an important pest of soybean (*Glycine max* (L.) Merr.) in Japan and Korea (Panizzi et al. 2000). Adults and nymphs of the bean bug can cause considerable yield loss and damage to seed quality of legume crops by sucking sap from developing pods and seeds (Sehgal and Ujagir 1988). Kang et al. (2003) reported that bean bug caused damage to young pods of legume crops such as soybean, mungbean and cowpea from flowering time to pod ripening period. Bean bug feeding can also transmit pathogens to the legume plants. Bean bug develops into adult stage and reproduces successively by feeding on dried legume seeds and water (Kadosawa and Santa 1981). Although there has been no report on genetic improvement for the resistance to bean bug in mungbean, there are reports on genetics and genome analysis of the resistance to this insect. The wild mungbean TC1966 and cultivated mungbean V2709 which possessing the *Br* gene conferring resistance to bruchids are also resistant to bean bug (Ishimoto and Kitamura 1993; Jeong et al. 2015; Hong et al. 2015). The resistance is due to the biochemical in seeds that inhibits nymphal growth of the bean bug (Ishimoto and Kitamura 1991). Jeong et al. (2015) isolated an active compound that shows inhibitory growth effect against bean bug nymphs from bruchid resistance mungbean. The active compound has 4–5 carbonyl functional groups. The *Br* gene possibly confers the resistance to bean bug too (Ishimoto and Kitamura 1993; Kaga and Ishimoto 1998). However, Hong et al. (2015) demonstrated that the resistance in mungbean Jagannogdu that obtained the *Br* gene from V2709 is controlled by two genes with dominant suppression epistasis and that the resistance genes for bean bug and bruchid are different but closely linked to each other. Nonetheless, QTL analysis in 420 F_2 plants of the cross between Jangannogdu and Sunhwa using a partial linkage map of SSR, STS and CAP markers revealed that the resistance to bean bug is controlled by a single major QTL between markers RP and COPU06. This QTL explains 42% of the resistance variation. This QTL co-localized with a major QTL for *C. chinensis* resistance. However, the QTL for the bean bug has not yet been confirmed. In addition, due to a partial linkage map that was used in the QTL analysis, other QTLs for the resistance, if exists, are not detected in this analysis. Co-localization between QTLs for the bean bug and bruchid resistance is interesting. *VrPGIP1* and *VPGIP2* are the candidate genes for bruchid resistance in mungbean (Chotechung et al. 2016; Kaewwongwal et al. 2017). PGIPs play roles in the feeding behavior of heteropteran insects. PvPGIP3 and PvPGIP4 in common bean (*Phaseolus vulgaris* L.) inhibit PGs of mirid bugs (Frati et al. 2006). *VrPGIP1* and *VrPGIP2* in the resistance mungbean may also confer resistance to the bean bug by preventing the digestion of plant tissue and seeds by insect PGs.

10.7 New Genomics Approaches for Resistance to Disease and Insect Resistance in Mungbean

Forward genetics, especially QTL mapping, is a major approach in genome analysis of reaction to biotic stresses in mungbean. Although several QTLs have been identified for resistance to powdery mildew disease, Cercospora leaf spot disease, yellow mosaic disease, bruchids and pod sucking bug in mungbean, the QTL intervals are usually large. Conventional QTL mapping requires genotyping of many individuals in a segregating population using a large number of suitable DNA markers such as SNPs and SSRs. Not many markers of these types have been developed for mungbean as compared to the other legumes with similar economic importance. In addition, SSR polymorphism in mungbean is low (Chankaew et al. 2011, 2013; Gwag et al. 2006; Isemura et al. 2012; Kasettranan et al. 2010; Seehalak et al. 2009; Somta et al. 2008b, 2009; Tangphatsornraung et al. 2009). Nonetheless, since the number of genes/QTL controlling resistance to the biotic stresses mentioned in this chapter is only one or two genes (Alam et al. 2014a; Chankaew et al. 2011, 2013; Hong et al. 2015; Ishimoto and Kitamura 1993; Jeong et al. 2015; Kitamura et al. 1988; Reddy et al. 1994; Reddy 2009; Somta et al. 2007; Thakur et al. 1977), it would not be difficult to identify the candidate causative gene for the resistance. This can be achieved by combining a bulked segregant analysis and a whole-genome resequencing (WGR) by next-generation sequencing (NGS), known as Bulk-Seq (Zou et al. 2016) or QTL-Seq (Takagi et al. 2013). BSA-Seq can be used to rapidly genotype the resistance versus susceptible DNA or RNA pools at thousands of SNPs spreading across the genome. The marker data generated from BSA-Seq can reveal a narrow genomic region harboring the resistance gene or even the lesion of the resistance gene (Zou et al. 2016). BSA-Seq has been used to identify the causative genes for qualitative and quantitative traits in plant species. For disease and insect resistance, BSA coupled with WGR has been used successfully in identifying a candidate gene for tomato yellow leaf curl virus resistance in tomato (*Solanum lycopersicum* L.) (Wang et al. 2018), major QTL for blast disease resistance in rice (*Oryza sativa* L.) (Takagi et al. 2013) and the candidate genes for aphid (*Aphis gossypii* Glover) resistance in cucumber (*Cucumis sativus* L.) (Liang et al. 2016).

Due to a large difference in genome size of mungbean cultivars and the reorganization in the reference mungbean genome of VC1973A as clearly shown in genome resequencing and genome mapping of the bruchid resistance in mungbean (Liu et al. 2016; Schafleitner et al. 2016). Additional reference genomes of the mungbean with higher quality and more completeness should be developed for a more precise genome analysis of important traits including biotic stress resistance.

10.8 Conclusion

Genetic improvement for resistance to powdery mildew disease, Cercospora leaf spot disease, yellow mosaic disease, bruchids and bean bug has become permanent goals in breeding for biotic stress resistance in mungbean. Resistance germplasm and information on genetics of the resistance in these biotic stresses are available. However, a bottleneck to incorporate genes conferring the resistance and screen for the resistant genotypes is that evaluation for the resistance is environmental-dependent. In most cases, the resistance is controlled by one or two major genes/QTLs providing highly resistance, application of a marker-assisted breeding, especially marker-assisted backcross selection, has a high potential for improving the biotic stress resistance in mungbean. Information on the resistance at the genome level and genomic tools is well developed and thus can be readily used for the development of bruchid-resistant genotypes. In addition, the *Br* gene for the bruchid resistance is interesting as it confers the broad-spectrum resistance to bruchid biotypes and species and possibly to bean bug as well. Gene cloning of the *Br* gene can be useful for the breeding of bruchid and insect resistance in other crops. The genomics resources developed and the candidate genes identified for the bruchid resistance from the recent studies should facilitate the identification of the *Br* gene in the near future. In cases of powdery mildew disease, Cercospora leaf spot disease and yellow mosaic disease, progress in their genome analyses may be achieved by comparative genomics approaches, especially the powdery mildew in which *MLO* genes can be used as candidate genes for the resistance. To facilitate identification of the QTLs and causative genes for biotic stress resistance in mungbean, array-based high-throughput DNA markers and genotyping platforms should be developed possibly by a consortium of mungbean research laboratories. However, to conduct a marker-assisted selection effectively, suitable genotyping platforms such as kompetitive allele-specific PCR (KASP) for SNP genotyping assay must be developed for the mungbean because most mungbean research and breeding are in developing countries.

References

Afzal AJ, Wood AJ, Lightfoot DA (2008) Plant receptor-like serine threonine kinases: roles in signaling and plant defense. Mol Plant Microbe Interact 21:507–517

Alam AM, Somta P, Srinives P (2014a) Generation mean and path analyses of reaction to mungbean yellow mosaic virus (MYMV) and yield-related traits in mungbean (*Vigna radiata* (L.) Wilczek). SABRAO J Breed Genet 46:150–159

Alam AM, Somta P, Srinives P (2014b) Identification and confirmation of quantitative trait loci controlling resistance to mungbean yellow mosaic disease in mungbean [*Vigna radiata* (L.) Wilczek]. Mol Breed 34:1497–1506

Ali MM, Sabir IA, Ahmed HM (1997) The mungbean green revolution in Pakistan. Asian Vegetable Research and Development Centre, Shanhua

Ammavasai S, Phogat DS, Solanki IS (2004) Inheritance of resistance to mungbean yellow mosaic virus (MYMV) in green gram (*Vigna radiata* (L.) Wilczek). Indian J Genet 64:145–146

Anton KW, Halperin J, Calderon M (1997) An annotated list of the Bruchidae (Coleoptera) of Israel and adjacent areas. Isr J Entomol 31:59–96

Appiano M, Pavan S, Catalano D, Zheng Z, Bracuto V, Lotti C et al (2015) Identification of candidate MLO powdery mildew susceptibility genes in cultivated *Solanaceae* and functional characterization of tobacco *NtMLO1*. Transgenic Res 24:1–12

Arora GL (1977) Bruchidae of Northwest India. Orient Insects 7:1–132

AVRDC (1974) Progress report 1974. Asian Vegetable Research and Development Center, Shanhua, Taiwan, Republic of China

AVRDC (1978) Progress report, 1978. Asian Vegetable Research and Development Center, Shanhua, Taiwan, Republic of China

AVRDC (1980) Progress report 1980. Asian Vegetable Research and Development Center, Shanhua, Taiwan, Republic of China

AVRDC (1984) Progress report 1984. Asian Vegetable Research and Development Center, Shanhua, Taiwan, Republic of China

Bai Y, Pavan S, Zheng Z, Zappel NF, Reinstädler A, Lotti C et al (2008) Naturally occurring broad-spectrum powdery mildew resistance in a Central American tomato accession is caused by loss of *Mlo* function. Mol Plant Microbe Interact 21: 30–39

Bhanu AN, Kumar P, Singh MN, Srivastava K, Hemantaranjan A (2017) Assessment of genetic purity of inter-specific F_1 hybrids involving *Vigna radiata* and *Vigna umbellata*. J Exp Biol Agric Sci 5:5

Bhanu AN, Singh MN, Srivastava K (2019) Genetic analysis of gene-specific resistance to mungbean yellow mosaic virus in mungbean (*Vigna radiata*). Plant Breed 138:202–206

Boutin SR, Young ND, Olson TC, Yu ZH, Shoemaker RC, Vallejos CE (1995) Genome conservation among three legume genera detected with DNA markers. Genome 38:928–937

Büschges R, Hollricher K, Panstruga R, Simons G, Wolter M, Frijters A et al (1997) The barley *Mlo* gene: a novel control element of plant pathogen resistance. Cell 88:695–705

Cayalvizhi BS, Nagarajan P, Raveendran M, Rabindran R, Kannan BJR, Senthil N (2017) Understanding the inheritance of mungbean yellow mosaic virus (MYMV) resistance in mungbean (*Vigna radiata* L. Wilczek). Mol Breed 37:63

Chaitieng B, Kaga A, Han OK, Wang X, Wongkaew S, Laosuwan P, Tomooka N, Vaughan DA (2002) Mapping a new source of resistance to powdery mildew in mungbean. Plant Breed 121:521–525

Chankaew S, Somta P, Sorajjapinun W, Srinives P (2011) Quantitative trait loci mapping of *Cercospora* leaf spot resistance in mungbean, *Vigna radiata* (L.) Wilczek. Mol Breed 28:255–264

Chankaew S, Somta P, Isemura T, Tomooka N, Kaga A, Vaughan DA, Srinives P (2013) Quantitative trait locus mapping reveals conservation of major and minor loci for powdery mildew resistance in four sources of resistance in mungbean [*Vigna radiata* (L.) Wilczek]. Mol Breed 32:121–130

Chathiranrat N, Nitisit S, Chaiyapan C, Wansuriwong N, Papan P, Tantasawat PA (2018) Selection of mungbean resistant to powdery mildew in BC_1F_1 progenies based on ISSR and ISSR-RGA markers. IJASEAT 6:73–77

Chen KC, Lin CY, Kuan CC, Sung HY, Chen CS (2002) A novel defensin encoded by a mungbean cDNA exhibits insecticidal activity against bruchid. J Agric Food Chem 50:7258–7263

Chen HM, Liu CA, Kuo CG, Chien CM, Sun HC, Huang CC, Lin YC, Ku HM (2007) Development of a molecular marker for a bruchid (*Callosobruchus chinensis* L.) resistance gene in mungbean. Euphytica 157:113–122

Chen HM, Ku HS, Schafleitner R, Bains TS, Kuo GC, Liu CA, Nair RM (2013) The major quantitative trait locus for mungbean yellow mosaic Indian virus resistance is tightly linked in repulsion phase to the major bruchid resistance locus in a cross between mungbean [*Vigna radiata* (L.) Wilczek] and its wild relative *Vigna radiata* ssp. sublobata. Euphytica 192:205–216

Chotechung S, Somta P, Chankaew S, Srinives P, Somta P (2011) Identification of DNA markers associated with bruchid resistance in mungbean. Khon Khan Agri J 39:221–226

Chotechung S, Somta P, Chen J, Yimram T, Chen X, Srinives P (2016) A gene encoding a polygalacturonase-inhibiting protein (PGIP) is a candidate gene for bruchid (Coleoptera: Bruchidae) resistance in mungbean (*Vigna radiata*). Theor Appl Genet 129:1673–1683

Clarry S (2016) The rise and rise of mungbeans. In: GroundCover™ Supplement Issue 125 November–December. Grains Research and Development Corporation. Available online at: https://grdc.com.au/resources-and-publications/groundcover/ground-cover-supplements/ground-cover-issue-125-pulse-breeding-advances/the-rise-and-rise-of-mungbeans. Accessed 5 Dec 2018

Collard BCY, Mackill DJ (2008) Marker-assisted selection: an approach for precision plant breeding in the twenty-first century. Philos Trans R Soc Lond B Biol Sci 12(363):557–572

Consonni C, Humphry ME, Hartmann HA, Livaja M, Durner J, Westphal L, Vogel J, Lipka V, Kemmerling B, Schulze-Lefert P, Somerville SC, Panstruga R (2006) Conserved requirement for a plant host cell protein in powdery mildew pathogenesis. Nat Genet 38:716–720

Das G, Patra JK, Baek KH (2017) Insight into MAS: a molecular tool for development of stress resistant and quality of rice through gene stacking. Front Plant Sci 8:985

Deshmukh R, Singh VK, Singh BD (2014) Comparative phylogenetic analysis of genome-wide Mlo gene family members from *Glycine max* and *Arabidopsis thaliana*. Mol Genet Genomics 289:345–359

Desroches P, Elshazly E, Mandon N, Duc G, Huignard J (1995) Development of *Callosobruchus chinensis* (L) and *C. maculatus* (F) (Coleoptera, Bruchidae) in seeds of *Vicia faba* L. differing in their tannin, vicine and convicine contents. J Stored Prod Res 31:83–89

Devoto A, Hartmann HA, Piffanelli P, Elliott C, Simmons C, Taramino G et al (2003) Molecular phylogeny and evolution of the plant-specific seven-transmembrane MLO family. J Mol Evol 56:77–88

Dhole VJ, Reddy KS (2013) Development of a SCAR marker linked with a MYMV resistance gene in mungbean (*Vigna radiata*). Plant Breed 132:127–132

Duangsong U, Kaewwongwal A, Somta P, Chankaew S, Srinives P (2016) Identification of a major QTL for resistance to *Cercospora* leaf spot disease in cowpea (*Vigna unguiculata* (L.) Walp.) revealed common genomic region with that for the resistance to angular leaf spot in common bean (*Phaseolus vulgaris* L.). Euphytica 209:199–207

Fernandez GCJ, Shanmugasundaram S (1988) The AVRDC mungbean improvement programme: the past, present and future. In: Proceedings of the 2nd international mungbean sypmposium. Asian Vegetable Research and Development Center, Shanhua, Tainan, Taiwan, pp 58–70

Frati F, Galletti R, De Lorenzo G, Salerno G, Conti E (2006) Activity of endo-polygalacturonases in mirid bugs (Heteroptera: Miridae) and their inhibition by plant cell wall proteins (PGIPs). Eur J Entomol 103:515–522

Fujii K, Miyazaki S (1987) Infestation resistance of wild legumes (*Vigna sublobata*) to azuki bean weevil, *Callosobruchus chinensis* (L.) (Coleoptera: Bruchidae) and its relationship with cytogenetic classification. Appl Entomol Zool 22:229–230

Fujii K, Ishimoto M, Kitamura K (1989) Patterns of resistance to bean weevils (Bruchidae) in *Vigna radiata mungo sublobata* complex inform the breeding of new resistant varieties. Appl Ent Zool 24:126–132

Gawande VL, Patil JV (2003) Genetics of powdery mildew (*Erysiphe polygoni* D.C.) resistance in Mungbean (*Vigna radiata* (L.) Wilczek). Crop Prot 22:567–571

Gill AS, Verma MM, Dhaliwal HS, Sandhu TS (1983) Interspecific transfer of resistance to mungbean yellow mosaic virus from *Vigna mungo* to *Vigna radiata*. Current Sci 52:31–33

Grewal JS (1978) Disease of mungbean in Indian. Proceeding of the 1st mungbean symposium. University of the Philippines, Los Banos, pp 165–168

Grewal JS, Machendra P, Kulshrestha DP (1980) Control of *Cercospora* leaf spot of green gram by spraying Bavistin. Indian J Agric Sci 50:707–711

Gupta SK, Gopalakrishna T (2010) Development of unigene-derived SSR markers in cowpea (*Vigna unguiculata*) and their transferability to other *Vigna* species. Genome 53:508–523

Gupta S, Gupta DS, Anjum TK, Pratap A, Kumar J (2013) Inheritance and molecular tagging of MYMIV resistance gene in blackgram (*Vigna mungo* L. Hepper). Euphytica 193:27–37

Gwag JG, Chung JW, Chung HK, Lee JH, Ma KH, Dixit A, Park YJ, Cho EG, Kim TS, Lee SH (2006) Characterization of new microsatellite markers in mung bean, *Vigna radiata* (L.). Mol Ecol Notes 6:1132–1134

Han OK, Kaga A, Isemura T, Wang XW, Tomooka N, Vaughan DA (2005) A genetic linkage map for azuki bean [Vigna angularis (Willd.) Ohwi & Ohashi]. Theor Appl Genet 111(7):1278–1287

Hartman GL, Wang TC, Kim D (1993) Field evaluation of mungbeans for resistance to Cercospora leaf spot and powdery mildew. Int J Pest Manag 39:418–421

Hong MG, Kim KH, Ku JH, Jeong JK, Seo MJ, Park CH et al (2015) Inheritance and quantitative trait loci analysis of resistance genes to bruchid and bean bug in mungbean (*Vigna radiata* L. Wilczek). Plant Breed Biotechnol 3:39–46

Humphry ME, Magner T, McIntyre CL, Aitken EA, Liu CL (2003) Identification of major locus conferring resistance to powdery mildew (*Erysiphe polygoni* D. C.) in mungbean (*Vigna radiata* L. Wiczek) by QTL analysis. Genome 46:738–744

Humphry M, Reinstädler A, Ivanov S, Bisseling T, Panstruga R (2011) Durable broad-spectrum powdery mildew resistance in pea *er1* plants is conferred by natural loss-of-function mutations in *PsMLO1*. Mol Plant Pathol 12:866–878

Hussain M, Qazi J, Mansoor S, Iram S, Bashir M, Zafar Y (2004) First report of mungbean yellow mosaic India virus on mungbean in Pakistan. New Disease Rep 9:13

Ilyas M, Qazi J, Mansoor S, Briddon RW (2010) Genetic diversity and phylogeography of begomoviruses infecting legumes in Pakistan. J Gen Virol 91:2091–2101

Imrie BC, Lambrides CJ (1998) Marker-assisted selection for resistance to bruchids. In: Libas EM, Lopez KS (eds) International consultation workshop on mungbean: proceedings of the mungbean workshop, 7–11 September 1997, New Delhi, India. Asian Vegetable Research and Development Center, Tainan, Taiwan, pp 135–140

Iqbal SM, Ghafoor A, Bashir M, Malik BA (1995) Estimation of losses in yield components of mugbean due to Cercospora leaf spot. Pakistan J Phytopath 7:80–81

Isemura T, Kaga A, Konishi S, Ando T, Tomooka N, Han OK, Vaughan DA (2007) Genome dissection of traits related to domestication in azuki bean (*Vigna angularis*) and comparison with other warm season legumes. Ann Bot 100:1053–1071

Isemura T, Kaga A, Tabata S, Somta P, Srinives P, Shimizu T, Jo U, Vaughan DA, Tomooka N (2012) Construction of a genetic linkage map and genetic analysis of domestication related traits in mungbean (*Vigna radiata*). PLoS ONE 7:e41304

Ishimoto M, Kitamura K (1989) Growth inhibitory effects of an α-amylase inhibitor from kidney bean, *Phaseolus vulgaris* (L.) on three species of bruchids (Coleoptera: Bruchidae). Appl Ent Zool 24:281–286

Ishimoto I, Kitamura K (1991) Biochemical basis of bruchid resistance in TC1966 (*Vigna radiata* var. *sublobata*). Japan J Breed 41:408–409

Ishimoto I, Kitamura K (1993) Inhibitory effects of adzuki bean weevil-resistant mungbean seeds on growth of bean bug. Japan J Breed 43:75–80

Ishimoto M, Sato T, Chrispeels MJ, Kitamura K (2006) Bruchid resistance of transgenic azuki bean expressing seed α-amylase inhibitor of common bean. Entomol Exp Appl 79:309–315

Jeong JK, Kim YH, Hwang TY, Lee IJ, Seo BY, Moon CK, Kim CS, Kim YT, Lee HE (2015) Identification of genes and active compounds related to insect-resistance from mungbean, and analysis of their function. Final report, National Institute of Food Science and Technology, Rural Development Administration, Korea, pp 1–50

Kadosawa T, Santa H (1981) Growth and reproduction of soybean pod bugs (Heteroptera) on seeds of legumes. Bull Chugoku Natl Agric Exp Stn Ser E 19:75–97

Kaewwongwal A, Chen J, Somta P, Kongjaimun A, Yimram T, Chen X, Srinives P (2017) Novel alleles of two tightly linked genes encoding polygalacturonase-inhibiting proteins (*VrPGIP1* and *VrPGIP2*) associated with the *Br* locus that confer bruchid (*Callosobruchus* spp.) resistance to mungbean (*Vigna radiata*) accession V2709. Front Plant Sci 8: 1692

Kaga A, Ishimoto M (1998) Genetic localization of a bruchid resistance gene and its relationship to insecticidal cyclopeptide alkaloids, the vignatic acids in mungbean (*V. radiata* L. Wilczek). Mol Gen Genet 258:378–384

Kaga A, Teraishi M, Iijima N, Sugawara F, Ishimoto M (2000) Progresses in identification of the bruchid resistance gene in mungbean (*Vigna radiata* (L.)). In: Abstract in plant and animal genome VIII conference, SanDiego, CA

Kaga A, Isemura T, Shimizu T, Somta P, Srinives P, Tabata S, Tomooa N, Vaughan DA (2011) Asian Vigna genome research. In: Tomooka N, Vaughan DA (eds) Proceedings of the 14th NIAS international workshop on genetic resources: genetic resources and comparative genomics of legumes (*Glycine* and *Vigna*). National Institute of Agrobiological Sciences, Tsukuba, Japan, pp 33–39

Kang CH, Huh HS, Park CG (2003) Review on true bugs infesting tree fruits, upland crops, and weeds in Korea. Korean J Appl Entomol 42:269–277

Kang YJ, Kim S, Kim MY, Lestari P, Kim KH, Ha BK et al (2014) Genome sequence of mungbean and insights into evolution within *Vigna* species. Nat Commun 5:5443

Karthikeyan K, Shobhana VG, Sudha M, Raveendran M, Senthil N, Pandiyan M, Nagarajan P (2014) Mungbean yellow mosaic virus (MYMV): a threat to green gram (*Vigna radiata*). Int J Pest Manag 60:314–324

Kasettranan W, Somta P, Sirnives P (2009) Genetics of the resistance to powdery mildew disease in mungbean (*Vigna radiate* (L.) Wilczek). J Crop Sci Biotech 12:37–42

Kasettranan W, Somta P, Srinives P (2010) Mapping of quantitative trait loci controlling powdery mildew resistance in mungbean (*Vigna radiata* (L.) Wilczek). J Crop Sci Biotechnol 13:155–161

Kashiwaba K, Tomooka N, Kaga A, Han OK, Vaughan DA (2003) Characterization of resistance to three bruchid species (*Callosobruchus* spp., Coleoptera, Bruchidae) in cultivated rice bean, (*Vigna umbellata* (Thunb.) Ohwi and Ohashi). J Econ Entomol 96:207–213

Khajudparn P, Wongkaew S, Thipyapong P (2007) Mungbean powdery mildew resistance identification of genes for powdery mildew resistance in mungbean. Afr Crop Sci Conf Proc 8:743–745

Khan MMK, Khan A, Ishimoto M, Kitamura K, Komatsu S (2003) Proteome analysis of the relationship between bruchid-resistant and -susceptible mungbean genotypes. Plant Genet Resou 1:115–123

Khattak GSS, Haq MA, Ashraf M, Elahi T (2000) Genetics of mungbean yellow mosaic virus (MYMV) in mungbean (*Vigna radiata* (L.) Wilczek). J Genet Breed 54:237–243

Khunti JP, Bhoraniya MF, Vora VD (2002) Management of powdery mildew and Cercospora leaf spot of mungbean by some systemic fungicides. J Mycol Pl Pathol 32:103

Kitamura K, Ishimoto M, Sawa M (1988) Inheritance of resistance to infestation with azuki bean weevil in *Vigna sublobata* and successful incorporation to *V. radiata*. Jap J Breed 38:459–464

Kitamura K, Ishimoto M, Ishii S (1990) Bruchid resistance factors in *Phaseolus* and *Vigna* legumes. In: Fujii K, Gatehouse AMR, Johnson CD, Mitchel R, Yoshida T (eds) Bruchids and legumes: economics, ecology and coevolution. Kluwer Academic Publishers, Dordrecht, pp 229–240

Kitsanachandee R, Somta P, Chatchawankanphanich O, Akhtar KP, Shah TM, Nair RM, Bains TS, Sirari A, Kaur L, Srinives P (2013) Detection of quantitative trait loci for mungbean yellow mosaic India virus (MYMIV) resistance in mungbean (*Vigna radiata* (L.) Wilczek) in India and Pakistan. Breed Sci 63:367–373

Kongjaimun A, Kaga A, Tomooka N, Somta P, Shimizu T, Shu Y, Isemura T, Vaughan DA, Srinives P (2012) An SSR-based linkage map of yardlong bean (*Vigna unguiculata* (L.) Walp. subsp. *unguiculata* Sesquipedalis Group) and QTL analysis of pod length. Genome 55:81–92

Kusch S, Pesch L, Panstruga R (2016) Comprehensive phylogenetic analysis sheds light on the diversity and origin of the MLO family of integral membrane proteins. Genome Biol Evol 8:878–895

Lambrides CJ, Godwin I (2007) Mungbean. In: Kole C (ed) Genome mapping and molecular breeding in plants pulses, sugar and tuber crops, vol 3. Springer, Berlin, pp 69–90

Lambrides CJ, Imrie BC (2000) Susceptibility of mungbean varieties to the bruchid species *Callsobruchus maculatus* (F.), *C. phaseoli* (Gyll.), *C. chinensis* (L.), and *Acanthoscelides obtectus* (Say.) (Coleoptera: Chrysomelidae). Aust J Agri Res 51:85–89

Lambrides CJ, Diatloff AL, Liu CJ, Imrie BC (1999) Molecular studies in mungbean *Vigna radiata*. In: Proceedings of the 11th Australasian plant breeding conference, 19–23 April, Adelaide, Australia

Lawrence PK, Koundal KR (2002) Plant protease inhibitors in control of phytophagous insects. Electron J Biotechnol 5(1):3

Leabwon U, Oupadissakoon S (1984) Inheritance of resistance to *Cercospora* leaf spot in mungbean. Kasetsart J (Nat Sci) 18:14–19

Lee YH, Moon JK, Park KY, Ku JH, Yun HT, Chung WK, Kim SD, Kim HS, Kim DH, Chung MN (2000) A new mungbean cultivar with bruchid resistance, 'Jangannogdu'. Korean J Breed 32:296–297

Lekhi P, Gill RK, Kunal Kaur S (2018) Identification of molecular marker linked to mungbean yellow mosaic virus (MYMV) resistance in *Vigna radiata* (L.) Wilczek. Electr J Plant Breed 9:839–845

Li Y, Chen X, Chen Z, Cai R, Zhang H, Xiang Y (2016) Identification and expression analysis of BURP domain-containing genes in *Medicago truncatula*. Front Plant Sci 7:485

Liang D, Chen M, Qi X, Xu Q, Zhou F, Chen X (2016) QTL mapping by SLAF-seq and expression analysis of candidate genes for aphid resistance in cucumber. Front Plant Sci 7:1000

Lin WJ, Ko CY, Liu MS, Kuo CY, Wu DC, Chen CY et al (2016) Transcriptomic and proteomic research to explore bruchid-resistant genes in mungbean isogenic lines. J Agric Food Chem 64:6648–6658

Liu YJ, Cheng CS, Lai SM, Hsu MP, Chen CS, Lyu PC (2006) Solution structure of the plant defensin VrD1 from mung bean and its possible role in insecticidal activity against bruchids. Proteins: Struct, Funct, Bioinf 63:777–786

Liu MS, Kuo TCY, Ko CY, Wu DC, Li KY, Lin WJ, Lin CP, Wang YW, Schafleitner R, Lo HF, Chen CY, Chen LF (2016) Genomic and transcriptomic comparison of nucleotide variations for insights into bruchid resistance of mungbean (*Vigna radiata* [L.] R. Wilczek). BMC Plant Biol 16:46

Liu CY, Su QZ, Fan BJ, Cao ZM, Zhang ZX, Wu J, Cheng XZ, Tian J (2018) Genetic mapping of bruchid resistance gene in mungbean V1128. ACTA Agron Sinica 44:1875–1881

Maiti S, Basak J, Kundagrami S, Kundu A, Pal A (2011) Molecular marker-assisted genotyping of mungbean yellow mosaic India virus resistant germplasm of mungbean and urdbean. Mol Biotechnol 47:95–104

Maiti S, Paul S, Pal A (2012) Isolation characterization and structural analysis of non-TR-NBS-LRR encoding candidate gene from MYMIV-resistant *Vigna mungo*. Mol Biotechnol 52:217–233

Malathi VG, John P (2009) Mungbean yellow mosaic viruses. In: Mahy BWJ, Van Regenmortal MHV (eds) Desk encyclopedia of plant and fungal virology. Academic Press, London, pp 217–226

Malik LA, Sarwar G, Ali Y (1986) Genetic studies in mungbean (*Vigna radiata* (L.) Wilczek). I: Inheritance of tolerance to mungbean yellow mosaic virus and some morphological characters. Pak J Bot 18:189–198

Mandhare VK, Suryawanshi AV (2008) Dual resistant against powdery mildew and yellow mosaic virus in greengram. Agric Sci Digest 28:39–41

Marimuthu T, Subramanian CL, Mohan R (1981) Assessment of yield losses due to yellow mosaic infection in mungbean. Pulse Crop News Lett 1:104

Mariyammal I, Seram D, Samyuktha SM, Karthikeyan A, Dhasarathan M, Murukarthick J, Kennedy JS, Malarvizhi D, Yang TJ, Pandiyan M, Senthil N (2019) QTL mapping in *Vigna radiata* × *Vigna umbellata* population uncovers major genomic regions associated with bruchid resistance. Mol Breed 39:110

Mathivathana MK, Murukarthick J, Karthikeyan A, Jang W, Dhasarathan M, Jagadeeshselvam N, Sudha M, Vanniarajan C, Karthikeyan G, Yang TJ, Raveendran M, Pandiyan M, Senthil N (2019) Detection of QTLs associated with mungbean yellow mosaic virus (MYMV) resistance using the interspecific cross of *Vigna radiata* × *Vigna umbellata*. J Appl Genet. https://doi.org/10.1007/s13353-019-00506-x

Mei L, Cheng XZ, Wang SH, Wang LX, Liu CY, Sun L et al (2009) Relationship between bruchid resistance and seed mass in mungbean based on QTL analysis. Genome 52:589–596

Menancio-Hautea D, Kumar L, Danesh D, Young ND (1993) A genome map for mungbean [*Vigna radiata* (L.) Wilczek] based on DNA genetic markers (2 N = 2X = 22). In:O'Brien SJ (ed) Genetic maps: locus maps of complex genomes, 6th edn. Cold Spring Harbor Laboratory Press, Cold Spring Harbor, New York, pp 6.259–6.261

Mishra SP, Asthana AN, Yadav L (1988) Inheritance of *Cercospora* leaf spot resistance in mungbean, *Vigna radiata* (L.) Wilczek. Plant Breed 100:228–229

Mishra SK, Macedo MLR, Panda SK, Panigrahi J (2018) Bruchid pest management in pulses: past practices, present status and use of modern breeding tools for development of resistant varieties. Ann Appl Biol 72:4–19

Miyagi M, Humphry M, Ma ZY, Lambrides CJ, Bateson M, Liu CJ (2004) Construction of bacterial artificial chromosome libraries and their application in developing PCR-based markers closely linked to a major locus conditioning bruchid resistance in mungbean (*Vigna radiata* L. Wilczek). Theor Appl Genet 110:151–156

Nair RM, Schafleitner R, Kenyon L, Srinivasan R, Easdown W, Ebert A, Hanson P (2012) Genetic improvement of mungbean. SABRAO J Breed Genet 44:177–190

Nair RM, Götz M, Winter S, Giri RR, Boddepalli VN, Sirari A, Bains TS, Taggar GK, Dikshit HK, Aski M, Boopathi M, Swain D, Rathore A, Kumar VA, Lii EC, Kenyon L (2017) Identification of mungbean lines with tolerance or resistance to yellow mosaic in fields in India where different begomovirus species and different *Bemisia tabaci* cryptic species predominate. Eur J Plant Pathol 149:349–365

Nogueira FCS, Silva CP, Alexandre D, Samuels RI, Soares EL, Aragão FJL et al (2012) Global proteome changes in larvae of *Callosobruchus maculatus* (Coleoptera: Chrysomelidae: Bruchinae) following ingestion of a cysteine proteinase inhibitor. Proteomics 12:2704–2715

Oliveira AEA, Sales MP, Machado OLT, Fernandes KVS, Xavier-Filho J (1999) The toxicity of Jack bean (*Canavalia ensiformis*) cotyledon and seed coat proteins to the cowpea weevil (*Callosobruchus maculatus*). Entomol Eexp Appl 92:249–255

Pandey AK, Burlakoti RR, Kenyon L, Nair RM (2018) Perspectives and challenges for sustainable management of fungal diseases of mungbean [*Vigna radiata* (L.) Wilczek var. *radiate*]: a review. Front Environ Sci 6: 53

Panizzi AR, Schaefer CW, Natuhara Y (2000) Broad-headed bugs (Alydidae). In: Schaefer CW, Panizzi AR (eds) Heteroptera of economic importance. CRC, Boca Raton, FL, pp 321–336

Pauchet Y, Wilkinson P, Chauhan R, Ffrench CRH (2010) Diversity of beetle genes encoding novel plant cell wall degrading enzymes. PLoS ONE 5:e15635

Pavan S, Jacobsen E, Visser RGF, Bai Y (2009) Loss of susceptibility as a novel breeding strategy for durable and broad-spectrum resistance. Mol Breed 25:1–12

Pavan S, Schiavulli A, Appiano M, Marcotrigiano AR, Cillo F, Visser RGF et al (2011) Pea powdery mildew er1 resistance is associated to loss-of-function mutations at a MLO homologous locus. Theor Appl Genet 123:1425–1431

Pedra JHF, Brandt A, Westerman R, Lobo N, Li HM, Romero SJ et al (2003) Transcriptome analysis of the cowpea weevil bruchid: identification of putative proteinases and alpha-amylases associated with food breakdown. Insect Mol Biol 12:405–412

Pham A-T, Harris DK, Buck J, Hoskins A, Serrano J, Abdel-Haleem H, Cregan P, Song Q, Boerma HR, Li Z (2015) Fine mapping and characterization of candidate genes that control resistance to *Cercospora sojina* K. Hara in two soybean germplasm accessions. PLoS ONE 10(5):e0126753. https://doi.org/10.1371/journal.pone.0126753

Poolsawat O, Kativat C, Arsakit K, Tantasawat PA (2017) Identification of quantitative trait loci associated with powdery mildew resistance in mungbean using ISSR and ISSR-RGA markers. Mol Breed 37:150

Rani A, Kumar V, Gill BS, Rathi P, Shukla S, Singh RK, Husain SM (2017) Linkage mapping of mungbean yellow mosaic India virus (MYMIV) resistance gene in soybean. Breed Sci 67:95–100

Reddy KS (2009) A new mutant for yellow mosaic virus resistance in mungbean (*Vigna radiata* L Wilczek) variety SML-668 by recurrent gamma-ray irradiation. In: Shu QY (ed) Induced plant mutation in the genomics era. Food and Agriculture Organization of the United Nations, Rome, pp 361–362

Reddy KS, Pawar SE, Bhatia CR (1987) Screening for powdery mildew (*Erysiphe polygoni* D.C.) resistance in mungbean (*Vigna radiata*(L.) Wilczek) using excised leaves. Proc Indian Acad Sci (Plant Sci) 97:365–369

Reddy KS, Pawar SE, Bhatia CR (1994) Inheritance of powdery mildew (*Erysiphe polygoni* D.C.) resistance in mungbean [*Vigna radiata*(L.) Wilczek]. Theor Appl Genet 88:945–948

Rispail N, Rubiales D (2016) Genome-wide identification and comparison of legume *MLO* gene family. Sci Rep 6:32673

Sales MP, Gerhardt IR, Grossi-de-Sa MF, Xavier J (2000) Do legume storage proteins play a role in defending seeds against bruchids? Plant Physiol 124:515–22

Sarkar S, Ghosh S, Chatterjee M, Das P, Lahari T, Maji A, Mondal N, Pradhan KK, Bhattacharyya S (2011) Molecular markers linked with bruchid resistance in *Vigna radiata* var. *sublobata* and their validation. J Plant Biochem Biotechnol 20:155–160

Schafleitner R, Huang S, Chu S, Yen J, Lin C, Yan M, Krishnan B, Liu M, Lo H, Chen C, Chen O, Wu D, Bui TGT, Ramasamy S, Tung C, Nair R (2016) Identification of single nucleotide polymorphism markers associated with resistance to bruchids (*Callosobruchus* spp.) in wild mungbean (*Vigna radiata* var. *sublobata*) and cultivated *V. radiata* through genotyping by sequencing and quantitative trait locus analysis BMC Plant Biol 16:159. https://doi.org/10.1186/s12870-016-0847-8

Seehalak W, Somta P, Sommanas W, Srinives P (2009) Microsatellite markers for mungbean developed from sequence database. Mol Ecol Resour 9:862–864

Sehgal VK, Ujagir R (1988) Insect pests and pest management of mungbean in India. In: Shanmugasundaram S, McLean BT (eds) Mungbean: proceedings of the second international symposium. AVRDC, Tainan, Taiwan, pp 335–328

Selvi R, Muthiah AR, Manivannan N (2006) Tagging of RAPD marker for MYMV resistance in mungbean (*Vigna radiata* (L.) Wilczek). Asian J Plant Sci 5:277–280

Shade RE, Schroeder HE, Pueyo JJ, Tabe LL, Murdock TJV, Higgins MJ, Chrispeels MJ (1994) Transgenic pea seeds expressing the α-amylase inhibitor of the common bean are resistant to bruchid beetles. Nature Biotech 12:793–796

Shao Y, Wei G, Wang L, Dong Q, Zhao Y, Chen B et al (2011) Genome-wide analysis of BURP domain-containing genes in *Populus trichocarpa*. J Integr Plant Biol 53:743–755

Shukla GP, Pandya BP (1985) Resistance to yellow mosaic in greengram. SABRAO J Breed Genet 17:165–171

Singh G, Brar JS, Sharma YR, Kuar L (1998) Enhancement of mungbean yellow mosaic virus resistance through inter/intraspecific hybridization and mutation. In: Libas EM, Lopez KS (eds) International consultation workshop on mungbean: proceedings of the mungbean workshop, 7–11 September 1997, New Delhi, India. Asian Vegetable Research and Development Center, Tainan, Taiwan, pp 86–91

Somta P, Ammaranan C, Ooi PAC, Srinives P (2007) Inheritance of seed resistance to bruchids in cultivated mungbean (*Vigna radiata*, L. Wilczek). Euphytica 155:47–55

Somta C, Somta P, Tomooka N, Ooi PAC, Vaughan DA, Srinives P (2008a) Characterization of new sources of mungbean (*Vigna radiata* (L) Wilczek) resistance to bruchids, *Callosobruchus* spp. (Coleoptera: Bruchidae). J Stored Prod Res 44:316–321

Somta P, Musch W, Kongsamai B, Chanprame S, Nakasathien S, Toojinda T, Sorajjapinun W, Seehalak W, Tragoonrung S, Srinives P (2008b) New microsatellite markers isolated from mungbean (*Vigna radiata* (L.) Wilczek). Mol Ecol Resour 8:1155–1157

Somta P, Seehalak W, Srinives P (2009) Development, characterization and cross-species amplification of mungbean (*Vigna radiata*) genic microsatellite markers. Conserv Genet 10:1939

Sorajjapinun W, Rewthongchum S, Koizumi M, Srinives P (2005) Quantitative inheritance of resistance to powdery mildew disease in mungbean (*Vigna radiata* (L.) Wilczek). SABRAO J Breed Gent 37:91–96

Southgate BJ (1979) Biology of the *Bruchidae*. Annu Rev Entomol 24:449–473

Srinives P, Somta P, Somta C (2007) Genetics and breeding of resistance to bruchids (*Callosobruchus* spp.) in *Vigna* crops: a review. NU Int J Sci 4:1–17

Sudha M, Anusuya P, Mahadev NG, Karthikeyan A, Nagarajan P, Raveendran M, Senthil N, Pandiyan M, Angappan K, Balasubramanian P (2013a) Molecular studies on mungbean (*Vigna radiata* (L.) Wilczek) and ricebean (*Vigna umbellata* (Thunb.)) interspecific hybridization for mungbean yellow mosaic virus resistance and development of species-specific SCAR marker for ricebean. Arch Phytopath Plant Protect, 503–517

Sudha M, Karthikeyan A, Anusuya P, Ganesh NM, Pandiyan M, Senthil N, Raveendran M, Nagarajan P, Angappan K (2013b) Inheritance of resistance to mungbean yellow mosaic virus (MYMV) in inter and intra specific crosses of mungbean (*Vigna radiata*). Am J Plant Sci 4:1924–1927

Sudha M, Karthikeyan A, Shobhana VG, Nagarajan P, Raveendran M, Senthil N, Pandiyan M, Angappan K, Balasubramanian P, Rabindran R, Bharathi M (2015) Search for *Vigna* species conferring resistance to mungbean yellow mosaic virus in mungbean. Plant Genetic Resourc 13:162–167

Sugawara F, Ishimoto M, Le-Van N, Koshino H, Uzawa J, Yoshida S, Kitamura K (1996) Insecticidal peptide from mungbean: a resistant factor against infestation with azuki bean weevil. J Agric Food Chem 44:3360–3364

Sun L (2007) Allelism test of bruchid resistant genes and inheritance and microsatellite tagging of resistance to bruchid in a cultivar mungbean (*Vigna radiata* (L.) Wilczek) V2709. Master thesis, Chinese Academy of Agricultural Sciences, China, pp 1–50

Sun L, Cheng XZ, Wang SH, Wang LX, Liu CY, Mei L, Xu N (2008) Heredity analysis and gene mapping of bruchid resistance of a mungbean cultivar V2709. Agric Sci China 7:672–677

Takagi H, Abe A, Yoshida K, Kosugi S, Natsume S, Mitsuoka C, Uemura A, Utsushi H, Tamiru M, Takuno S, Innan H, Cano LM, Kamoun S, Terauchi R (2013) QTL-seq: rapid mapping of quantitative trait loci in rice by whole genome resequencing of DNA from two bulked populations. Plant J 74:174–183

Talekar NS (1988) Biology, damage and control of bruchid pests of mungbean. In: Shanmugasundaram S, McLean BT (eds) Mungbean: proceedings of the second international symposium. AVRDC, Tainan, Taiwan, pp 329–342

Talekar NS, Lin CP (1992) Characterization of *Callosobruchus chinensis* (Coleoptera: Bruchidae) resistance in mungbean. J Econ Entomol 85:1150–1153

Tang Y, Cao Y, Qiu J, Gao Z, Ou Z, Wang Y et al (2014) Expression of a vacuole-localized BURP-domain protein from soybean (SALI3-2) enhances tolerance to cadmium and copper stresses. PLoS ONE 9:e98830

Tangphatsornruang S, Somta P, Uthaipaisanwong P, Chanprasert J, Sangsrakru D, Seehalak W, Sommanas W, Tragoonrung S, Srinives P (2009) Characterization of microsatellites and gene contents from genome shotgun sequences of mungbean (*Vigna radiata* (L.) Wilczek). BMC Plant Biol 9:137

Thakur RP, Patel PN, Verma JP (1977) Genetical relationships between reactions to bacterial leaf spot, yellow mosaic and *Cercospora* leaf spot diseases in mungbean (*Vigna radiata*). Euphytica 26:765–774

Tomooka N, Kashiwaba K, Vaughan DA, Ishimoto M, Egawa Y (2000) The effectiveness of evaluating wild species: searching for sources of resistance to bruchids beetles in the genus *Vigna* subgenus *Ceratotropis*. Euphytica 115:27–41

Tomooka N, Vaughan DA, Moss H, Maxted N (2002) The Asian *Vigna*: genus *Vigna* subgenus *Ceratotropis* genetic resources. Kluwer Academic Publishers, Dordrecht, Netherlands, 270 pages

Tuda M, Chou LY, Niyomdham C, Buranapanichpan S (2005) Tateishi Y (2005) Ecological factors associated with pest status in *Callosobruchus* (Coleoptera: Bruchidae): high host specificity of non-pests to *Cajaninae* (Fabaceae). J Stored Prod Res 41:31–45

Várallyay É, Giczey G, Burgyán J (2012) Virus-induced gene silencing of Mlo genes induces powdery mildew resistance in *Triticum aestivum*. Arch Virol 157:1345–1350

Vinod JD, Kandali SR (2012) Genetic analysis of resistance to mungbean yellow mosaic virus in mungbean (*Vigna radiata*). Plant Breed 131:414–417

Wang X, Kaga A, Tomooka N, Vaughan DA (2004) The development of SSR markers by a new method in plants and their application to gene flow studies in azuki bean [*Vigna angularis* (Willd.) Ohwi & Ohashi]. Theor Appl Genet 109:352–360

Wang L, Wu C, Zhong M, Zhao D, Mei L, Chen H, Wang S, Liu C, Chen X (2016) Construction of an integrated map and location of a bruchid resistance gene in mung bean. Crop J 4:360–366

Wang Y, Jiang J, Zhao L, Zhou R, Yu W, Zhao T (2018) Application of whole genome resequencing in mapping of a tomato yellow leaf curl virus resistance gene. Sci Rep 8:9592

Watanasit A, Pichitporn S (1996) Improvement of mungbean for resistance to bruchids. In: Srinives P, Kitbamroong C, Miyazaki S (eds) Mungbean germplasm: collection, evaluation and utilization for breeding program. Japan International Research Center for Agricultural Sciences, Tsukuba, Japan, pp 67–71

Yao Y, Cheng X, Ren G (2015) A 90-day study of three bruchid-resistant mung bean cultivars in Sprague-Dawley rats. Food Chem Toxicol 76:80–85

Young N, Kumar L, Menancio HD, Danesh D, Talekar NS, Shanmugasundaram S, Kim DH (1992) RFLP mapping of a major bruchid resistance gene in mungbean (*Vigna radiata*, L. Wilczek). Theor Appl Genet 84:839–844

Young ND, Danesh D, Menancio HD, Kumar L (1993) Mapping oligogenic resistance to powdery mildew in mungbean with RFLPs. Theor Appl Genet 87:243–249

Zhang MC, Wang DM, Zheng Z, Humphry M, Liu CJ (2008) Development of PCR-based markers for a major locus conferring powdery mildew resistance in mungbean (*Vigna radiata*). Plant Breed 127:429–432

Zhang H, GaoZ Zheng X, Zhang Z (2012) The role of G-proteins in plant immunity. Plant Signal Behav 7:1284–1288

Zhao D, Cheng XZ, Wang LX, Wang SH, Ma YL (2010) Construction of mungbean genetic linkage map. Acta Agron Sin 36:932–939

Zheng Z, Nonomura T, Appiano M, Pavan S, Matsuda Y, Toyoda H, Wolters AMA, Visser RGF, Bai Y (2013) Loss of function in *Mlo* orthologs reduces susceptibility of pepper and tomato to powdery mildew disease caused by *Leveillula taurica*. PLoS ONE 8:e70723

Zhu K, Huesing JE, Shade RE, Bressan RA, Hasegawa PM, Murdock LL (1996) An insecticidal N-acetylglucosamine-specific lectin gene from *Griffonia simplicifolia* (Leguminosae). Plant Physiol 110:195–202

Zou C, Wang P, Xu Y (2016) Bulked sample analysis in genetics, genomics and crop improvement. Plant Biotechnol J 14:1941–1955

11 Genomic Approaches to Abiotic Stresses in Mungbean

Thomas J. Noble, Brett Williams, Thi My Linh Hoang, Sudipta Shekhar Das Bhowmik, Grace Zi Hao Tan and Sagadevan Mundree

Abstract

Mungbean (*Vigna radiata* (L.) R. Wilczek var. *radiata*) is an important legume crop widely produced and consumed throughout Southeast Asia, cultivated on more than 6 million hectares worldwide. Minimizing the impact climate variability has on production is vital to smallholder farmers rely on mungbeans as a source of income and nutrition Abiotic stress factors such as drought, water availability, heat and salinity pose a major risk to global food security. Variability in the climate and the increasing demand for food crops means innovative approaches must be implemented now to secure the food of tomorrow. Conventional breeding programs lead by the World Vegetable Centre and the Australian National Mungbean Improvement Program have dramatically increased the yields, reliability and sustainability of mungbean crops worldwide. Breeders and researchers are building on that foundational work through the implementation of genomic technologies. Sequencing the genomes of large diverse sets of mungbean germplasm aims to quantify how the genetic diversity present among the world's mungbean collections and to identify genes associated with agronomically important traits. By combining sequence and phenotyping data regions of the genome associated with important traits link to, the maintenance of photosynthetic pathways and water-use efficiency can be targeted. Once identified, those pathways can be directly manipulated using genome-editing tools reduce current breeding times by more than half. Although abiotic stressors pose an immediate and extensive risk, fortunately the technologies and researchers needed to address the issues exist today.

11.1 Introduction

The world population is increasing dramatically, and from 1993 to 2017, the population grew by two billion reaching 7.6 billion in mid-2017 (United Nations 2017). By 2050, the world population is predicted to rise to 10.2 billion people (United Nations 2017). Producing enough food for the increasing population means food production needs to increase by 70% (FAO 2009a). This is a challenge because there is very little potential for future expansion of arable lands due to climate variability and environmental stresses compounding in severity (Eckardt 2009; FAO 2009b, 2012; Cominelli et al. 2013). Drought, water availability, heat, and

T. J. Noble · B. Williams · T. M. L. Hoang · S. S. Das Bhowmik · G. Z. H. Tan · S. Mundree (✉)
Centre for Tropical Crops and Biocommodities, Queensland University of Technology, Brisbane, QLD, Australia
e-mail: sagadevan.mundree@qut.edu.au

R. M. Nair et al. (eds.), *The Mungbean Genome*, Compendium of Plant Genomes,
https://doi.org/10.1007/978-3-030-20008-4_11

salinity significantly affect crop production worldwide. A global synthesis of drought effects on crop production of legume, cereal, tuber and root by Daryanto et al. (2017) reported varying yield reductions between crop species. When subjected to a water deficit, yield loss in legumes ranged from 20 to 80%. A similar loss of 25–70% was recorded for root/tuber groups; however, cereals were not as severely affected with losses of 25–40% (Daryanto et al. 2017). Changes in the global environment due to climate change are predicted to cause further yield losses (Eynard et al. 2005).

Mungbean (*Vigna radiata*) is an important tropical pulse crop produced and consumed widely throughout Southeast Asia including India, China, Vietnam, Pakistan and Korea. More than 6 million ha of cultivated land are used worldwide for mungbean production, accounting for approximately 8.5% of global pulses area (Kim et al. 2015a). With high protein content up to 27% in dry beans, rich in dietary fibre, vitamins and essential amino acids including methionine and lysine, mungbean is one of the most nutritious and economical sources of protein (Day 2013). Furthermore, low levels of oligosaccharides make mungbean easily digestible providing a good source of protein for a wide range of consumers including infants and the elderly (Ihsan et al. 2013). Mungbean also plays a role in the productivity of other crops due to increased soil health through reduced use of nitrogen fertilizers and breaking disease chains (Fernandez et al. 1988). Although mungbeans possess many beneficial properties, the crop is susceptible to abiotic stresses, especially waterlogging, salinity and extreme temperatures, see review by Hanumantharao et al. (2016). For example, in mungbean, a level of 50 mM NaCl is reported to cause up to a 60% yield loss (Hanumantharao et al. 2016).

In this chapter, we review the effects of abiotic stresses on mungbean, current approaches and resources for enhancing abiotic stress tolerance and future directions for mungbean breeding in the twenty-first century.

11.2 Abiotic Stress in Mungbean

Mungbean is considered an important crop for global food security providing essential dietary protein, carbohydrate, vitamins and minerals to smallholder farmers. However, environmental stresses mentioned earlier are severely limiting the productivity of mungbean (Sarkar et al. 2017; Hasanuzzaman et al. 2012). Abiotic stresses such as drought, salinization of arable lands, flash floods due to unseasonal rainfall, extreme temperature fluctuations and soil degradation (acidity and aluminium toxicity) are the main factors that critically impact mungbean growth and yield.

A fast-growing, warm-season legume crop requiring as little as 90 days to reach maturity (Ranawake et al. 2011) also performs well under low soil moisture due to its well-developed root system (Kole 2007). Although considered to withstand drought, mungbean yield is significantly affected by drought stress conditions, with reports indicating adequate water supply as the critical environmental factor (Kramer et al. 1997; Pandey et al. 1984). The most susceptible stages of development are flowering and pod filling resulting in high overall yield loss (Ghanbari and Javan 2015). The rapid expansion of drought-stressed areas around the world and increasing population are exacerbating the issues (Postel 2000). Many studies have reported the negative effects of drought stress on the growth, physiology and yield components of mungbean (Ahmed et al. 2002; Rafiei and Shirvan 2009; Ranawake et al. 2011).

Salinity is also a major problem globally estimated to cause ~50% loss of arable land by 2050 (Hasanuzzaman et al. 2012). Salinity stress induces ionic and osmotic imbalances leading to secondary oxidative stress that has a detrimental effect on the growth and metabolism by affecting the integrity of the plant membrane, pigment content, osmotic adjustments, water retention capacity and photosynthetic activity (Munns and Tester 2008; Pang et al. 2010; Saha et al. 2010). Production of mungbean has been greatly affected due to salinity stresses worldwide (Sehrawat et al. 2013). The adverse effect of salt stress on mungbean grain yield is most pronounced during reproductive stages of growth (Ayers and Westcot 1985; Minhas et al. 1990). High yield losses at low concentrations of NaCl have been reported by (Abd-Alla et al. 1998), caused by a reduction in seed germination, fresh and dry biomass, shoot and root lengths, photosynthesis and yield attributes (Ghosh et al. 2015).

Mungbean cannot withstand water logging, particularly during the early stages of growth (Singh and Singh 2011). Water logging reduces oxygen concentrations around the roots of mungbean under submerged conditions restricting gas exchange and nodule activity (Singh and Singh 2011). Photosynthesis is the primary physiological process inhibited as a result of water logging leading to extensive grain yield loss (Ahmed et al. 2002, 2006). Thus, mungbean is not suited to the wet tropics with annual precipitation above 1000 mm (Fernandez and Shanmugasundaram 1987).

Extreme fluctuations in temperature have a direct effect on flower preservation and pod formation in mungbean, particularly at temperatures above 40 °C (Kumari and Varma 1983). Terminal heat stress can severely affect mungbeans in spring/summer causing a drastic reduction in seed yield due to pollen sterility, lack of fertilization and complete flower shedding (HanumanthaRao et al. 2016; Kaur et al. 2015; Sinha 1977; Tickoo et al. 1996).

Degradation of arable land mainly due to soil acidity is a major factor that affects root growth and nutrient availability mainly nitrogen, phosphorus and potassium (Meena and Varma 2016). Acidic soils make up approximately 30% of the world's total land area and more than 50% of the world's potentially arable lands, particularly in the tropics and subtropics (Kochian et al. 2004; Meena and Yadav 2014). Aluminium (Al) toxicity is a common problem in many productivity areas of the world, and of the world's arable soils, 30–40% are affected by Al toxicity (Kochian et al. 2015). Physiological and metabolic disorders, growth inhibition and cell death are common symptoms caused by aluminium toxicity (Panda et al. 2009). Mungbean being a short maturity plant with well-developed root systems is greatly affected by aluminium toxicity (Nahar et al. 2017).

11.3 Current Approaches for Enhancing Abiotic Stress Tolerance in Mungbean

To minimize the effect of abiotic stress on mungbean, several strategies have been employed. Among these, water and soil management practices have facilitated agricultural production on soils marginalized by salinity and drought. Achieving additional gains while continuing to use these approaches appear problematic. In recent years crop improvement strategies based on molecular markers and biotechnology techniques are overcoming the limitations of traditional breeding efforts.

11.3.1 Conventional Breeding

Conventional breeding by means of selecting desirable genotypes based on visual phenotypes and crossing the elite lines is the most foundational, accessible and easiest method of crop improvement. The underlying principle is simple —parents with desired traits are mated to generate superior offspring bearing the desirable traits of both parents. It is a method tried, tested and true since the dawn of agriculture and has given rise to most modern crops and livestock (Borlaug 1983).

Prior to the 1970s, mungbean was considered unpopular amongst growers due to poor yields and agronomic properties. Efforts at crop improvement only began upon recognition of its potential as a protein source and cash crop for smallholder farmers, resulting in the launch of an improvement program by the World Vegetable Centre (AVRDC) within Asia (Shanmugasundaram et al. 2009). Despite this, progress was slow, hampered by the lack of financial support to support R&D efforts. Some components of the original plan could only be started in 1997 despite the plan having been laid more than two decades prior (Shanmugasundaram et al. 2009).

In parallel, yet independent to the work in Asia were the breeding efforts advancing in Australia. Kick-started by the introduction of commercial varieties provided by the AVRDC in the late 1960s and 70s mungbean breeding in

Australia has grown exponentially (AMA 2015). Improved varieties such as crystal generated from the Australian breeding program revolutionized the industry. Production doubled from 10,000 to 20,000 tonnes between the 1980s and 1996 to about 45,000 tonnes by the mid-2000s, reaching a pinnacle of 150,000 tonnes in 2016 (AMA 2015).

Despite being independent ventures, the Asian and Australian breeding programs share a fundamental aim to develop a crop more appealing to growers (ACIAR 2016). Improvement in yield, seed quality, synchronous maturity and disease resistance remains core targets for breeders (Shanmugasundaram et al. 2009) though increasingly pressing challenges such as climate change have led to a growing interest in abiotic stress tolerance. Despite this growing interest, there has been little advancement in terms of actual outcomes. Much of this can be attributed to the limitations of the existing mungbean breeding programs as well as the inherent complexity of the trait.

As an orphan crop, mungbeans have not attracted as much R&D funding compared to major crops like soybean or cereals. Genetic information has been lacking in comparison, and crop development has been primarily limited to conventional breeding. While conventional breeding is simple in principle, it is laborious and time consuming to execute. Undesirable traits from the parents are inherited along with the desirable ones in a process known as linkage drag, requiring several generations of selection and backcrossing before a cultivar can be presented for release (Flint-Garcia et al. 2003).

The ease through which a trait can be bred in or out depends on the number of genes governing it (Agrawal 1998). Qualitative traits, such as some disease resistance, can be coded for by a single gene or a small group of genes (Moody et al. 2003). On the other hand, quantitative traits like abiotic stress tolerance are governed by multiple genes and produce a range of phenotypic variations (Hill 2010). This is made even more variable by the influence exerted by environmental factors. Predictably, breeding for a quantitative trait is more difficult compared to its qualitative counterpart (Moody et al. 2003).

Where abiotic stress tolerance is concerned, however, the complexity involved renders conventional breeding next to impossible. Unlike disease resistance which is dependent a small number of resistance genes (R-genes), abiotic stress tolerance is a summation of morphological and biochemical interactions. Breeding and selection based on phenotype alone would be insufficient and will require the support of more discriminating markers.

11.3.2 Marker-Assisted Selection

One way through which precise trait selection can be done is marker-assisted selection (MAS). Markers associated with a desired trait are used to assist the selection process. Generally, markers can be morphological, cytological, biochemical or molecular/DNA—the term MAS, however, is exclusively associated with molecular/DNA markers (Jiang 2013).

MAS has several benefits, namely efficiency and specificity. With DNA being used for screening, marker detection is not dependent on plant maturity, and there is less interference from environmental factors. Advances in molecular biology and the supporting technologies have also enabled economical high-throughput screening (Wetterstrand 2018). The capacity to distinguish between alleles in genotypes allows breeders to make selections based purely on genetics prior to phenotypic evaluation (Ota et al. 2007). This means large breeding populations can be quickly narrowed down prior to selection in the field, speeding up the time it takes to integrate the desired traits.

Five types of DNA markers are commonly used for MAS—restriction fragment length polymorphism (RFLP), amplified fragment length polymorphism (AFLP), random amplified polymorphic DNA (RAPD), microsatellites or simple sequence repeat (SSR) and single-nucleotide polymorphism (SNP). Each has their unique advantages and disadvantages, the specifics of which as well as the general methodology and considerations can be found in the review by Jiang (2013).

11.3.3 Mungbean Genome Sequencing and Molecular Markers

Understanding the genetic basis of important agronomic traits such as abiotic stress tolerance has been dramatically advanced through the use of next-generation sequencing (NGS) (Varshney et al. 2009). In recent years, nutritionally dense legume crops such as mungbean have gained global interest with major efforts made to unlock the crops' true potential through genomic tools (Kim et al. 2015a). Although mungbean is primarily grown by smallholder farmers in developing countries with limited funding, the importance of mungbean worldwide has spurred many researchers to explore genomic tools for crop improvement (Noble et al. 2018; Schafleitner et al. 2015). The first draft whole-genome sequence (WGS) of mungbean published by Kang et al. (2014) has facilitated genomic research in mungbean with the aim of accelerating molecular breeding. Two significant association mapping populations have used the mungbean WGS as a basis to anchor markers to a physical position in the genome (Schafleitner et al. 2015; Noble et al. 2018). Both populations employed the same genotype-by-sequencing (GBS) technology to identify thousands of highly polymorphic single-nucleotide polymorphisms (SNPs) markers to be used in genome-wide association studies (GWAS). Further advances in sequencing technology have made it economical and feasible to re-sequence large numbers of lines ushering in a new age of genotyping by re-sequencing (Zhou et al. 2015). Mungbean geneticists are taking advantage of this cost-effective technology and are re-sequencing many genetically distinct accessions from the above populations as well as wild relatives. Using this data to unravel the domestication history of mungbeans will provide an opportunity to identify the genetic basis of adaptation affecting agronomically important traits.

Crop wild relatives (CWR) have been underutilized in most breeding programs due to the inherent difficulty of integrating only the desired traits and a lack of accessions collected and conserved in seed banks worldwide (Castañeda-Álvarez et al. 2016). Wild relatives have become a major interest globally to increase the productivity of crops as a result of limited genetic diversity in domesticated cultivars (Castañeda-Álvarez et al. 2016). Gene editing is proving to be a pivotal method to introduce traits from wild species and other closely related species which cannot be achieved through conventional means (Zsögön et al. 2017). Mungbeans are in an excellent position to take advantage of CWR due to Australian researchers gathering, evaluating and conserving extensive collections of wild relatives (Lawn and Rebetzke 2006). Development of a pan-genome including CWR genome sequences will be integral to understanding the genetic diversity present within the crop's gene pool. Characterizing the core set of genes present in all accessions and most importantly the dispensable genome where many abiotic tolerance traits are found but have been unwittingly lost during domestication (Li et al. 2014).

11.3.4 QTL Mapping and Genome-Wide Association Studies (GWAS) of Abiotic Stress Tolerance in Mungbean

Abiotic stress tolerance is a quantitative trait with complex inheritance that is difficult to measure, and unstable over different environments (Deshmukh et al. 2014). Researchers have been using DNA markers and linkage maps since the early 1990s to identify quantitative trait locus (QTL) related to essential agronomic traits (Kim et al. 2015b). The first studies used RFLP technology to analyse bruchid resistance (Young et al. 1992) and seed weight (Fatokun et al. 1992), and since then, numerous QTLs have been identified for mungbean traits such as flower, leaf, pod and seed characteristics (Isemura et al. 2012; Kim et al. 2015a). However, the outdated marker technology and traditional linkage analysis used in these analyses lack the resolution and allele richness of association mapping (Huang and Han 2014). Proxy phenotypic traits such as plant height, maximum leaf area, biomass and relative water content have been used to identify potential drought tolerance QTLs indicating possible candidate regions of the genome (Liu et al. 2017).

Genome-wide association studies (GWAS) take advantage of historic recombination events of large diverse mapping populations to dissect the genetic source of complex traits such as abiotic stress tolerance (Huang and Han 2014). Utilizing the World Vegetable Centre mini-core and Australian diversity panel populations to study complex traits such as drought and nutritional content will provide greater precision during analyses, helping identify false positives and narrow down the genomic region associated with the trait (Schafleitner et al. 2015; Noble et al. 2018). The advantages of GWAS have been demonstrated in other legume species to more precisely identify QTLs controlling polygenic traits such as drought and heat tolerance; soybean (Dhanapal et al. 2015), chickpea (Thudi et al. 2014) and protein and oil content soybean (Hwang et al. 2014), chickpea (Upadhyaya et al. 2016).

Nested association mapping (NAM) is a technique used to dissect the genetic architecture of complex traits by combining the advantages of two traditional methods linkage analysis and association mapping (Yu et al. 2008). The maize NAM has been used to identify QTLs in flowering time (McMullen et al. 2009) and southern leaf blight (Kump et al. 2011). Researchers found rather than a few large-effect QTLs controlling a trait, it was the accumulative effect of numerous small-effect QTLs that impacted these traits. To enhance the productivity and resilience of mungbeans, a NAM population has been developed as a resource for the mungbean community. Consisting of twenty-six exotic founder lines identified as the most genetically and phenotypically distinct genotypes from the Australian diversity panel, these were then backcrossed to a single elite commercial cultivar (Noble et al. 2018). To further increase the genetic diversity within the population, four sub-populations were made with wild relatives to capture unique alleles not present in domesticated accessions. Overall, the mungbean NAM is made up of 2000 RILs all of which have been genotyped-by-sequenced but are yet to be phenotype and analysed.

Marker-assisted selection (MAS) established from QTLs which account for a large proportion of phenotypic variance is effective for simple Mendelian traits but becomes particularly challenging for complex traits (Deshmukh et al. 2014). The prediction model used in genomic selection overcomes the shortcomings of MAS and avoids marker bias by accounting for all phenotypic and marker data to calculate a breeding value for lines (Heffner et al. 2009). Genomic selection is a prevailing approach that will considerably speed up breeding times by more reliably integrating complex polygenic traits through selecting lines based on marker information alone (Heffner et al. 2009).

11.3.5 Mungbean Transformation

Mungbean improvement through conventional breeding has significantly improved overall yield in recent years. However, efforts to develop robust abiotic stress tolerance are lacking, leaving the crop vulnerable. Many of the abiotic stress tolerance traits such as salinity and drought are polygenic, and often undesirable traits are inherited through linkage drag (Turan et al. 2012; Vikram et al. 2015). Genetic modification can be used as a complementary tool in traditional breeding programmes by providing a platform to functionally characterize genes. Functional characterization of the stay-green gene from rice and sorghum (Hörtensteiner 2009; Jiang et al. 2007; Paterson 2008) can be targeted through genome editing. A high-throughput, efficient, transformation system is essential to deliver genome-edited plants. Mungbean like other legumes is very challenging to transform due to a recalcitrant responses to in vitro regeneration (Lambrides and Godwin 2007). Limited reports are available on successful generation of transgenic mungbean using *Agrobacterium*-mediated transformation with varying rates of efficiency (Sahoo et al. 2016; Sahoo and Jaiwal 2009; Saini et al. 2007; Vijayan et al. 2006). Further research is required to develop a high efficiency, reliable and repeatable mungbean transformation system to generate transgenic mungbean for detail genetic studies.

Recent reports of genetic improvements for abiotic stress tolerance in mungbean through genome editing allowed the transfer of foreign genes from distantly related species (Baloda and Madanpotra 2017; Kumar et al. 2017; Sahoo et al. 2016). The team led by Kumar and his co-workers (Kumar et al. 2017), co-expressed Arabidopsis *NHX1* and *bar* genes through gene stacking and significantly improved mungbean tolerance to salinity, oxidative stress and herbicide resistance (Kumar et al. 2017). Similar gene-editing approaches for enhanced abiotic stress tolerance in mungbean could complement current breeding programs.

11.3.6 Genome Editing and Other Molecular Technologies to Bring Mungbean Breeding in the Twenty-First Century

Genome editing is the generic term for a group of technologies that enable the direct modification of target genomes through the creation of double-stranded DNA breaks (DSBs) and subsequent "editing" by host DNA repair machinery (Shah et al. 2018). The power of genome editing is unprecedented. To put it into perspective, the use of molecular breeding approaches has accelerated the generation of new crop varieties from 25 years to as little as seven (Zhang et al. 2018). Depending on the regulatory restrictions imposed by the host country and the availability of efficient plant transformation and regeneration systems, genome editing may cut the time taken to produce new crop varieties by as much as half.

Genome editing has already been used to generate elite crop varieties, namely in grains such as rice. To improve resistance to rice blast, CRISPR/Cas was used to mutate the OSERF922 gene vital to disease progression (Xu et al. 2015; Wang et al. 2016). In soybean, Huang and Han (2014) used the TALEN genome-editing system to mutate the fatty acid desaturase 2 genes, FAD2-1A and FAD2-1B to increase the proportion of monounsaturated fatty acids by up to 80%. The mungbean reference sequence, NAM and other association mapping populations will be integral to taking advantage of genome-editing tools now and into the future (Kang et al. 2014; Noble et al. 2018).

A potential target for mungbean improvement is the development of powdery mildew resistant genotypes. The *mlo* gene family is a highly conserved group of proteins that are grouped into seven phylogenetic clades and have been linked to disease susceptibility, plant reproduction and root thigmomorphogenesis (Lyngkjær et al. 2000; Pavan et al. 2009). In particular, MLO proteins from clades IV and V are important genetic drivers of crop susceptibility to powdery mildew (Rispail and Rubiales 2016). Plants that are homozygous for recessive alleles at the *mlo* locus are more resistant to penetration by powdery mildew (Rispail and Rubiales 2016). The relationship between *mlo* recessive alleles and powdery mildew remains one of few examples of monogenic traits that provide resistance in the field. The monogenic nature of the *mlo* alleles makes it an ideal target for genome-editing strategies. Current studies have shown that mungbean contains 1 and 6, clade IV and V *mlo* genes, and therefore, further work is required before this approach can be implemented into breeding programmes.

Recent genome-wide association studies of the mungbean diversity set identified nine SNPs that play a significant role in seed colour (Noble et al. 2018). Given that seed coat colour is a key trait for the grading and evaluation of mungbean seeds, these SNPs could be further investigated to determine whether genome-editing strategies could be utilized to produce shiny green mungbean varieties into backgrounds with abiotic stress tolerance. Genome-editing strategies also provide a pathway to overcome incompatible hybridisation during crossing. For example, when cultivated mungbean (*V. radiata*) is hybridised with black gram (*V. mungo*) or wild mungbean (*V. radiata* ssp sublobata) sterility of the progeny and poor morphonology characteristics have historically prevented their use in breeding programming (Lambrides et al. 1999). There is potential to utilize genome-editing strategies to introduce the traits from black gram into cultivated mungbean without the need to perform crosses. These are just a few examples; however, it is becoming apparent that genome engineering will be a powerful tool for the improvement of mungbean in the near future.

In recent years, we have witnessed the huge impact abiotic stresses have had on agricultural production across the world, due to significant variability in climatic conditions. Mungbean, an essential food security crop for many developing nations, has been equally vulnerable to these stresses and until recently, has received little attention and support from the major R&D funding agencies. During the past decade, conventional breeding has been leapfrogged by significant advancements in sequencing technologies and genome manipulation tools such as genome editing. These advancements and technology platforms will accelerate the development of more resilient and nutritious crops, including mungbean.

References

Abd-Alla M, Vuong T, Harper J (1998) Genotypic differences in dinitrogen fixation response to NaCl stress in intact and grafted soybean. Crop Sci 38:72–77

ACIAR (2016) Establishing the international mungbean improvement network [Online]. Available: https://www.aciar.gov.au/project/CIM/2014/079. Accessed 11 Feb 2018]

Agrawal RL (1998) Fundamentals of plant breeding and hybrid seed production. Science Publishers, Inc

Ahmed S, Higuchi H, Nawata E, Sakuratani T (2002) Effects of exogenous ABA and ethylene application and waterlogging on photosynthesis in mungbean (*Vigna radiata* (L.) Wilczak). Japanese J Trop Agric (Japan)

Ahmed S, Nawata E, Sakuratani T (2006) Changes of endogenous ABA and ACC, and their correlations to photosynthesis and water relations in mungbean (*Vigna radiata* (L.) Wilczak cv. KPS1) during waterlogging. Environ Exp Bot 57:278–284

AMA (2015) Australian Mungbean Association about us [Online]. http://www.mungbean.org.au/about-us.html. Australian Mungbean Association (AMA). Available: http://www.mungbean.org.au/about-us.html#industry-size-and-value. Accessed 15 Oct 2018

Ayers RS, Westcot DW (1985) Water quality for agriculture. Food and Agriculture Organization of the United Nations Rome

Baloda A, Madanpotra S (2017) Transformation of Mungbean plants for salt and drought tolerance by introducing a gene for an osmoprotectant glycine betaine. J Plant Stress Physiol 3:5–11

Borlaug NE (1983) Contributions of conventional plant breeding to food production. Science 219:689

Castañeda-Álvarez NP, Khoury CK, Achicanoy HA, Bernau V, Dempewolf H, Eastwood RJ, Guarino L, Harker RH, Jarvis A, Maxted N, Müller JV, Ramirez-Villegas J, Sosa CC, Struik PC, Vincent H, Toll J (2016) Global conservation priorities for crop wild relatives. Nature Plants 2:16022

Cominelli E, Conti L, Tonelli C, Galbiati M (2013) Challenges and perspectives to improve crop drought and salinity tolerance. New Biotechnol 30:355–361

Daryanto S, Wang L, Jacinthe PA (2017) Global synthesis of drought effects on cereal, legume, tuber and root crops production: a review. Agric Water Manag 179:18–33

Day L (2013) Proteins from land plants—potential resources for human nutrition and food security. Trends Food Sci Technol 32:25–42

Deshmukh R, Sonah H, Patil G, Chen W, Prince S, Mutava R, Vuong T, Valliyodan B, Nguyen HT (2014) Integrating omic approaches for abiotic stress tolerance in soybean. Front Plant Sci 5:244. https://doi.org/10.3389/fpls.2014.00244

Dhanapal AP, Ray JD, Singh SK, Hoyos-Villegas V, Smith JR, Purcell LC, Andy King C, Cregan PB, Song Q, Fritschi FB (2015) Genome-wide association study (GWAS) of carbon isotope ratio (δ13C) in diverse soybean [*Glycine max* (L.) Merr.] genotypes. Theor Appl Genet 128:73–91

Eckardt NA (2009) The future of science: food and water for life. Plant Cell 21:368–372

Eynard A, Lal R, Wiebe K (2005) Crop response in salt-affected soils. J Sustain Agric 27:5–50

FAO (2009a) Declaration of the world summit on food security. World Summit on Food Security Rome

FAO (2009b) Declaration of the world summit on food security. 16–18 November, Rome. http://www.fao.org/wsfs/world-summit/wsfs-challenges/en/

FAO (2012) World water day 2012 celebration. UN Conference Centre, Bangkok, 22 March 2012. http://www.fao.org/asiapacific/rap/home/meetings/list/detail/en/?meetings_id=637&year=2012

Fatokun CA, Menancio-Hautea DI, Danesh D, Young ND (1992) Evidence for orthologous seed weight genes in cowpea and mung bean based on RFLP mapping. Genetics 132:841–846

Fernandez G, Shanmugasundaram S (1987) The AVRDC mungbean improvement program. The past, present and future, pp 58–70
Fernandez G, Shanmugasundaram S, Mclean B (1988) The AVRDC mungbean improvement program: the past, present and future. In: Mungbean. Proceedings of the second international symposium, Bangkok, Thailand. Asian Vegetable Research and Development Center, Tainan, Taiwan, pp 58–70
Flint-Garcia SA, Thornsberry JM, Buckler EST (2003) Structure of linkage disequilibrium in plants. Annu Rev Plant Biol 54:357–374
Ghanbari M, Javan SM (2015) Study the response of mung bean genotypes to drought stress by multivariate analysis. Int J Agric Innov Res 3:1298–1302
Ghosh S, Mitra S, Paul A (2015) Physiochemical studies of sodium chloride on mungbean (*Vigna radiata* L. Wilczek) and its possible recovery with spermine and gibberellic acid. The Sci World J, 858016. https://doi.org/10.1155/2015/858016
Hanumantharao B, Nair RM, Nayyar H (2016) Salinity and high temperature tolerance in mungbean [*Vigna radiata* (L.) Wilczek] from a physiological perspective. Front Plant Sci 7:957
Hasanuzzaman M, Hossain MA, Da Silva JAT, Fujita M (2012) Plant response and tolerance to abiotic oxidative stress: antioxidant defense is a key factor. Crop stress and its management: perspectives and strategies. Springer
Heffner EL, Sorrells ME, Jannink JL (2009) Genomic selection for crop improvement. Crop Sci 49:1–12
Hill WG (2010) Understanding and using quantitative genetic variation. Philos T R Soc B 365:73–85
Hörtensteiner S (2009) Stay-green regulates chlorophyll and chlorophyll-binding protein degradation during senescence. Trends Plant Sci 14:155–162
Huang X, Han B (2014) Natural variations and genome-wide association studies in crop plants. Ann Rev Plant Biol 65:531–551
Hwang EY, Song Q, Jia G, Specht JE, Hyten DL, Costa J, Cregan PB (2014) A genome-wide association study of seed protein and oil content in soybean. BMC Genom 15:1
Ihsan MZ, Shahzad N, Kanwal S, Naeem M, Khaliq A, El-Nakhlawy FS, Matloob A (2013) Potassium as foliar supplementation mitigates moisture induced stresses in mung bean (*Vigna radiata* L.) as revealed by growth, photosynthesis, gas exchange capacity and Zn analysis of shoot. Int J Agronom Plant Prod 4:3828–3835
Isemura T, Kaga A, Tabata S, Somta P, Srinives P, Shimizu T, Jo U, Vaughan DA, Tomooka N (2012) Construction of a genetic linkage map and genetic analysis of domestication related traits in mungbean (*Vigna radiata*). PLoS ONE 7:e41304
Jiang GL (2013) Molecular markers and marker-assisted breeding in plants. Plant breeding from laboratories to fields. InTech
Jiang H, Li M, Liang N, Yan H, Wei Y, Xu X, Liu J, Xu Z, Chen F, Wu G (2007) Molecular cloning and function analysis of the stay green gene in rice. The Plant J 52:197–209
Kang YJ, Kim SK, Kim MY, Lestari P, Kim KH, Ha BK, Jun TH, Hwang WJ, Lee T, Lee J, Shim S, Yoon MY, Jang YE, Han KS, Taeprayoon P, Yoon N, Somta P, Tanya P, Kim KS, Gwag J-E, Moon J-K, Lee Y-H, Park B-S, Bombarely A, Doyle JJ, Jackson SA, Schafleitner R, Srinives P, Srinives RK, Lee S-H (2014) Genome sequence of mungbean and insights into evolution within *Vigna* species. Nat Commun 5:5443
Kaur R, Bains T, Bindumadhava H, Nayyar H (2015) Responses of mungbean (*Vigna radiata* L.) genotypes to heat stress: effects on reproductive biology, leaf function and yield traits. Scientia Horti 197:527–541
Kim SK, Nair RM, Lee J, Lee SH (2015) Genomic resources in mungbean for future breeding programs. Front Plant Sci 6:626
Kochian LV, Hoekenga OA, Pineros MA (2004) How do crop plants tolerate acid soils? Mechanisms of aluminum tolerance and phosphorous efficiency. Annu Rev Plant Biol 55:459–493
Kochian LV, Piñeros MA, Liu J, Magalhaes JV (2015) Plant adaptation to acid soils: the molecular basis for crop aluminum resistance. Ann Rev Plant Biol 66:571–598
Kole (2007) Genome mapping and molecular breeding in plants. Springer, Heidelberg
Kramer PJ, Boyer JS, Carlson W (1997) Water relations of plants and soils. Forest Sci 43:151–152
Kumar S, Kalita A, Srivastava R, Sahoo L (2017) Co-expression of Arabidopsis NHX1 and bar improves the tolerance to salinity, oxidative stress, and herbicide in transgenic mungbean. Front Plant Sci 8:1896
Kumari P, Varma S (1983) Genotypic differences in flower production/shedding and yield in mungbean (*Vigna radiata*). Indian J Plant Physiol 26:402–405
Kump KL, Bradbury PJ, Wisser RJ, Buckler ES, Belcher AR, Oropeza-Rosas MA, Zwonitzer JC, Kresovich S, Mcmullen MD, Ware D, Balint-Kurti PJ, Holland JB (2011) Genome-wide association study of quantitative resistance to southern leaf blight in the maize nested association mapping population. Nature Genet 43:163
Lambrides C, Godwin I (2007) Mungbean. Pulses, sugar and tuber crops. Springer
Lambrides CJ, James AT, Lawn RJ, Williams RW (1999) Cross fertility of australian accessions of wild mungbean (*Vigna radiata* ssp. *sublobata*) with green gram (*V. radiata* ssp. *radiata*) and black gram (*V. mungo*). Australian J Bot 47:601–610
Lawn RJ, Rebetzke GJ (2006) Variation among Australian accessions of the wild mungbean (*Vigna radiata* ssp sublobata) for traits of agronomic, adaptive, or taxonomic interest. Australian J Agric Res 57:119–132
Li YH, Zhou G, Ma J, Jiang W, Jin LG, Zhang Z, Guo Y, Zhang J, Sui Y, Zheng L, Zhang S, Zuo Q, Shi X, Li Y, Zhang, W, Hu Y, Kong G, Hong, H, Tan B, Song J, Liu Z, Wang Y, Ruan H, Yeung CKL, Liu J,

Wang H, Zhang L, Guan R, Wang K, Li W, Chen S, Chang R, Jiang Z, Jackson SA, Li R, Qiu LJ (2014) De novo assembly of soybean wild relatives for pan-genome analysis of diversity and agronomic traits. Nature Biotechnol 32:1045

Liu C, Wu J, Wang L, Fan B, Cao Z, Su Q, Zhang Z, Wang Y, Tian J, Wang S (2017) Quantitative trait locus mapping under irrigated and drought treatments based on a novel genetic linkage map in mungbean (*Vigna radiata* L.). Theor Appl Genet 130:2375–2393

Lyngkjær MF, Newton AC, Atzema JL, Baker SJ (2000) The Barley mlo-gene: an important powdery mildew resistance source. Agronomie 20:745–756

Mcmullen MD, Kresovich S, Villeda HS, Bradbury P, Li H, Sun Q, Flint-Garcia S, Thornsberry J, Acharya C, Bottoms C, Brown P, Browne C, Eller M, Guill K, Harjes C, Kroon D, Lepak N, Mitchell SE, Peterson B, Pressoir G, Romero S, Rosas MO, Salvo S, Yates H, Hanson M, Jones E, Smith S, Glaubitz JC, Goodman M, Ware D, Holland JB, Buckler ES (2009) Genetic properties of the maize nested association mapping population. Science 325:737–740

Meena RS, Yadav RS (2014) Phenological performance of groundnut varieties under sowing environments in hyper arid zone of Rajasthan, India. J Appl Nat Sci 6:344–348

Meena RS, Varma D (2016) Mungbean yield and nutrient uptake performance in response of NPK and lime levels under acid soil in Vindhyan region, India. J Appl Nat Sci 8:860–863

Minhas P, Sharma D, Khosla B (1990) Mungbean response to irrigation with waters of different salinities. Irrigation Sci 11:57–62

Moody D, Emebiri L, Spackman M, Mekuria G (2003) Breeding for qualitative and quantitative traits using markers. In: Proceedings of a joint meeting of the 11th Australian barley technical symposium and the 53rd Australian cereal chem conference, Glenelg, SA

Munns R, Tester M (2008) Mechanisms of salinity tolerance. Annu Rev Plant Biol 59:651–681

Nahar K, Hasanuzzaman M, Suzuki T, Fujita M (2017) Polyamines-induced aluminum tolerance in mung bean: a study on antioxidant defense and methylglyoxal detoxification systems. Ecotoxicol 26:58–73

Noble TJ, Tao Y, Mace ES, Williams B, Jordan DR, Douglas CA, Mundree SG (2018) Characterization of linkage disequilibrium and population structure in a mungbean diversity panel. Front Plant Sci 8:2102

Ota M, Fukushima H, Kulski JK, Inoko H (2007) Single nucleotide polymorphism detection by polymerase chain reaction-restriction fragment length polymorphism. Nat Protoc 2:2857

Panda SK, Baluška F, Matsumoto H (2009) Aluminum stress signaling in plants. Plant Signal Behav 4:592–597

Pandey R, Herrera W, Villegas A, Pendleton J (1984) Drought response of grain legumes under irrigation gradient: III. Plant growth 1. Agronomy J 76:557–560

Pang Q, Chen S, Dai S, Chen Y, Wang Y, Yan X (2010) Comparative proteomics of salt tolerance in *Arabidopsis thaliana* and *Thellungiella halophila*. J Proteome Res 9:2584–2599

Paterson AH (2008) Genomics of sorghum. Int J Plant Genomics Article ID 362451, 6 pages

Pavan S, Jacobsen E, Visser RGF, Bai Y (2009) Loss of susceptibility as a novel breeding strategy for durable and broad-spectrum resistance. Mol Breed 25:1

Postel SL (2000) Entering an era of water scarcity: the challenges ahead. Ecol Appl 10:941–948

Rafiei M, Shirvan AM (2009) Yield reaction and morphological characteristics of some mung bean genotypes to drought stress. J Mod Agric Knowl 5:67–76

Ranawake A, Dahanayaka N, Amarasingha U, Rodrigo W, Rodrigo U (2011) Effect of water stress on growth and yield of mung bean (*Vigna radiata* L.). Trop Agric Res Ext 14:76–79

Rispail N, Rubiales D (2016) Genome-wide identification and comparison of legume MLO gene family. Sci Rep 6:32673

Saha P, Chatterjee P, Biswas AK (2010) NaCl pretreatment alleviates salt stress by enhancement of antioxidant defense system and osmolyte accumulation in mungbean (*Vigna radiata* L. Wilczek). Indian J Exp Biol 48:593–600

Sahoo DP, Kumar S, Mishra S, Kobayashi Y, Panda SK, Sahoo L (2016) Enhanced salinity tolerance in transgenic mungbean overexpressing Arabidopsis antiporter (NHX1) gene. Mol Breed 36:144

Sahoo L, Jaiwal PK (2009) Asiatic beans. Compendium of Transgenic Crop Plants, pp 115–132

Saini R, Singh RP, Jaiwal PK (2007) Agrobacterium tumefaciens mediated transfer of Phaseolus vulgaris α-amylase inhibitor-1 gene into mungbean *Vigna radiata* (L.) Wilczek using bar as selectable marker. Plant Cell Rep 26:187–198

Sarkar M, Datta S, Kundagrami S (2017) Global climate change and mung bean production: a roadmap towards future sustainable agriculture. Sustaining future food security in changing environments

Schafleitner R, Nair RM, Rathore A, Wang YW, Lin CY, Chu SH, Lin PY, Chang JC, Ebert AW (2015) The AVRDC—the world vegetable center mungbean (*Vigna radiata*) core and mini core collections. BMC Genom 16:344

Sehrawat N, Jaiwal P, Yadav M, Bhat K, Sairam R (2013) Salinity stress restraining mungbean (*Vigna radiata* (L.) Wilczek) production: gateway for genetic improvement. Int J Agric Crop Sci 6:505

Shah T, Andleeb T, Lateef S, Noor MA (2018) Genome editing in plants: advancing crop transformation and overview of tools. Plant Physiol Biochem 131:12–21

Shanmugasundaram S, Keatinge J, Hughes J (2009) The mungbean transformation: diversifying crops, defeating malnutrition. IDEAS Working Paper Series from RePEc

Singh D, Singh B (2011) Breeding for tolerance to abiotic stresses in mungbean. J Food Leg 24:83–90

Sinha SK (1977) Food legumes: distribution, adaptability and biology of yield. FAO

Thudi M, Upadhyaya HD, Rathore A, Gaur PM, Krishnamurthy L, Roorkiwal M, Nayak SN, Chaturvedi SK, Basu PS, Gangarao NVPR, Fikre A, Kimurto P, Sharma PC, Sheshashayee MS, Tobita S, Kashiwagi J, Ito O, Killian A, Varshney RK (2014) Genetic dissection of drought and heat tolerance in chickpea through genome-wide and candidate gene-based Association mapping approaches. PLoS ONE 9:e96758

Tickoo J, Mahto GR, Manji C (1996) Plant type in mungbean (*Vigna radiata* (L.) Wilczek). In: Asthana AN, Kim DH (eds) Proceeding of recent Advances in mungbean research. Indian Society of Pulses Research, Kanpur, India, pp 197–213

Turan S, Cornish K, Kumar S (2012) Salinity tolerance in plants: breeding and genetic engineering. Aust J Crop Sci 6:1337

United Nations DOEASA, Population Division (2017) World population prospects: the 2017 revision, world population 2017 Wallchart. ST/ESA/SER.A/398

Upadhyaya HD, Bajaj D, Narnoliya L, Das S, Kumar V, Gowda CLL, Sharma S, Tyagi AK, Parida SK (2016) Genome-wide scans for delineation of Candidate genes regulating seed-protein content in chickpea. Front Plant Sci 7:302. https://doi.org/10.3389/fpls.2016.00302

Varshney RK, Nayak SN, May GD, Jackson SA (2009) Next-generation sequencing technologies and their implications for crop genetics and breeding. Trends Biotechnol 27:522–530

Vijayan S, Beena M, Kirti P (2006) Simple and effective regeneration of mungbean (Vigna radiata (L) Wilczek) using cotyledonary node explants. J Plant Biochem Biotechnol 15:131–134

Vikram P, Swamy BM, Dixit S, Singh R, Singh BP, Miro B, Kohli A, Henry A, Singh N, Kumar A (2015) Drought susceptibility of modern rice varieties: an effect of linkage of drought tolerance with undesirable traits. Sci Rep 5:14799

Wang F, Wang C, Liu P, Lei C, Hao W, Gao Y, Liu YG, Zhao K (2016) Enhanced rice blast resistance by CRISPR/Cas9-targeted mutagenesis of the ERF transcription factor gene OsERF922. PLoS ONE 11: e0154027

Wetterstrand K (2018) DNA sequencing costs: data from the NHGRI genome sequencing program (GSP) [Online]. Available: https://www.genome.gov/27541954/dna-sequencing-costs-data/. Accessed 11 Feb 2018

Xu RF, Li H, Qin RY, Li J, Qiu CH, Yang YC, Ma H, Li L, Wei PC, Yang JB (2015) Generation of inheritable and "transgene clean" targeted genome-modified rice in later generations using the CRISPR/Cas9 system. Sci Rep 5:11491

Young ND, Kumar L, Menancio-Hautea D, Danesh D, Talekar NS, Shanmugasundarum S, Kim DH (1992) RFLP mapping of a major bruchid resistance gene in mungbean (*Vigna radiata*, L. Wilczek). Theor Appl Genet 84:839–844

Yu J, Holland JB, Mcmullen MD, Buckler ES (2008) Genetic design and statistical power of nested association mapping in maize. Genetics 178:539–551

Zhang J, Zhang H, Botella JR, Zhu JK (2018) Generation of new glutinous rice by CRISPR/Cas9-targeted mutagenesis of the Waxy gene in elite rice varieties. J Integr Plant Biol 60:369–375

Zhou Z, Jiang Y, Wang Z, Gou Z, Lyu J, Li W, Yu Y, Shu L, Zhao Y, Ma Y, Fang C, Shen Y, Liu T, Li C, Li Q, Wu M, Wang M, Wu Y, Dong Y, Wan W, Wang X, Ding Z, Gao Y, Xiang H, Zhu B, Lee S-H, Wang H, Tian Z (2015) Resequencing 302 wild and cultivated accessions identifies genes related to domestication and improvement in soybean. Nat Biotechnol 33:408

Zsögön A, Cermak T, Voytas D, Peres LEP (2017) Genome editing as a tool to achieve the crop ideotype and de novo domestication of wild relatives: Case study in tomato. Plant Sci 256:120–130

12 Future Prospects and Challenges

Roland Schafleitner and Ramakrishnan M. Nair

Abstract

Legume crops play a key role for producing proteins for human and animal nutrition. Sustainable increase of plant protein production is essential to satisfy the rising demand of a growing world population. Breeding varieties with high and stable yields and with optimized nutritional value, which at the same time require less input in terms of energy and labor is one of the pathways for sustainable rise of mungbean productivity. Mungbean is mainly used in rotation with cereals. Therefore, producing an economically viable harvest in the short time window between two main crops, often under stressful conditions of a hot and dry season, is an important breeding aim for this crop. Breeding improved varieties requires access to the genetic diversity of the crop and crop wild relatives to source new traits. As natural plant populations are endangered by loss of habitats and climate change, ex situ collections have gained increased importance to conserve biodiversity for crop improvement. Effective screening methods for desired agronomical traits, including biotic and abiotic stress tolerances and pre-breeding technologies to introgress new traits from non-adapted materials into elite lines are facilitating breeding efforts. Often new traits have to be sourced from wild relatives. Crossing barriers between different *Vigna* species and the need of technologies to restore fertility add additional complexity when traits have to be sourced from wild species. Genomics methods such as quantitative trait mapping or pangenomics studies elucidate the genetic basis of traits of interest, and marker assisted or genomic selection are guiding breeding efforts. Well-coordinated phenotyping efforts to collect and analyze crop performance data across multiple locations are essential for effective breeding of a more productive, nutritious and resilient mungbean crop.

R. Schafleitner (✉)
The World Vegetable Center, Shanhua, Tainan, Taiwan
e-mail: roland.schafleitner@worldveg.org

R. M. Nair
World Vegetable Center, South Asia, ICRISAT Campus, Patancheru, Hyderabad, Telangana, India

12.1 Introduction

One of the major challenges of humanity for the next decades is producing sufficient food for a growing world population, which is estimated to reach 9.15 billion by 2050 (Alexandratos and Bruinsma 2012). To satisfy the demand of this number of people, an overall increase in agricultural output of about 60% is required. Proteins are essential macronutrient for the human diet. Legumes are an important protein source either for direct consumption by humans, or as feed for meat, poultry or fish production. Especially, in Asia, with a large vegetarian population, legumes

R. M. Nair et al. (eds.), *The Mungbean Genome*, Compendium of Plant Genomes,
https://doi.org/10.1007/978-3-030-20008-4_12

contribute to meeting the protein requirement of a growing population. Among the legumes, soybean has the largest protein content (33–45%) and mungbean and pea rank sixth with 21–33% of protein in the seed (Salunkhe et al. 1985).

Achieving food production increases in a sustainable manner will be critical to avoid further natural resource degradation and damage to the environment. Reducing the energy requirement of agriculture and lowering the release of greenhouse gases into the atmosphere and for food production will become increasingly important. Legume crops are an important component of sustainable production systems. They release 5–7 times less greenhouse gases such as CO_2 and nitrous oxide per unit area compared with other crops. Their capacity to establish symbiosis with microorganisms that fix nitrogen from the air save approximately 277 kg CO_2 ha^{-1} and per year, an amount which is otherwise emitted by processes for industrial nitrogen fixation (Stagnari et al. 2017). In addition, legume cultivation can increase the phosphorous availability at the rhizosphere level (Latati et al. 2016). Phosphorus is an essential macronutrient for plants. Rock phosphate required for producing phosphorous fertilizers is an increasingly scarce resource and reducing the phosphorus fertilizer amount for crop production makes agriculture more sustainable.

Increasing mungbean production can be achieved by three approaches: improving the yield of the crop by breeding, extend the cultivation surface into new areas, and integrate a short-duration mungbean crop into an existing cropping system like inter-cropping with maize or sorghum. For making work any of these approaches, mungbean cultivation must be attractive and profitable for farmers. This implies that production methods have to be adapted to the needs of the farmers. In developing countries, mungbean is hand harvested, with two or three times picking. Hand-picking requires labor resources that are becoming scarce, so mungbean varieties that mature synchronously and are amenable to machine harvest are the key for keeping mungbean production attractive for farmers. Access to good quality seed by farmers and the implementation of good agricultural practices is paramount for mungbean to compete in high-premium market segments like the sprout segment.

12.2 Access to Genetic Diversity for Breeding

Breeding needs access to genetic variation for the traits to be improved. Genetic variation that is not found in breeding populations, needs to be sourced from germplasm accessions stored in genebanks or found in the wild. Access to seed from genebanks and to passport and evaluation information of germplasm accessions is speeding up the identification of new genetic variation for breeding. Specialized populations for trait screening such as core collections or improved screening methods like focused identification of germplasm set (FIGS) improve the access to diversity. Core collection diversity sets for mungbean are available (Liu et al. 2008; Bisht et al. 1998; Barkley et al. 2008; Moe et al. 2012; Schafleitner et al. 2015; Noble et al. 2018). Introgression line populations containing fragments of wild genomes provide access to new traits not available in the cultivated genepool in materials that can be easily crossed with elite varieties. In various crops, such as tomato or wheat, available trait variability was strongly increased by such lines (Eshed and Zamir 1995; Pestsova et al. 2001). Nested association mapping (NAM) populations have a similar function and display the alleles of a biodiverse set of lines in the background of a selected line (McMullen et al. 2009). A NAM population created for mungbean is under construction (Col Douglas, personal communication). Multi-parent advanced generation inter-cross (MAGIC) populations combine alleles of biodiverse parents in a set of inbreeding lines in a novel manner (Pascual et al. 2015). The development of such resources will improve access to novel alleles and facilitate breeding and gene mapping. Developing resistant varieties in response to emerging diseases like dry root rot, bacterial diseases (halo blight and tan spot) and to growing infestation by stem fly and thrips will be essential.

12.3 Stabilize Yields and Adapt Mungbean to More Stressful Environments

Short-duration legumes such as mungbean fit into most cereal production systems in tropical and subtropical regions. As a rotation crop, mungbean increases overall productivity and sustainability of the system. Several studies demonstrated the beneficial effect of a short-term mungbean crop on the yield and quality of the following crops in terms of increased cereal yield and quality (Kayani et al. 2010).

Heat and drought tolerance make mungbean suitable for the time of the year when cereal crops would fail. The overall stress tolerance of the crop makes it also a good option for agricultural production in marginal areas with a high risk of drought and heat stress and low soil fertility. Its capacity to mature rapidly and produce yield in only two months lets it escape seasonal stress factors better than plants with a longer vegetative period.

Short-duration mungbean is often cultivated during the hot and dry season. Current varieties have relatively high levels of heat and drought tolerance. Investigation of a very small set of four lines suggested a linear relationship between leaf conductance and net photosynthetic or transpiration rates, suggesting that productivity is strongly correlated with water use (Hamid et al. 1990). Investigations on larger germplasm may yield lines with more variation in water use efficiency. Optimization of irrigation schedules is an alternative way of genetic improvement to increase mungbean production under water-limiting conditions (Pannu and Singh 1993; Nakhlawy et al. 2018).

Most varieties can cope with temperatures above 40 °C, but still higher heat tolerance is required for mungbean cultivation during the hot and dry season. Mungbean under field conditions gradually adapts to heat (Hanumantha Rao et al. 2016). Flower shattering is one of the consequences of excessive heat, but may also be associated with thrips infestation (Chhabra and Kooner 1985). Climate change is likely to make stressful environments even harsher, affecting crop production in various ways, including by drought and heat spells or by flooding. Stress-tolerant short-duration mungbean may be less affected by climate change than other more heat and drought-susceptible species. The elevated CO_2 level that contributes to climate change, however, may have a beneficial effect on plant growth. Elevated CO_2 concentrations resulted in taller mungbean plants with larger leaf area, root length and more dry matter, including pods and seed than ambient CO_2 grown plants at all growth stages (Srivastava et al. 2001). Response to high CO_2 was highest at early growth stages. The quest for genotypes with superior response to elevated CO_2 by producing more grain has not yet started for mungbean.

Saline soils are an increasing problem in arid regions, where salt of low-quality irrigation water or leaching from lower soil layers accumulate in top soils. Mungbean is known to be salt susceptible, and in addition, symbiotic microorganisms may be affected by salt stress, lowering the overall performance of the crop on saline soils (Hanumantha Rao et al. 2016). Variation in salt tolerance in *V. radiata* has been found, but not yet systematically analyzed and mobilized for breeding. Inoculation of mungbean seed with consortia of Rhizobia with plant growth-promoting bacteria showed good results in pot trials (Ahmad et al. 2012), but more research is needed to recommend this method for broad application on saline soils. The introduction of tolerance from wild relatives, as discussed further below, may help increasing salt tolerance of mungbean.

12.4 Exploit Variation in Phenology to Improve Yields of Short-Duration Mungbean

Introgression of mungbean as a catch crop in cereal systems requires short-duration varieties with high yields that are adapted to the local conditions. Short-duration varieties have less time for photosynthesis and generally yield less than long-duration varieties. Accelerating plant establishment and early growth can reduce the yield gap between short- and long-duration varieties. Seed priming by soaking the seed in water before planting has shown the potential to accelerate plant growth (Rashid et al. 2004). Early flower induction and early maturation are key traits of short-duration mungbean. Up to 6.5 days earlier flowering and up to 9.6 days earlier maturation compared to control lines was achieved by gamma-ray mutagenesis in comparison to un-mutated controls (Khan and Goyal 2009). Large seed is associated with more vigorous seeding growth (Khattak et al. 2003), consequently, combining earliness with large seededness may reduce the yield gap for short-duration varieties. Further, increases in yield could be achieved by better matching the crop duration to the available time window and the target environment (Chauhan and Williams 2018).

Phenology in mungbean shows high plasticity. Mungbean is a quantitative short-day plant with variation in optimum mean diurnal temperatures for flowering (Summerfield and Lawn 1987). Crop models are excellent tools to investigate and predict crop duration. Chauhan and Rachaputi (2014) used an APSIM model (Carberry et al. 1992) that measures phenology by separating plant development into different stages, such as sowing to emergence, emergence to the end of the basic vegetative phase, a photoperiod-induced phase which ends at floral initiation and a floral development phase which ends at 50% flowering. Three of these stages have been modeled as a function of temperature; only the photoperiod-induced phase ending at floral induction has been modeled as a function of temperature and photoperiod. The model has been found to be reasonably accurate for predicting mungbean phenology.

12.5 Improve Yield Potential of Mungbean

Dry matter accumulation is a function of leaf area index, light extinction coefficient, the duration of light interception and radiation use efficiency. The interaction between these factors under various environmental conditions can be investigated with crop models to define an ideotype that is well suited for specific target environments (Chauhan and Williams 2018). Modulating leaf area and form could adapt mungbean better to the irradiation and evaporation conditions of new target areas with the potential to augment light interception. Mungbean leaves with narrower leaflets would allow better light distribution within the canopy than the broad leaves of most accessions, leading to higher yield potential (Lee et al. 2004). Leaf formation and senescence rates affect solar radiation interception. Higher temperatures can result in quicker canopy development, but similar conditions can enhance senescence, especially under drought conditions. Genetic variation in both leaf appearance and senescence rates could be useful for breeding lines that maintain leaf area longer for more dry matter accumulation and grain filling (Chauhan and Williams 2018).

Harvest index for mungbean is around 0.3, which is low compared to about 0.5 in soybean and peanuts (Bushby and Lawn 1992; Robertson et al. 2004; Singh and Singh 2011). Non-synchronous flowering is a major contribution to the low harvest index. The ideal photoperiod for flower induction and harvest index may be different, leading to a trade-off between these traits (Chauhan and Williams 2018). Choosing an appropriate planting time that ensures a suitable photoperiod for ontogenetic development is not a suitable approach when the crop should fit in a short cropping window between two other crops. Searching variation in photoresponse and harvest index in mungbean germplasm, perhaps even in near relatives, producing lines with variation in this trait and subsequent model-based analysis of the traits for specific target environments will be a suitable approach to improve mungbean yields by increasing harvest index.

Hybrid breeding has been a successful strategy to increase crop yields and overall plant vigor. For grain legumes hybrids such as pigeon pea (*Cajanus cajanus*) yield advantages of 30–50% have been reached in hybrids (Saxena et al. 2016). Information on improved hybrid vigor of mungbean is already available for a long time (Ramanujam et al. 1974), but no attempts for commercial hybrid production were made. Mungean is a cleistogam species, which means that fertilization is taking place while the flower is closed, which hinders hybrid production. A chasmogamous mutant of mungbean has been developed (Sorajjapinun and Srinives 2011) and candidate genes for this trait were mapped (Chen et al. 2016), so breeding for non-cleistogamous mungbean is possible. However, successful hybrid production for field crops needs a workable cytoplasmic male sterility system and fertility restorers, both components are still not available for mungbean.

12.6 Biofortification and Consumption

Mungbean is rich in easily digestible protein with a concentration of about 24–28% in seed. When mungbeans are combined with cereals, the result is a complete protein. In addition, mungbeans contain minerals such as iron (up to 0.06 g kg^{-1}) and zinc (up to 0.04 g kg^{-1}) (reviewed by Nair et al. 2013). However, concerted efforts are required to improve protein quality (by increasing the sulfur-containing amino acids) of mungbean by conventional plant breeding. Increased bioavailability of micronutrients through improved processing, storage and cooking; improved varieties with increased density of micronutrients, decreased levels of inhibitors or increased levels of promoters of food absorption; promoting awareness about the quality of mungbean and developing customized recipes would all pave the way for better diets and higher mungbean consumption. There is a growing demand for the development of mungbean varieties suited for the premium sprout market segment and nutrient-dense varieties would help to add more value to the product. It is important to encourage the domestic consumption of mungbean in countries like Myanmar, where the market is predominantly skewed towards the export market. This strategy would help in tackling nutritional issues as well as reduce the risks of fluctuating export prices faced by farmers.

12.7 New Traits for Mungbean Breeding

Some breeder-desired traits may be absent in the cultivated genepool and have to be sourced from wild relatives. While some relatives of mungbean are easily crossed with cultivated lines, crossing barriers limit the use of more distant species in mungbean improvement. Crosses between cultivated mungbean and its presumptive wild ancestor *V. radiata* ssp. *sublobata* are easy to perform and offspring is fertile (Kitamura et al. 1988). Most of these hybrids were made to introgress bruchid resistance into cultivated mungbean. *V. mungo* can also be crossed with mungbean, but the fertility of the hybrid depends on the lines used for making the crosses. Mungbean yellow mosaic disease resistance and improved methionine content have been sourced from this species for mungbean breeding. More distant *Vigna* species are difficult to cross (Pandiyan et al. 2010). However, the development of a RIL population to map bruchid resistance by crossing *Vigna radiata* with *Vigna umbellata* is promising (Mariyammal et al. 2019). A few fertile hybrids were obtained in crosses with *V. vexillata* and a few other species, but in most cases, embryo rescue over two generations of backcrossing was required to restore fertility. In general, pollen from the wild (or distant) relative is harvested to fertilize *V. radiata*. Overcoming fertility barriers in wide crossings is one of the key challenges in using wild *Vigna* species as a source for new traits for improving mungbean.

12.8 Genomics for Mungbean Crop Improvement

Genomics tools have become essential components for plant science and crop improvement. Sequence information of crop plants and wild relatives is used to determine the available diversity and to map traits of interest for breeding. Genomics contributes to both the analysis of the genetic base of traits and the prediction of plant performance based on sequence information. Access to biodiverse plant material segregating for the trait(s) of interest, polymorphic markers for genotyping and high-quality phenotypic data are preconditions for meaningful genomic studies. Quantitative trait locus (QTL) mapping has revealed several loci associated with disease and pest resistances as well as with domestication and yield in mungbean (chapter "Molecular marker resources and their application" in this book). Genome-wide association studies to tackle traits in germplasm panels are just taking off in mungbean (Noble et al. 2018). The availability of various germplasm sets and coordinated efforts to phenotype these panels and the ease to get markers for genotyping will facilitate these studies. Markers associated with traits of interest can be used to combine these traits with other desirable properties. Genomic selection (Heffner et al. 2009) will be the next development step to accelerate mungbean breeding.

Getting markers to conduct genotype–phenotype associations are not limiting anymore. DNA sequencing has become quite affordable and technologies were put in place for genotyping populations with large numbers of single nucleotide polymorphisms (SNPs) at low cost. Genotyping by sequencing (GBS) and single primer enrichment technology (SPET) are highly cost-effective methods for simultaneous marker development and genotyping and improved technologies are continuously developed. For example, skim-based GBS (skimGBS) applying low-coverage whole-genome sequencing has been developed for high-resolution genotyping (Golicz et al. 2015). In SkimGBS, the parents of a population are submitted to whole-genome re-sequencing at about 30× sequencing depth and the offspring is sequenced at much lesser depth (approximately 1×). The parental reads are mapped to the reference genome and SNPs are called (Lorenc et al. 2012). The progeny reads are mapped to the same reference, SNPs are identified and the SNP density is enhanced by imputing missing genotypes. The resulting maps allow for accurate definition of cross-over breakpoints and significantly improve mapping compared to conventional GBS. SkimGBS has been successful for general marker discovery, haplotype identification, recombination characterization, QTL analysis, genome-wide association studies and genomic selection on a range of different plant species (Golicz et al. 2015; Bayer et al. 2015), but was not yet performed on mungbean.

High-quality reference sequences are a key for successful genotyping. Genotyping by sequencing as well as whole-genome re-sequencing depend on correctly assembled reference sequences. But the genomes of genotypes under investigation may show structural differences compared to the reference sequence. In mungbean, Liu et al. (2016) reported differences in genome size of up to 50 MB among four cultivated mungbean lines, including line VC1973A which was used for producing the mungbean reference sequence (Kang et al. 2014). This finding indicates that beside mutations leading to SNPs, major genome rearrangements have taken place during breeding of landraces and varieties. Like SNPs, the structural differences may be associated with phenotypes. Structural rearrangements may implicate copy number or presence–absence variations of genes. Many agronomically important traits are known to be associated with copy number variants or gene presence/absence variants, such as the response to photoperiod (Diaz et al. 2012) or nematode resistance in soybean (Cook et al. 2012), to name just a few.

If genotyping is based on mapping short reads of some 100 bp length coming from biodiverse germplasm to a reference sequence for subsequent SNP calling, highly polymorphic regions are not captured, as the reads cannot be mapped unambiguously to such sequences. In addition, structural changes larger than the mapped fragments remain undetected. Genetic mapping can detect chromosomal rearrangements provided that each rearranged part is tagged with markers, but the detection is specific for the mapping population. Consequently, sequencing and assembling of one or a few genomes clearly do not depict the genomic diversity of a species. In contrast, sequencing a set of biodiverse lines and de novo assembly of the sequencing reads to a pangenome would give a much more accurate picture of the available genomic diversity (Golicz et al. 2015).

Ideally, a pangenome covers the entire gene set of all strains of a species. With dropping sequencing costs establishing pangenomes, a concept that originally has been developed to analyze bacterial species becomes more and more applicable also for plants. Practical limitations are sampling constraints, the resources available for sequencing and sequence assembly and the computing capacity for sequence analysis. The choice of biodiverse material is most critical for the quality of pangenome studies (Golicz et al. 2015). A usual approach is to investigate the diversity in a large collection with conventional genotyping, and then select a diverse subset for pangenome assembly. Published pangenomes contain sequences of seven to many thousands of individuals (Li et al. 2014; Hirsch et al. 2014). Pangenome analysis results in estimating how many genes or gene families are present in all the investigated individuals, which genes are common in all lines, and which are specific for certain lineages (Golicz et al. 2015). Pangenome analysis, in addition to SNP information, also pinpoints structural rearrangements.

The major challenge of pangenome studies is the correct assembly and alignment of fragmented genome sequences obtained by the next-generation sequencing. Miss-alignments may mask structural variants; therefore, sequencing technologies that yield longer reads are preferred to improve the correctness of the assembly. Current long-read sequencing technologies suffer from low accuracy, but tools are becoming available to overcome this constraint (Rang et al. 2018). A number of pangenome analysis tools are becoming available improving ortholog detection, pangenome storage, analysis and visualization (Vinuesa et al. 2018; Zekic et al. 2018; Yang et al. 2018).

Pangenome analysis can provide unpreceded insight in crop evolution and diversity. A study on 66 accessions of cultivated and wild rice revealed introgression of "wild" DNA in cultivated rice species and genetic exchange among cultivated species and gave insight into functional allele diversity (Zhao et al. 2018). By following the evolutionary route of well-characterized quantitative trait nucleotides, information on the origin and selection point for these traits was obtained, showing that most of the naturally occurring variants in rice are of low frequency. Such low-frequency traits usually cannot be mapped in genome-wide association studies using usual sized germplasm panels. Such pangenome studies yield information on genepools where specific traits can be sourced plus markers associated with the traits. Knowledge gained about the frequency of the desired alleles can guide germplasm screening and pre-breeding activities.

Pangenomes have been constructed for the legume crop soybean (Lam et al. 2010; Li et al. 2014). The first soybean pangenome was established by DNA sequencing and de novo assembly of sequencing reads from seven *Glycine soja* accessions, the wild relative of cultivated soybean (*Glycine max*). Comparisons between the individual sequences identified lineage-specific genes, genes with copy number variation or large-effect mutations. Some variants showed evidence of positive selection and specific mutations may contribute to agronomic traits such as biotic stress resistance, seed composition, flowering and maturity time, organ size and biomass (Li et al. 2014).

A pangenome study in *V. radiata* and near relatives has not yet been conducted. Such a study could inform about the domestication history of mungbean and pinpoint genetic diversity that is useful for improving adaptation to diverse environments. Mapping of complex traits would be enhanced by pangenome studies through improved mapping accuracy and inclusion of copy number variants that cause trait variation. Association of traits with specific loci would become more accurate and reliable, and insight into the functional allele diversity for breeder-desired traits in mungbean would reveal new alleles for crop improvement.

12.9 Improving Trait Mapping

Trait mapping makes use of plant genotypes with contrasting traits. Bulked segregant analysis compares groups of lines with extreme opposite values of a specific trait of a bi-parental population and identifies markers that are associated with this trait types (Michelmore et al. 1991). QTL mapping tests for marker-trait associations in segregating populations. If the marker genotyping of the extreme pools is replaced by whole-genome re-sequencing, the technology is called QTL-Seq (Takagi et al. 2013). The technique has been proven to be a quick and cost-effective method for QTL identification. It has been applied on various legume species, including chickpea, pigeonpea and groundnut (reviewed by Varshney et al. 2018). The high-marker density and especially the precise phenotyping over various seasons allowed narrowing down the QTL loci for quantitative traits such as seed weight and plant biomass to small genomic regions with a few genes (Singh et al. 2016). Techniques like QTL-seq should be applied to map breeder-desired traits for mungbean. The crucial requirement for the success of such an approach is, as outlined above, high-quality phenotyping data.

12.10 Genomic Prediction and Genomic Selection

Genomic selection has been proposed as a superior method for the rapid selection of favorable genotypes that accelerates breeding (Crossa et al. 2017). Instead of markers associated with a trait of interest, genomic selection uses breeding values for selection. The breeding values are estimated by genomic-enabled prediction for each marker in a training population that was extensively phenotyped over locations and seasons and was also well genotyped (Heffner et al. 2009). The statistical complexities of genomics prediction models, especially genomic genotype by environment interactions impose great challenges to the technology.

Once available, the breeding values can be used to select favorable genotypes in large breeding populations, minimizing the need for phenotyping, saving time and resources for breeding, and enhancing the genetic gain per generation, especially for multigenic traits with low heritability such as yield and stress tolerance (Crossa et al. 2017). For legumes, only a few genomic selection studies were reported so far (reviewed by Varshney et al. 2018). Investment in legume breeding (except soybean) is lower than, for example, cereal breeding, consequently, there are less multi-location and multi-season evaluation data available for legume crops, which limits the estimation of breeding values. Less investment in legume breeding also means that legume breeding populations generally are small, reducing the scope of genomic selection. Nevertheless, genomic selection is also useful in small populations to increase the genetic gain of low heritability traits, especially when selection should be performed for multiple traits simultaneously. Methods for genomic prediction and calculating prediction accuracies are available (Crossa et al. 2017).

12.11 Conclusion

To cope with the expected rise of demand, mungbean production needs new improved high-yielding disease resistant and abiotic stress-tolerant varieties that are well adapted to their target environment and production system. To produce such varieties, the breeders need access to mungbean diversity. Access to diversity is enhanced through specific populations such as core collections, introgression line populations or lines that combine the diversity of founder lines such as NAM or MAGIC populations.

There are several different pathways for improving mungbean yield potential, like changes in phenology or light interception capacity. Crop growth models are helpful tools to estimate the impact of changes in plant phenotypes on yield development in specific environments. Once the traits to be targeted and genetic resources harboring these traits are identified, genomics-aided breeding can be conducted. These works are generally initiated by identifying molecular markers associated with the trait(s) of interest and pangenome analysis adds crucial information to such mapping studies. Improving complex multigenic traits and simultaneous selection for multiple traits may better be achieved by genomic selection than by selection with markers identified in QTL studies. Availability of multi-location and multi-seasonal phenotypic data for mungbean training populations would open the path for genomic selection in this species.

References

Ahmad M, Zahir ZA, Asghar HN, Arshad M (2012) The combined application of rhizobial strains and plant growth promoting rhizobacteria improves growth and productivity of mung bean (*Vigna radiata* L.) under salt-stressed conditions. Ann Microbiol 62(3):1321–1330

Alexandratos N, Bruinsma J (2012) World agriculture towards 2030/2050: the 2012 revision. ESA working paper. FAO, Rome

Barkley NA, Wang ML, Gillaspie AG, Dean RE, Pederson GA, Jenkins TM (2008) Discovering and verifying DNA polymorphisms in a mung bean [*V. radiata* (L.) R. Wilczek] collection by EcoTILLING and sequencing. BMC Res Notes 1:28. https://doi.org/10.1186/1756-0500-1-28

Bayer PE, Ruperao P, Mason AS, Stiller J, Chan CK, Hayashi S et al (2015) High-resolution skim genotyping by sequencing reveals the distribution of crossovers and gene conversions in *Cicer arietinum* and *Brassica napus*. Theor Appl Genet 128:1039–1048. https://doi.org/10.1007/s00122-015-2488-y

Bisht IS, Mahajan RK, Patel DP (1998) The use of characterisation data to establish the Indian mungbean core collection and assessment of genetic diversity. Genet Resour Crop Ev 45:127–133

Bushby H, Lawn R (1992) Accumulation and partitioning of nitrogen and dry matter by contrasting genotypes of mungbean (*Vigna radiata* (L.) Wilczek). Aust J Agric Res 43:1609–1628

Carberry P, Muchow R, Williams R, Sturtz J, McCown R (1992) A simulation model of kenaf for assisting fibre industry planning in northern Australia. I. General introduction and phenological model. Aust J Agric Res 43:1501–1513

Chauhan Y, Williams R (2018) Physiological and agronomic strategies to increase mungbean yield in climatically variable environments of Northern Australia. Agronomy 8(6):83

Chauhan Y, Rachaputi RC (2014) Defining agro-ecological regions for field crops in variable target production environments: a case study on mungbean in the northern grains region of Australia. Agric Forest Meteorol 194:207–217

Chen J, Somta P, Chen X, Cui X, Yuan X, Srinives P (2016) Gene mapping of a mutant mungbean (*Vigna radiata* L.) using new molecular markers suggests a gene encoding a yuc4-like protein regulates the chasmogamous flower trait. Front plant Sci 7:830

Chhabra KS, Kooner BS (1985) Problem of flower shedding caused by thrips, *Megalurothrips distalis* (Karny), on summer mungbean, *Vigna radiata* (L.) Wilczek, and its control. Int J Pest Manag 31(3):186–188

Cook DE, Lee TG, Guo X, Melito S, Wang K, Bayless AM, Wang J et al (2012) Copy number variation of multiple genes at Rhg1 mediates nematode resistance in soybean. Science 338:1206–1209

Crossa J, Pérez-Rodríguez P, Cuevas J, Montesinos-López O, Jarquín D, de los Campos G, Burgueño J, González-Camacho JM, Pérez-Elizalde S, Beyene Y, Dreisigacker S (2017) Genomic selection in plant breeding: methods, models, and perspectives. Trends Plant Sci 22(11):961–975

Dıaz A, Zikhali M, Turner AS, Isaac P, Laurie DA (2012) Copy number variation affecting the photoperiod-B1 and vernalization-A1 genes is associated with altered flowering time in wheat (*Triticum aestivum*). PLoS ONE 7:e33234

Eshed Y, Zamir D (1995) An introgression line population of *Lycopersicon pennellii* in the cultivated tomato enables the identification and fine mapping of yield-associated QTL. Genetics 141(3):1147–1162

Golicz AA, Bayer PE, Edwards D (2015) Skim-based genotyping by sequencing. InPlant genotyping. Humana Press, New York, NY, pp 257–270

Hamid A, Agata W, Kawamitsu Y (1990) Photosynthesis, transpiration and water use efficiency in four cultivars of mungbean, *Vigna radiata* (L.) Wilczek. Photosynthetica 24(1):96–101

Hanumantha Rao B, Nair RM, Nayyar H (2016) Salinity and high temperature tolerance in mungbean [*Vigna radiata* (L.) Wilczek] from a physiological perspective. Front Plant Sci 7:957

Heffner EL, Sorrells ME, Jannink JL (2009) Genomic selection for crop improvement. Crop Sci 49(1):1–2

Hirsch CN, Foerster JM, Johnson JM, Sekhon RS, Muttoni G, Vaillancourt B, Peñagaricano F, Lindquist E, Pedraza MA, Barry K, de Leon N (2014) Insights into the maize pan-genome and pan-transcriptome. Plant Cell 26:121–135

Kang YJ, Kim SK, Kim MY, Lestari P, Kim KH, Ha BK, Jun TH, Hwang WJ, Lee T, Lee J, Shim S (2014) Genome sequence of mungbean and insights into evolution within *Vigna* species. Nat Commun 5 (5443):6443

Kayani AK, Qureshi S, Kayani WK, Qureshi R, Waheed A, Arshad M, Gulfraz M, Laghai MK (2010) Assessment of wheat yield potential after cropping mungbean (*Vigna radiata* (L.) Wilczek). Pak J Bot 42(3):1535–1541

Khan S, Goyal S (2009) Mutation genetic studies in mungbean IV. Selection of early maturing mutants. Thai J Agri Sci 42(2):109–113

Khattak GS, Ashraf M, Elahi T, Abbas G (2003) Selection for large seed size at the seedling stage in mungbean (Vigna radiata (L.) Wilczek). Breed Sci 53 (2):141–143

Kitamura K, Ishimoto M, Sawa M (1988) Inheritance of resistance to infestation with azuki bean weevil in *Vigna sublobata* and successful incorporation to *V. radiata*. Japanese J Breed 38(4):459–464

Lam HM, Xu X, Liu X, Chen WB, Yang GH, Wong FL, Li MW et al (2010) Resequencing of 31 wild and cultivated soybean genomes identifies patterns of genetic diversity and selection. Nat Genet 42:1053–1059

Latati M, Bargaz A, Belarbi B, Lazali M, Benlahrech S, Tellah S (2016) The intercropping common bean with maize improves the rhizobial efficiency, resource use and grain yield under low phosphorus availability. Eur J Agron 72:80–90

Lee YS, Lee JY, Kim DK, Yoon CY, Bak GC, Park IJ, Bang GP, Moon JK, Oh YJ, Min KS (2004) A new high-yielding mungbean cultivar, "Samgang" with lobed leaflet. Kor Breed J 36:183–184

Li Y-H, Zhou G, Ma J, Jiang W, Jin L-G, Zhang Z et al (2014) De novo assembly of soybean wild relatives for pan-genome analysis of diversity and agronomic traits. Nat Biotech 32:1045–1052

Liu C, Wang S, Wang L, Sun L, Mei L, Xu N, Cheng X (2008) Establishment of candidate core collection in Chinese mungbean germplasm resources. Acta Agron Sin 34:700

Liu MS, Kuo TC, Ko CY, Wu DC, Li KY, Lin WJ, Lin CP, Wang YW, Schafleitner R, Lo HF, Chen CY (2016) Genomic and transcriptomic comparison of nucleotide variations for insights into bruchid resistance of mungbean (*Vigna radiata* [L.] R. Wilczek). BMC Plant Biol 16(1):46

Lorenc MT, Hayashi S, Stiller J, Lee H, Manoli S, Ruperao P, Visendi P, Berkman PJ, Lai K, Batley J, Edwards D (2012) Discovery of single nucleotide polymorphisms in complex genomes using SGSautoSNP. Biology 1(2):370–382

Mariyammal I, Seram D, Samyuktha SM, Karthikeyan A, Dhasarathan M, Murukarthick J, Kennedy JS, Malarvizhi D, Yang TJ, Pandiyan M, Senthil N (2019) QTL mapping in *Vigna radiata* × *Vigna umbellata* population uncovers major genomic regions associated with bruchid resistance. Mol Breed 39:110

McMullen MD, Kresovich S, Villeda HS, Bradbury P, Li H, Sun Q, Flint-Garcia S, Thornsberry J, Acharya C, Bottoms C, Brown P (2009) Genetic properties of the maize nested association mapping population. Science 325(5941):737–740

Michelmore RW, Paran I, Kesseli RV (1991) Identification of markers linked to disease-resistance genes by bulked segregant analysis: a rapid method to detect markers in specific genomic regions by using segregating populations. Proc Natl Acad Sci USA 88 (21):9828–9832

Moe KT, Gwag JG, Park YJ (2012) Efficiency of PowerCore in core set development using amplified fragment length polymorphic markers in mungbean. Plant Breed 131:110–117

Nair RM, Yang RY, Easdown WJ, Thavarajah D, Thavarajah P, Hughes JD, Keatinge JD (2013) Biofortification of mungbean (*Vigna radiata*) as a whole food to enhance human health. J Sci Food Agri 93(8):1805–1813

Nakhlawy FS, Ismail SM, Basahi JM (2018) Optimizing mungbean productivity and irrigation water use efficiency through the use of low water-consumption during plant growth stages. Legume Res. https://doi.org/10.18805/lr.v40i04.9014

Noble TJ, Tao Y, Mace ES, Williams B, Jordan DR, Douglas CA, Mundree SG (2018) Characterization of linkage disequilibrium and population structure in a mungbean diversity panel. Front Plant Sci 8:2102

Pandiyan M, Senthil N, Ramamoorthi N, Muthiah AR, Tomooka N, Duncan V, Jayaraj T (2010) Interspecific hybridization of *Vigna radiata* x 13 wild *Vigna* species for developing MYMV donor. Elect J Plant Breed 1(4):600–610

Pannu RK, Singh DP (1993) Effect of irrigation on water use, water-use efficiency, growth and yield of mungbean. Field Crops Res 31(1–2):87–100

Pascual L, Desplat N, Huang BE, Desgroux A, Bruguier L, Bouchet JP, Le QH, Chauchard B, Verschave P, Causse M (2015) Potential of a tomato MAGIC population to decipher the genetic control of quantitative traits and detect causal variants in the resequencing era. Plant Biotech J 13(4):565–577

Pestsova EG, Börner A, Röder MS (2001) Development of a set of *Triticum aestivum-Aegilops tauschii* introgression lines. Hereditas 135(2–3):139–143

Ramanujam S, Tiwari AS, Mehra RB (1974) Genetic divergence and hybrid performance in mung bean. Theor App Genet 45(5):211–214

Rang FJ, Kloosterman WP, de Ridder J (2018) From squiggle to basepair: computational approaches for improving nanopore sequencing read accuracy. Genome Biol 19(1):90

Rashid A, Harris D, Hollington PA, Rafiq M (2004) Improving the yield of mungbean (*Vigna radiata*) in the North West Frontier Province of Pakistan using on-farm seed priming. Exp Agri 40(2):233–244

Robertson MJ, Fukai S, Peoples MB (2004) The effect of timing and severity of water deficit on growth, development, yield accumulation and nitrogen fixation of mungbean. Field Crops Res 86:67–80

Salunkhe DK, Kadam SS, Chavan JK (1985) Post-harvest biotechnology of food legumes. CRC Press, Boca Raton, FL, p 160

Saxena KB, Sameer Kumar CV, Saxena RK, Vijay Kumar R, Singh IP, Hingane AJ, Mula MG, Patil SB, Varshney RK (2016) Hybrid pigeonpea: accomplishments and challenges for the next decade. Legume Persp 11:30–32

Schafleitner R, Nair RM, Rathore A, Wang YW, Lin CY, Chu SH, Lin PY, Chang JC, Ebert AW (2015) The AVRDC–the world vegetable center mungbean (*Vigna radiata*) core and mini core collections. BMC Genom 16(1):344

Singh VK, Khan AW, Jaganathan D, Thudi M, Roorkiwal M, Takagi H, Garg V, Kumar V, Chitikineni A, Gaur PM, Sutton T (2016) QTL-seq for rapid identification of candidate genes for 100-seed weight and root/total plant dry weight ratio under rainfed conditions in chickpea. Plant Biotech J14(11):2110–2119

Singh D, Singh B (2011) Breeding for tolerance to abiotic stresses in mungbean. J Food Legumes 24:83–90

Sorajjapinun W, Srinives P (2011) Chasmogamous mutant, a novel character enabling commercial hybrid seed production in mungbean. Euphytica 181:217–222

Srivastava AC, Pal M, Das M, Sengupta UK (2001) Growth, CO_2 exchange rate and dry matter partitioning in mungbean (*Vigna radiata* L.) grown under elevated CO_2. Indian J Exp Bio 39:572–577

Stagnari F, Maggio A, Galieni A, Pisante M (2017) Multiple benefits of legumes for agriculture sustainability: an overview. Chem Biol Tech Agri 4(1):2

Summerfield R, Lawn R (1987) Environmental modulation of flowering in mung bean (*Vigna radiata*): a reappraisal. Exp Agric 23:461–470

Takagi H, Abe A, Yoshida K, Kosugi S, Natsume S, Mitsuoka C, Uemura A, Utsushi H, Tamiru M, Takuno S, Innan H, Cano LM, Kamoun S, Terauchi R (2013) QTL-seq: rapid mapping of quantitative trait loci in rice by whole genome resequencing of DNA from two bulked populations. Plant J 74:174–183

Varshney RK, Pandey MK, Bohra A, Singh VK, Thudi M, Saxena RK (2018) Toward the sequence-based breeding in legumes in the post-genome sequencing era. Theor App Genet. https://doi.org/10.1007/s00122-018-3252-x

Vinuesa P, Ochoa-Sánchez LE, Contreras-Moreira B (2018) GET_PHYLOMARKERS, a software package to select optimal orthologous clusters for phylogenomics and inferring pan-genome phylogenies, used for a critical geno-taxonomic revision of the genus *Stenotrophomonas*. Front Microbiol 9:771. https://doi.org/10.3389/fmicb.2018.00771

Yang HL, Le DO, Hui WA, Liu CL, Fang LI, Xie CX (2018) A simple way to visualize detailed phylogenetic tree of huge genome-wide SNP data constructed by SNPhylo. J Integ Agri 17(9):1972–1978

Zekic T, Holley G, Stoye J (2018) Pan-genome storage and analysis techniques. Comparative genomics. Humana Press, New York, NY, pp 29–53

Zhao Q, Feng Q, Lu H, Li Y, Wang A, Tian Q, Zhan Q, Lu Y, Zhang L, Huang T, Wang Y (2018) Pan-genome analysis highlights the extent of genomic variation in cultivated and wild rice. Nat Genet 50 (2):278